Oxide-Based Materials and Structures

Oxide-Based Materials and Structures

Fundamentals and Applications

Edited by
Rada Savkina
Larysa Khomenkova

CRC Press is an imprint of the
Taylor & Francis Group, an **Informa** business

First edition published 2020
by CRC Press
6000 Broken Sound Parkway NW, Suite 300,
Boca Raton, FL 33487-2742

and by CRC Press
2 Park Square, Milton Park, Abingdon, Oxon, OX14 4RN

First issued in paperback 2021

CRC Press is an imprint of Taylor & Francis Group, an Informa business

ISBN 13: 978-1-03-224149-4 (pbk)
ISBN 13: 978-0-367-25239-7 (hbk)

Publisher's Note
The publisher has gone to great lengths to ensure the quality of this reprint but points out that some imperfections in the original copies may be apparent.

Library of Congress Cataloging-in-Publication Data

Names: Khomenkova, Larysa, editor. | Savkina, Rada, editor.
Title: Oxide-based materials and structures : fundamentals and applications / edited by Larysa Khomenkova & Rada Savkina.
Description: First edition. | Boca Raton : CRC Press, 2020. | Includes bibliographical references and index.
Identifiers: LCCN 2019059106 | ISBN 9780367252397 (hardback) | ISBN 9780429286728 (ebook)
Subjects: LCSH: Oxides. | Oxides–Surfaces. | Materials science.
Classification: LCC QD181.O1 O947 2020 | DDC 620.1/6–dc23
LC record available at https://lccn.loc.gov/2019059106

Visit the Taylor & Francis Web site at
www.taylorandfrancis.com

and the CRC Press Web site at
www.crcpress.com

Dedication

To our families for their endless love, guiding, and patience.
To our friends and colleagues for their kind support and friendly critics.

Contents

Preface

Oxide materials are becoming of increasing interest due to their potential applications in advanced nanodevices and nanosystems toward energy, environmental, and sensing sectors. As we know, oxide materials can be conductive, semiconductive, or dielectric, as well as they can exhibit magnetic or multiferroic properties. They are used in field effect devices, capacitors for nonvolatile memories, transparent electronics, tunable devices for wireless communications, photocatalysis and environment protection, and so on. This book provides a review of the original works and the patents received over the last years.

This book covers a wide range of materials, techniques, and approaches that could be interesting for both experienced researchers and young scientists.

This book consists of nine chapters. Each of them is written individually. **Chapter 1** concerns general considerations of the oxide-based materials and some important fields of their application. **Chapter 2** focuses on the optical modeling of oxides and oxide-based structures using spectroscopic ellipsometry. The parameterization of the dielectric function is discussed for single-phase and composite multilayer materials, as well as for the vertical, lateral, and optical anisotropy. Modeling of low-dimensional oxide structures such as nanotubes and nanodots embedded in oxides is also presented. Some applications of these materials in optical sensing are also mentioned in this chapter. **Chapter 3** addresses the impedance spectroscopy of oxides and oxide-based structures. **Chapter 4** gives an overview of the basic principles and applications of synchrotron experimental methods for analyzing the local environment of atoms and/or ions along with their electronic structure reorganization: X-ray spectroscopy (including XAS, XES, and RIXS) and X-ray photoelectron spectroscopy (including RPES).

Chapter 5 aims to review and highlight the recent developments and applications of low-dimensional structures based on SnO_2, TiO_2, and In_2O_3. These wide-bandgap oxides have attained growing attention during the last decade as they have become key materials in several applications of technological interest including microelectronic, energy harvesting and storage, optoelectronic, catalysis or gas sensing. However, challenges still remain in the understanding of their fundamental properties and their functionalization in order to face the advent of modern devices. In **Chapter 6**, the latest advances in the field of magnetic properties and applications of transition metal oxides are discussed. In particular, an interesting class of materials, known as magnetoelectric multiferroics, is considered. Isostructural and isovalent iron and chromium oxides are the subjects of active experimental and theoretical investigations. Recent advances in the use of transition metal oxides for water splitting are presented.

The development of different express methods for the investigation of materials and structures is very important. Infrared spectroscopy, being one of the nondestructive, fast, and sensitive methods, allows the investigation of vibrational properties of different materials. In **Chapter 7**, main attention is paid to theoretical consideration of elementary excitations such as surface polaritons and surface plasmon–phonon polaritons in anisotropic materials, as well as to the effects of magnetic field on these collective interactions. An application of this approach for the determination of electrophysical parameters of anisotropic materials is demonstrated. As an example, undoped and doped ZnO single crystals and thin films are addressed. It is shown that developed theoretical models can be implemented for the determination of the optical and electrophysical parameters of different anisotropic materials and, especially, for textured thin films grown on semiconductor and dielectric substrates.

Given the increasing importance of atomic layer deposition in present semiconductor industry, the basic objective of **Chapter 8** is to highlight the versatility of zinc oxide grown by this method. Depending on growth conditions, atomic layer deposition is able to provide epitaxial or polycrystalline ZnO films with various orientations and diverse surface morphologies.

Among the different oxide materials, Al_2O_3, HfO_2, and ZrO_2 as well as their silicates are considered mainly as alternatives to SiO_2 gate oxide and they are now mostly used in microelectronic devices. However, all these materials have unique optical and chemical properties. Being doped with different ions, they can offer novel applications in light-emitting structures and photovoltaic cells. In **Chapter 9**, some recent applications Al_2O_3, HfO_2, and ZrO_2 along with their silicates doped with rare-earth elements are presented.

We hope to present to the reader a recent survey of the topic of oxide-based composite materials and structures as well as to depict the potential and broadness of the application of such materials.

We would like to express our gratitude to all the contributors for their willingness to share their experiences and in-depth insights into this field with the readers and for taking the time off their schedules to write excellent chapters. We would also like to thank all our current and former colleagues who helped and advised us in conducting some research described in this book. Also, we are thankful to the staff of Taylor & Francis/CRC Press who invited us to write this book and helped immensely in its preparation for publication. Last but not least, we are extremely grateful to our families for their support, patience, and understanding over the many busy years of our scientific career.

Rada Savkina and **Larysa Khomenkova**

Acknowledgments

This book is a collective contribution of the authors from different countries: Hungary, Poland, Spain, and Ukraine. The authors would like to acknowledge all the organizations that supported their research in different years. Among them are the National Research, Development and Innovation Office of Hungary, the Polish National Science Centre, the Polish Academy of Sciences, the Department of Chemistry of the University of Warsaw (Poland), the National Academy of Sciences of Ukraine (NASU), the Ministry of Education and Science of Ukraine, the French National Research Agency, the National Research Centre of France, and the European Union Seventh Framework Programmeas well as the Ministry of Economy and Competitiveness (MINECO), Spain. Most of the results described by the authors in their chapters have been obtained while carrying out national and international projects such as:

- The project OTKA K131515 titled "Low-dimensional nanomaterials for the optical sensing of organic molecules on liquid and gas interfaces" financed by the National Research, Development and Innovation Office of Hungary;
- The project DEC-2018/07/B/ST3/03576 titled "Electrical conductivity and defect clusters in zinc oxide formed as a result of intentional and unitentional doping" financed by the Polish National Science Centre;
- Ukrainian–Polish joint research project "Characterization of the hybrid system of the nanometric layers based on transition metals oxides useful for spintronic" (Project # 23, 2018–2020) under the agreement on scientific cooperation between the Polish Academy of Sciences and the National Academy of Sciences of Ukraine.
- The project "Affine sensor system based on 'smart' hybrid nanocomplexes for determination of specific nucleic acid sequences" in the framework of target programme of the NASU, "'Intelligent' sensor devices of a new generation based on modern materials and technologies" (2018–2022, Ukraine);
- The project "Effect of the doping on structural, optical and electron-phonon properties and stability of anisotropic crystals" (No.: 89452, 2018–2020, Ukraine);
- The project "Development of technology and equipment for the creation of infrared, THz and sub-THz radiation based on low-dimensional structures" in the framework of the programme "Search and creation of promising semiconductor materials and functional structures for nano- and optoelectronics" (No.: III-41-17, 2017–2022,Ukraine);
- The project "Study and nanoanalysis of irradiation effects group" within the "Investments for the Future Programme" and the "GENESIS EQUIPEX Programme" (PIA, ANR, (ANR-11-EQPX-0020) and Normandy Region) (2014–2018, France);
- The project "High-k oxides doped with Si or Ge nanoclusters for microelectronic and photonic applications" in the framework of the Ukrainian–French bilateral programme "Dnipro", 2015–2016 edition (Nos. M/115-2015 and M/27-2016 (Ukraine), No. 34820QH (France));
- The project "High-k dielectric structures with modulated architecture for non-volatile memory application" in the framework of the *Programme Investissements d'Avenir* (ANR-11-IDEX-0002-02; ANR-10-LABX-0037-NEXT) (2015, France);
- The project 008-ABELCM-2013, project MAT2015-65274-R, Project MAT2016-81720-REDC, Project PCIN-2017-106, and RTI2018 -097195-B-100 supported by NILS program EEA Grants/MINECO/FEDER/M-ERA.NET COFUND. M. T. The FPI grant holder, M.T., thanks the financial support from MINECO. The Postdoctoral "Atracción del talento" Grant holder, J.B., Thanks the financial support from the Comunidad de Madrid.

- The European Union Seventh Framework Programme under Grant Agreement 312483 – ESTEEM2 (Integrated Infrastructure Initiative–I3) (2015, France)
- For the support via as well as for partial support of the scientific research. We also acknowledge the support from Elettra Sincrotrone (Italy, project number 20150152) and Helmholtz-Zentrum Dresden-Rossendorf (Germany) as well as Dr. Renata Ratajczak of National Centre for Nuclear Research (Poland) and Dr. Yevgen Melikhov of Cardiff University (UK) for their invaluable help and insightful discussions during RBS/XPS experiments and/or data interpretation.

Editors

Dr. Rada Savkina is senior staff researcher of V. Lashkaryov Institute of Semiconductor Physics at National Academy of Sciences of Ukraine (Kyiv, Ukraine) as well as Associate Professor of the Faculty of Natural Sciences at National University of "Kyiv-Mohyla Academy." She obtained M.Sc. degree in Solid State Physics in 1993 from T. Shevchenko Kyiv University; Ph.D. degree in Solid State Physics in 2002 and academic status of Senior Researcher in Solid State Physics in 2013, both from V. Lashkaryov Institute of Semiconductor Physics. She has gained valuable experience in developing new ideas from fundamental understanding to the manufacturing of multifunctional composite materials and devices on their basis. She is an expert in the fabrication of materials by physical methods (laser deposition, ion beam processing, and ultrasonic approach) as well as their characterization with optical and electrical methods. She has elaborated infrared and terahertz detectors and vision systems, which is covered by 5 patents. She has more than 70 papers in refereed journals and 2 book chapters in her name. She is a member of the editorial board of the international research journal "Recent Advances in Electrical & Electronic Engineering." Her novel field of activity concerns the elaboration of the concept of a new biocompatible photovoltaic cell by sonochemical method. She is the editor of two books.

Dr. Larysa Khomenkova is senior staff researcher of V. Lashkaryov Institute of Semiconductor Physics at National Academy of Sciences of Ukraine (Kyiv, Ukraine) as well as Associate Professor of the Faculty of Natural Sciences at National University of "Kyiv-Mohyla Academy." She obtained M.Sc. degree in Technology and Materials of Microelectronics in 1992 from T. Shevchenko Kyiv University; Ph.D. degree in Solid State Physics in 1999 and academic status of Senior Researcher in Solid State Physics in 2003, both from V. Lashkaryov Institute of Semiconductor Physics at National Academy of Sciences of Ukraine (Kyiv, Ukraine). Her work is centered around the multifunctional composite materials, in particular their fabrication with physical methods (solid state reaction, physical vapor deposition, and magnetron sputtering) and materials' characterization with optical, luminescent, and electrical methods. Her interest is also focused on the elaboration of novel technological approaches for creation of semiconductor and dielectric structures aiming at their photonic and microelectronic applications as well as on the instability and degradation phenomena in the devices. In addition, she focuses on the materials for alternative energy sources as solid oxide fuel cells and hydrogen storage. Novel field of her activity concerns the elaboration of the materials by atomic layer deposition for photovoltaic and photonic applications. She is the author of more than 150 research papers and 4 book chapters, and also the editor of 4 books.

Contributors

Peter Petrik
Institute of Technical Physics and Materials Science
Centre for Energy Research
Budapest, Hungary

Ana Cremades
Departamento de Física de Materiales
Facultad de Ciencias Físicas
Universidad Complutense de Madrid
Madrid, Spain

David Maestre
Departamento de Física de Materiales
Facultad de Ciencias Físicas
Universidad Complutense de Madrid
Madrid, Spain

Javier Bartolomé
Departamento de Física de Materiales
Facultad de Ciencias Físicas
Universidad Complutense de Madrid
Madrid, Spain

María Taeño
Departamento de Física de Materiales
Facultad de Ciencias Físicas
Universidad Complutense de Madrid
Madrid, Spain

Antonio Vázquez-López
Departamento de Física de Materiales
Facultad de Ciencias Físicas
Universidad Complutense de Madrid
Madrid, Spain

Félix del Prado
Departamento de Física de Materiales
Facultad de Ciencias Físicas
Universidad Complutense de Madrid
Madrid, Spain

Department of Neuroscience and Biomedical Engineering
Engineered Nanosystems
Aalto University – Micronova
Espoo, Finland

Miguel García-Tecedor
Departamento de Física de Materiales
Facultad de Ciencias Físicas
Universidad Complutense de Madrid, Spain

Institute of Advanced Materials (INAM)
Universitat Jaume I, Castelló
Madrid, Spain

G. Cristian Vásquez
Departamento de Física de Materiales
Facultad de Ciencias Físicas
Universidad Complutense de Madrid
Madrid, Spain

Department of Physics
Centre for Materials Science and Nanotechnology
University of Oslo
Oslo, Norway

Julio Ramírez-Castellanos
Departamento de Química Inorgánica
Facultad de Ciencias Químicas
Universidad Complutense de Madrid

Iraida N. Demchenko
Laboratory of Electrochemistry
Division of Physical Chemistry
Faculty of Chemistry
University of Warsaw
Warsaw, Poland

Elżbieta Guziewicz
Division of Physics and Technology of Wide Band Gap Semiconductor Nanostructures
Institute of Physics, Polish Academy of Science
Warsaw, Poland

Aleksej Smirnov
Department of Physics and Technology of Low-Dimensional Systems
V. Lashkaryov Institute of Semiconductor Physics NASU
Kyiv, Ukraine

Oleksandr Melnichuk
Mykola Gogol Nizhyn State University
Nizhyn, Ukraine

Lyudmyla Melnichuk
Mykola Gogol Nizhyn State University
Nizhyn, Ukraine

Evgen Venger
Department of Semiconductor Heterostructures
V. Lashkaryov Institute of Semiconductor Physics NASU
Kyiv, Ukraine

Abbreviations

AC	alternating current
AFM	antiferromagnetic
ALD	atomic layer deposition
ATR	attenuated total reflection
BEOL	back-end-of-the line
BST	$Ba_xSr_{1-x}TiO_3$
CB (CBM)	conduction band minimum
CL	cathodoluminescence
CMSA	chromatography-mass spectrometric analysis
CPE	constant phase element
CSE	combinatorial substrate epitaxy
CVD	chemical vapor deposition
DC	direct current
DEZn	$Zn(C_2H_5)_2$
DFT	density functional theory
DOS	density of states
DSSC	dye sensitized solar cell
EIS	electrochemical impedance spectroscopy
EL	electroluminescence
EMA	effective medium approximation
ENZ	epsilon-near-zero
EPR	electron paramagnetic resonance
EXAFS	extended X-ray absorption fine structure
FE	ferroelectricity
FET	field-effect transistor
FM	ferromagnetism
FMR	ferromagnetic resonance
FP	Fabry–Pérot
FTO	fluorine tin oxide
FWHM	full width at half maximum
GO	graphene oxide
GS	solution of graphene oxide
HEV (EV)	hybrid electric vehicles
HRXRD	high-resolution x-ray diffraction
IR	infrared
IRR	infrared reflection
IS	impedance spectroscopy
ITO	indium tin oxide
ITRS	International Technology Roadmap for Semiconductors
IUPAC	International Union of Pure and Applied Chemistry
LIB	lithium-ion batteries
LMAS	light modulated absorption spectroscopy
LSPR	localized surface plasmon resonance
MBE	molecular beam epitaxy
MDF	model dielectric function

ME	magneto-electrics
MIEC	mixed conductivity oxides
MIR	mid infrared
MOD	metal organic deposition
MOCVD	metal organic chemical vapor deposition
MOSFETs	metal-oxide semiconductor field-effect transistor
NMC	nickel manganese cobalt oxide
NXES	nonresonant X-ray emission spectroscopy
NWs	nanowires
ODC	oxygen deficient centers
ODR	open dielectric resonators
OER	oxygen evolution reaction
PEC	photoelectrochemical
PEFC	polymer electrolyte fuel cell
PBS	phosphate-buffered saline
PL	photoluminescence
PLD	pulsed laser deposition
PMAS	potentially modulated absorption spectroscopy
PMMA	polymethylmethacrylate
PMMA	polymethylmethacrylate
QSS-PC	quasi steady state photoconductance
RBS	Rutherford backscattering spectrometry
RE	rare earth
RIXES	resonant inelastic X-ray emission spectroscopy
RIXS	resonant inelastic X-ray scattering
RMS	root mean square
RPES	resonant photoelectron spectroscopy
RT	room temperature
RTP	rapid thermal processing
SIMS	secondary ion mass spectroscopy
SoC	state of charge
SOFC	solid oxide fuel cells
SP	surface plasmon
SPM	scanning probe microscopy
SPP	surface plasmon–polariton
SPPP	surface plasmon–phonon polariton
SSRT	solid-state redox transitions
SSSRT	solid surface state redox transitions
TCO	transparent conductive oxides
TEM	transmission electron microscopy
TFT	thin-film transistors
TIR	total internal reflections
TL	Tauc–Lorentz
TM	transition metal
UV	ultraviolet
VB (VBM)	valence band maximum
VLS	vapor–liquid–solid
VS	vapor–solid
WGM	whispering gallery modes

XANES X-ray absorption near edge structure
XAS X-ray absorption spectroscopy
XES X-ray emission spectroscopy
XPS X-ray photoelectron spectroscopy
XRD X-ray diffraction
YSZ yttria-stabilized zirconia

1 Oxide-Based Materials and Structures

General Remarks

Rada Savkina and Larysa Khomenkova

Oxide-based materials are characterized by a wide range of properties that make them appropriate for multiple applications in the field of novel electronic devices, such as for sensing, environment and human health diagnostics, use in nonvolatile memories memory, and so on (Lorenz et al., 2016). Among them are oxide-based electronic materials for resistive switching and thin film transistors (TFT) applications; perovskite-type materials for photocatalysis, environment protection and energy conversion; silicon photonics enhanced with functional oxides; complex-oxide field-effect transistors; topological oxide electronics; magnetoelectric multiferroic, superconductor and giant magnetoelectric materials and structures for magnetic and energy applications, and many others. As we see, all important fields of electronics and photonics, magnetics, and energetics are being based on achievements of oxides material science.

As we know, oxide materials can be conductive, semiconductive, and dielectric, as well as they can exhibit magnetic or multiferroic properties. In recent references, it is pointed out novel functionalities at oxide (multi)ferroic 2D states (Li et al., 2015), magnetoelectric coupling in multiferroic epitaxial thin film composites (Vaz, Hoffman, Ahn, & Ramesh, 2010) and self-assembled multiferroic nanocomposites (Kim, Ning, & Ross, 2019), and new oxide microcavities applications (Kalusniak, Sadofev, & Hennberger, 2014; Tyborski et al., 2015). The tremendous interest in interfaces between insulating oxides with different electrical polarizations has resulted in the study of the detailed atomic structure of the ferroelectric domain wall, electrostatic boundary conditions as well as the role of ionic point defects (Stemmer & Allen, 2018). These studies are directed at increasing the current density at the walls and manipulation of their position.

Many works have been conducted on the fundamental understanding, fabrication processes, and applications of magnetoelectric composite material systems, which has brought the technology closer to realization in practical devices (Srinivasan, Nan, Ramachandra Rao, & Sun, 2019). Efforts are focused on various electrically and magnetically cross-coupled devices, in particular, using exchange or strain-coupled heterostructures. Besides, biotechnological aspects (Madhumitha, Elango, & Roopan, 2016) and solar cell applications of ZnO-based structures (Boro, Gogoi, Rajbongshi, & Ramchiary, 2018) as well as a novel approach of oxide films epitaxy and bipolar oxide devices are discussed. There is increasing interest in microelectronics for oxides with a high dielectric constant (so-called high-k dielectrics) for use in integrated capacitors and different type transistors, in non-volatile and dynamic random access memories (DRAM). In this regard, different oxides such as Al_2O_3, HfO_2, ZrO_2, La_2O_3, and other rare-earth oxides, as well as their ternary and quaternary compositions have become to be very attractive for microelectronic application. At the same time, all these oxides have also unique optical and chemical properties that offer their application in photovoltaic and photonic devices. For instance, rare-earth-doped Al_2O_3 materials demonstrate intense light emission in specific spectral range and are attractive alternatives to rare-earth-doped Si-based materials because their crystalline structure allows higher solubility of rare-earth ions without their clustering at high content. Similar properties are demonstrated by rare-earth-doped HfO_2 and ZrO_2 materials. The fabrication of different waveguides for optical communication and frequency conversion layers (both down and up-conversion, down shifting) for solar cells were already demonstrated.

These materials, in powder or ceramic forms, along with TiO_2 are used for catalytic applications and water splitting. The development of solid oxide fuel cells and materials for hydrogen storage continues today, extending a hand for the elaboration of environmental friendly energetic solution.

Continuous down-scaling of electronic devices forces the elaboration of different nanomaterials that demonstrate high quality with tunable microstructure and morphology by means of cost-effective technology in comparison with traditional high-tech microelectronics fabrication. Among different oxides, wide bandgap SnO_2, TiO_2, and In_2O_3 and their solid solutions become key materials for photovoltaics, microelectronics, energy harvesting and storage, optoelectronic, catalysis, or gas sensing (Kumar, Kumari, Karthik, Sathish, & Shankar, 2017; Tran et al., 2017). In the particular case of In_2O_3, there is a substantial interest in developing devices with a low content of In, considered as a critical element (Critical Raw Materials, 2017), without losing its high performance characteristics, which also could be achieved by using nanomaterials. The combination of low-dimensional In_2O_3, SnO_2, or TiO_2 structures with organic materials such as graphene, carbon nanotubes, or conductive polymers results in a new generation of hybrid materials that are of low cost, easily processed and scalable, and show the ability to be used in flexible substrates.

Discovering new materials remains as the top priority in defining the future research directions. For example, room-temperature piezoforce microscopy and macroscopic polarization measurements have revealed that ε-Fe_2O_3 films display a typical ferroelectric switching with significant polarization (~1 $\mu C/cm^2$) and low switching voltages (Gich et al., 2014). It means that ε-Fe_2O_3 is a new and unique intrinsic room temperature multiferroic material that could find novel applications. New possibilities are opened by application of the graphene oxide, which is a unique material viewed as a single monomolecular layer of graphite with various oxygen-containing functionalities such as epoxide, carbonyl, carboxyl, and hydroxyl groups. It is worth pointing out also new non-van der Waals 2D transitional metal-based oxides – hematene and chromene. The nature of their magnetic ordering was investigated by Bandyopadhyay, Frey, Jariwala, and Shenoy (2019), who have shown that tuning the magnetic ordering in these materials controls the transport properties by modulating the band gap, which may be of use in spintronic or catalytic applications.

Integration of various oxide-based nanostructures, having specific electrical, optical, chemical, and magnetic properties as well as showing their biocompatibility, with widespread silicon-based CMOS technology allows production of novel multifunctional silicon-based structures and devices, with unique properties offering their application for energy storage, catalysis, microelectronic, photonic, and biomedical applications.

Discussing oxides, one cannot but say a few words about the methods of growth and study of these materials. To fully exploit the richness of oxide-based materials, it is paramount to control their synthesis with subnanometric scale precision. Well-known high-vacuum thin-film deposition techniques such as molecular beam epitaxy, pulsed laser deposition, and sputtering have rich opportunities. However, the continuous need for device miniaturization has led to a demand of even thinner films, uniform coatings, vertical processing, higher aspect ratios, and complex 3D structures with strict requirements for quality and functionality. Atomic layer deposition (ALD) is a low-temperature (<400°C) and low-vacuum (10^{-2} to 10 mbar) chemical gas-phase deposition technique that uniquely relies on the alternate pulsing of precursors, separate in time, that react with the surface in a self-limiting manner. Review of the rapid progress of atomic layer–deposited multicomponent oxides from doped to complex oxides focusing on single perovskite, spinel, delafossite, and scheelite structures and highlighting their potential technological applications was presented by Coll and Napari (2019). They say that the ALD method gives access to homogeneous ultrathin coatings not only on planar substrates but especially on complex features including nanorods and nanowires, nanoparticles, and mesoporous and nanoporous surfaces. ALD also enables the preparation of pinhole-free thin layers with atomically sharp interfaces for a wide variety of heterostructures offering, for example, unprecedented opportunities to study the novel physics arising at the oxide interface. ALD-grown films of wide band-gap high-k oxides as well as new applications in organic photovoltaics, electronics, and optoelectronics are demonstrated by Godlewski et al. (2017). A review of recent research and development advances in oxide TFTs fabricated by ALD processes is presented by Sheng, Park, Shong, and Park (2017).

Reducing the thickness of the layer down to a few unit cells is also interesting since novel properties can emerge at the interfaces between two oxides. The main approach to metastable compositions design and to generation of the novel electronic properties of the oxide films is to use strain engineering where little changes can drastically modify the electronic properties. It is possible, for example, to tune phase formation and preferred epitaxial orientation by film-substrate pairs tuning. Polycrystalline substrates are widely used for industrial coatings, but rarely used in epitaxial stabilization studies. Yet, they provide a large spread of surface orientations in a single experiment, and offer a high throughput method to study the role of orientation on film growth. Havelia et al. (2013) call this high-throughput method combinatorial substrate epitaxy, a combinatorial synthesis method in which a large number of template combinations are used to stabilize a precise composition in a specific structure.

And, finally, a few words about the research experimental techniques. Raman and photoluminescence spectroscopy investigations, ellipsometry and, in particular, spectroscopic ellipsometry are very capable optical methods used for oxides' characterization due to their numerous advantages. The high sensitivity of these methods combined with the non-destructive nature results in rapid and non-destructive qualification of the samples qualified as well as mapping on large surfaces. X-ray diffraction technique should also be noted, the *in-situ* application of which gives the possibility to follow the transformation of oxide during growth or deposition process. Besides, a combination of optical and X-ray methods with charge carriers transport and electron states density study techniques provides precise characterization of oxide-based materials. Local nanoscale domain structure, polarization switching, and piezoresponse in ferroelectric oxides can be studied by using AFM equipped with the piezoresponse force microscopy contact mode set up.

REFERENCES

Bandyopadhyay, A., Frey, N. C., Jariwala, D., & Shenoy, V. B. (2019). Engineering magnetic phases in two-dimensional non-van der Waals transition-metal oxides. *Nano Letters, 19*(11), 7793–7800.

Boro, B., Gogoi, B., Rajbongshi, B. M., & Ramchiary, A. (2018). Nano-structured TiO_2/ZnO nanocomposite for dye-sensitized solar cells application: A review. *Renewable and Sustainable Energy Reviews, 81*, 2264–2270.

Coll, M., & Napari, M. (2019). Atomic layer deposition of functional multicomponent oxides. *APL Materials, 7*(11), 110901.

Critical Raw Materials. (2017). Retrieved from http://criticalrawmaterials.org/european-commission-publishes-new-critical-raw-materials-list-27-crms-confirmed/.

Gich, M., Fina, I., Morelli, A., Sánchez, F., Alexe, M., Gàzquez, J., & Roig, A. (2014). Multiferroic Iron Oxide thin films at room temperature. *Advanced Materials, 26*(27), 4645–4652.

Godlewski, M., Pietruszka, R., Kaszewski, J., Witkowski, B. S., Gierałtowska, S., Wachnicki, Ł., … Gajewski, Z. (2017). Oxide-based materials by atomic layer deposition. *Proc SPIE 10105, Oxide-Based Materials and Devices, VIII*, 101050L.

Havelia, S., Wang, S., Balasubramaniam, K. R., Schultz, A. M., Rohrer, G. S., & Salvador, P. A. (2013). Combinatorial substrate epitaxy: A new approach to growth of complex metastable compounds. *CrystEngComm, 15*(27), 5434–5441.

Kalusniak, S., Sadofev, S., & Hennberger, F. (2014). ZnO as a tunable metal: New types of surface plasmon polaritons. *Physical Review Letters, 112*, 137401.

Kim, D. H., Ning, S., & Ross, C. A. (2019). Self-assembled multiferroic perovskite–spinel nanocomposite thin films: Epitaxial growth, templating and integration on silicon. *Journal of Materials Chemistry C, 7*, 9128–9148.

Kumar, D. P., Kumari, V. D., Karthik, M., Sathish, M., & Shankar, M. V. (2017). Shape dependence structural, optical and photocatalytic properties of TiO_2 nanocrystals for enhanced hydrogen production via glycerol reforming. *Solar Energy Materials and Solar Cells, 163*, 113–119.

Li, C., Liu, Z., Lü, W., Wang, X. R., Annadi, A., Huang, Z., … Venkatesan, T. (2015). Tailoring the two dimensional electron gas at polar $ABO_3/SrTiO_3$ interfaces for oxide electronics. *Scientific Reports, 5*, 13314.

Lorenz, M., Ramachandra Rao, M. S., Venkatesan, T., Fortunato, E., Barquinha, P., Branquinho, R., … Pryds, N. (2016). The 2016 oxide electronic materials and oxide interfaces roadmap. *Journal of Physics D: Applied Physics, 49*(43), 433001.

Madhumitha, G., Elango, G., & Roopan, S. M. (2016). Biotechnological aspects of ZnO nanoparticles: Overview on synthesis and its applications. *Applied Microbiology and Biotechnology, 100*(2), 571–581.

Sheng, J., Park, E. J., Shong, B., & Park, J. S. (2017). Atomic layer deposition of an indium gallium oxide thin film for thin-film transistor applications. *ACS Applied Materials & Interfaces, 9*(28), 23934–23940.

Srinivasan, G., Nan, C., Ramachandra Rao, M. S., & Sun, N. X. (2019). Special issue on magnetoelectrics and their applications. *Journal of Physics D Applied Physics, 52*(10), 100301.

Stemmer, S., & Allen, S. J. (2018). Non-Fermi liquids in oxide heterostructures. *Reports on Progress in Physics, 81*(6), 062502.

Tran, V. A., Truong, T. T., Phan, T. A. P., Nguyen, T. N., Van Huynh, T., Agresti, A., … Le, S. N. (2017). Application of nitrogen-doped TiO_2 nano-tubes in dye-sensitized solar cells. *Applied Surface Science, 399*, 515–522.

Tyborski, T., Kalusniak, S., Sadofev, S., Henneberger, F., Woerner, M., & Elsaesser, T. (2015). Ultrafast nonlinear response of bulk plasmons in highly doped ZnO layers. *Physical Review Letters, 115*(14), 147401.

Vaz, C. A. F., Hoffman, J., Ahn, C. H., & Ramesh, R. (2010). Magnetoelectric coupling effects in multiferroic complex oxide composite structures. *Advanced Materials, 22*, 2900–2918.

2 Optical Characterization of Oxide-Based Materials Using Ellipsometry

Peter Petrik

2.1 INTRODUCTION

Since ellipsometry measures surface and interface properties with sub-nanometer precision, an inherent requirement is that the surfaces and interfaces must be of optical quality, i.e. flat on the scale of nanometers or at most tens of nanometers. This is a challenge in many cases even today, but back to the beginnings of the method only very few surfaces could be investigated – they were typically polished metal surfaces (Rothen, 1945). As the technology for microelectronics and thin film preparation developed, the precision, speed, and non-destructive nature of ellipsometry made sense, and it could be applied very effectively for the testing and the development of materials and structures for microelectronics (Irene, 1993). Since the energies of the electronic transition in silicon and other semiconductors are in the photon energy range that could easily be covered by standard instrumentation, ellipsometry was proved to be a powerful tool for structural characterization of semiconductors (Petrik et al., 2014) except the first applications being the measurement of thickness of dielectric materials, such as the thickness of thermal oxide layers on silicon wafers (Irene, 2006).

For ellipsometry, as for any measurement method, the development of evaluation and instrumentation goes hand in hand. In case of optical modeling, the direction of development in ellipsometry is from well-defined dielectric layers with simple dispersions toward complex systems both in terms of dispersion models (semiconductors with multiple oscillators; see for example Petrik et al. (2009)) and layer structures (inhomogeneous multi-layer structures (Petrik et al., 2005)). Because of the versatile structures and properties of oxides, the modeling of all major categories of the dielectric functions will be discussed from dielectric to semiconducting features (Fried et al., 2011; Petrik et al., 2001).

In case of instrumentation, the main driving forces are the capabilities to measure in a broader wavelength range (Petrik, et al., 2019), in different environments (vacuum chamber (Collins, 1990) or flow cells (Kalas, et al., 2019)), to measure anisotropic materials or large surfaces (Fried et al., 2011; Shan et al., 2014).

In this chapter, the aspects of modeling and instrumentation will be discussed for oxide-based materials. Since the field is already huge and covers a broad range of characteristics (dielectrics, semiconductors or even metallic properties – as for transparent conductive oxides (TCOs)), this summary is far from being comprehensive, only discussing some important issues, mainly demonstrated by our own results, for those who are interested in oxides but who are no experts of ellipsometry.

2.2 OPTICAL CHARACTERIZATION BY ELLIPSOMETRY

Although the method of ellipsometry has been known since the end of the 19th century (Drude, 1889), the word "ellipsometer" has only been coined in the middle of the 20th century (Rothen, 1945), and the first major applications of ellipsometry are connected to the microelectronics technology (Irene, 1993) later on. The number of publications in ellipsometry increased rapidly in the years from ≈1980 to ≈2000, reaching a kind of saturation by the year of 2010, as shown, for example, in Figure 2.2 of Kalas et al. (2019). At the beginning, most publications have dealt with the development of the method, which gradually turned to a situation in which most of the publications are about the application of ellipsometry, rather than its development. Of course, it is not because the activities on development decreased, but because the number of papers on the application increased rapidly.

Compared to the numerous spectroscopic measurement methods, ellipsometry has some important features. (i) First of all, it doesn't require references, because it compares the reflection of two different polarizations of the same beam. Therefore, it is also very robust in terms of instrumentation, because it is not sensitive to intensity fluctuations from the lamp or vibrations. (ii) In spite all of these factors, it features a sub-nanometer sensitivity that can even be achieved using simple optical components like regular lamps and film polarizers. (iii) But most importantly, ellipsometry provides many data (e.g. as a function of wavelength and/or angles of incidence) which can be fitted by optical models. This fact ensures fully quantitative analysis as soon as proper optical models exist for the measured system. Consequently, although it is an indirect method, it is highly sensitive, quantitative, robust, fast, simple – in terms of instrumentation – and non-destructive. The latter feature is very important, because with proper hardware extensions, such as vacuum chambers (Fried et al., 2011), flow cells (Kalas, et al., 2017) or environmental cells (Petrik et al., 2019), real-time measurements can be performed with high sensitivity and good time resolution.

The literature on the theory and applications of ellipsometry is already huge, covering most areas of the field from the mathematical basics (Azzam & Bashara, 1977) and applications (Tompkins & Irene, 2005 or Fried, et al., (2001)) to a comprehensive practical guide with many excellent graphs and visualizations (Fujiwara, 2007). Therefore, in this chapter only some "oxides-related" aspects of hardware and evaluation will briefly be discussed, primarily based on first-hand experiences.

2.2.1 HARDWARE

The general components and working principle of an ellipsometer is shown in the example of a mapping ellipsometer in Figure 2.1 (Fried et al., 2011). The minimum requirement is

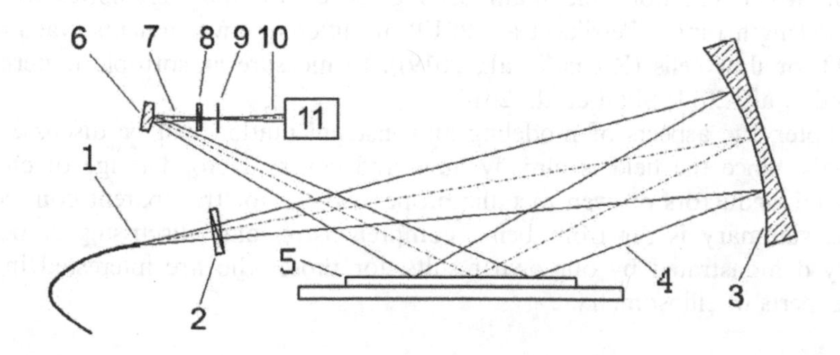

FIGURE 2.1 Beam path of a divergent light source mapping ellipsometer using components of a (1) light source, (2) film polarizer, (3) spherical mirror, (4) convergent beam, (5) sample, (6) cylindrical mirror, (7) corrected beam, (8) analyzer, (9) pinhole, (10) divergent beam, (11) spectrograph.

a light source, two polarizers, and a detector. The light source doesn't have to be coherent. One of the polarizers is in the beam path of the light source, serving as a polarization state modulator, sometimes combined with an additional compensator that changes the phase shift (Lee, Rovira, An, & Collins, 1998). The light reflected from the sample changes its polarization state, which can be measured by having another polarizer (called as analyzer in this case) in the detector path, sometimes also combined with a compensator to change the phase shift, increase the sensitivity, and allow measuring all the Muller matrix parameters. There are many configurations from null ellipsometry to rotating component ellipsometry (Aspnes & Studna, 1975; Collins, 1990), the latter of which utilizes one or two rotating polarizers or compensators. The most advanced Muller matrix ellipsometry uses simultaneously rotating compensators in both arms of the ellipsometer to allow full Muller matrix determination (Hauge, 1978; Lee et al., 2001). This kind of modulation ellipsometry results in a sinusoidal intensity signal in the detector. The analysis of this line shape provides the measured ellipsometric angles or elements of the Muller matrix (Fujiwara, 2007).

Our recent development of high-speed mapping ellipsometry (Figure 2.1) is only one of the many instrumental possibilities. There are also micro imaging ellipsometry (Wurstbauer et al., 2010), versatile *in situ* configurations in vacuum chambers (Cobet et al., 2018; Lee et al., 1998; Mikhaylova et al., 2007; Petrik et al., 2001), flow cells (Arwin, 2005; Kalas et al., 2017; Kozma et al., 2011; Poksinski & Arwin, 2004), or environmental cells (Petrik et al., 2019; Welch et al., 2017). The mapping or macro imaging configuration in Figure 2.1 also has numerous versions with or without a mirror after the sample (see also Figure 2.2), a line-scan spectroscopic version and many configurations to integrate in vacuum chambers or in a roll-to-roll environment. The advantage is (as in multichannel ellipsometry – see Collins et al. (1998)) that utilizing a detector matrix, both the spectral and lateral points can be measured simultaneously, by analyzing the line shapes of the signal at each pixel of the detector, replacing a one-by-one measurement with a simultaneous acquisition. In this way, the measurement speed of a typical (30-points by 30-points) matrix can be reduced from minutes to seconds, revealing an orders of magnitude improvement in terms of speed – but still maintaining a sensitivity of thickness in the range of nanometers.

One of the most important advantages of ellipsometry is its speed and non-destructive character, which results in applications such as *in situ* ellipsometry (Collins et al., 2000; Tompkins & Irene, 2005). First, the integration of ellipsometry in vacuum chambers was a major *in situ* application, but later, bioellipsometry (Arwin, 2005; Kalas et al., 2019), electrochemical *in situ* ellipsometry, and ellipsometry in environmental cells gained increasing interest. The reason is that the sub-nanometer sensitivity is usually possible even at high acquisition speeds (seconds or even better). Advanced configurations of environmental cells (Figure 2.3) even allow multiple angles of incidence or mapping measurements on heated samples and in controlled ambient conditions (both liquid and gas are possible).

Finally, besides sophisticated configurations, many developments of ellipsometry expand the wavelength range from the shortest ones (synchrotron – BESSY) through vacuum UV, near UV, and extending the visible range towards the near infrared, mid infrared and terahertz regions. A crucial problem in these fields is the availability of reliable references for the refractive indices. In case of the *in situ* ellipsometry during annealing, the problem is even larger due to the need of high temperature references. There are databases provided by hardware manufacturers; however, global databases are currently lacking.

2.2.2 Evaluation

From the analysis of the sinusoidal function generated on the detector as a function of the angle of rotating element(s) (polarizer, analyzer or compensator), the change of polarization

FIGURE 2.2 Prototype of a roll-to-roll compatible mapping ellipsometer using a configuration schematically plotted in Figure 2.1. The optical components of the ellipsometer are located on the left-hand side of the picture, except for the mirror on the right-hand side at the back, partly covered by an iron block. The mirror serves for imaging the divergent source on the pinhole camera and spectrograph, both located at the right-hand side. The apparatus only uses the central line of the image for mapping – the other index on the CCD detector is used for spectral information. This means that spectroscopic mapping can be done by moving the sample (i.e. the line with spectroscopic information in each point). The advantage of this arrangement is that it can easily be integrated in a vacuum chamber with only the mirror located in the chamber – all optical components are outside.

as a result of the reflection from the sample surface is calculated, described by the complex reflection coefficient

$$\rho = \tan \Psi e^{i\Delta} \equiv \frac{r_p}{r_s} \tag{1}$$

where ψ and Δ are the ellipsometry angles related to the amplitude and the phase, respectively, whereas r_p and r_s denote the reflection coefficients of light polarized parallel and perpendicular to the plane of incidence. The evaluation is only possible if one has an idea about the structure of the sample. It is required in order to construct an optical model that is good enough to supply a set of initial parameters from which spectra of ψ and Δ can be calculated. They should be close enough to the measured one, so that a proper parameter search or gradient fit algorithm can find suitable model parameters by minimizing the difference between the measured and the fitted spectra.

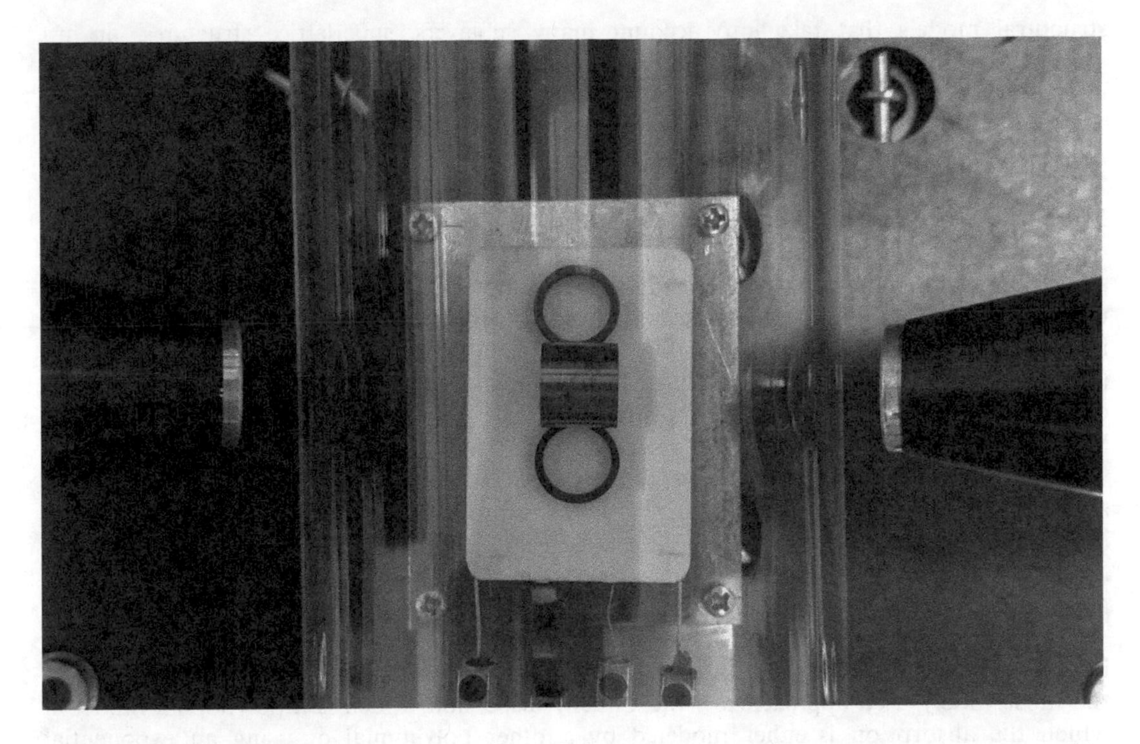

FIGURE 2.3 Environmental cell to study high-temperature processes using real-time ellipsometry in controlled ambient temperatures up to 600 °C.

There are three major evaluation issues of ellipsometry: (1) calibration of the instrument, (2) modeling the structure in terms of layers, lateral and vertical features, components, and phases, and (3) modeling the wavelength dependence of the refractive index of layers and components. The most important task of instrument calibration is the determination of the offsets (optical zero position regarding the plane of incidence) or the rotating components, and the polarization dependence of either the light source or the detector, depending on whether the source or detection part contains the rotating element. The structural modeling is getting more and more relevant as samples of ever-increasing complexity can be studied – due to the increasing computational power and experiences in complex modeling using many model parameters. The modeling of the dispersion (wavelength dependence of the refractive index) is mainly the materials science part of ellipsometry. From these parameters numerous material properties can indirectly be detected, such as the band structure, crystallinity, conductance, carrier density, physical density, porosity, band gap, and many more, limited only by the physical models, the wavelength region, and the sensitivity in the given optical configuration. Most importantly, the dispersion is a fingerprint of the materials in terms of whether it is a conductor, a semiconductor, or an isolator. Since the properties of oxides cover all of these categories, the modeling of these will briefly be discussed below.

2.3 UNIFORM STRUCTURES

From the three major evaluation issues of calibration, structural models, and dispersion, the two latter ones will be discussed in the next sections, regarded as the most important aspects of ellipsometry on complex oxides. First, we consider structurally uniform materials, and only deal with the dispersion. Then these dispersions will be included in more complex structural models that take into account many sublayers, superlattice structures, or just unknown vertical non-uniformities, multicomponent materials, and other complexities in the lateral and vertical structure.

2.3.1 Models for Dielectric Oxides

Most oxides are dielectric materials. At the dawn of microelectronics, the primary application of ellipsometry was the accurate measurement of the SiO_2 layer on silicon wafers. The accuracy, speed, and non-destructive capabilities were unprecedented, and ellipsometry has been used for quality checking and structural analysis for microelectronics until today – now used for many more kinds of oxides and thin films. The refractive index of thermally grown SiO_2 (Paulson, et al., 2002) is one of the most accurate references besides the single-crystalline silicon even today, thc latter being the most studied and cited of all optical references (Aspnes & Studna, 1983; Humlíček & Šik, 2015; Jellison, 1992; Paulson et al., 2002).

In the case of thermal oxide studies, the refractive index of oxide can usually be taken from a database or literature reference, because it is very reproducible. However, this is unfortunately exceptional, because most thin film properties largely depend on the parameters of the preparation. Even for thermal oxide, the properties deviate from the bulk value when its thickness is approaching the low-dimensional case, as discussed below (in section 2.4.2).

Being wide band gap materials, the oxides can be described by simple dispersion models in the visible wavelength range used by the majority of commercial ellipsometers. One of the most widely used approaches is the Cauchy dispersion, which is a polynomial model in which the absorption is either modeled by another polynomial or using an exponential dependence on the photon energy for the absorption. The real part of the refractive index (n) is described by the equation

$$n_C(\lambda) = A_C + \frac{Bc^2}{\lambda^2} + \frac{Cc^4}{\lambda^4} \tag{2}$$

where A_C, B_C and C_C are the Cauchy parameters, and λ denotes the wavelength. The extinction coefficient (k) is usually described by an exponential function:

$$k_C(E) = A_k e^{B_k(E-E_g)} \tag{3}$$

where A_k and B_k are the fitted parameters. In order to obtain physically relevant values, the real and imaginary parts of the optical functions must be Kramers-Kronig consistent. Consequently, the real part of the dielectric function ($\varepsilon_1 = n^2 - k^2$) must satisfy the following equation:

$$\varepsilon_1(E) = \frac{2}{\pi} P \int_{E_g}^{\infty} \frac{\xi \varepsilon_2(\xi)}{\xi^2 - E^2} d\xi \tag{4}$$

where ε_2 denotes the imaginary part of the dielectric function. The Cauchy equations are not Kramers-Kronig (KK) consistent, and therefore, in many cases rather the KK-consistent Sellmeier equation is used:

$$ns^2(\lambda) = \varepsilon_\infty + \frac{A_s \lambda^2}{\lambda^2 - B_s} \tag{5}$$

where A_S, B_S, and ε_∞ are the three fit parameters of the Sellmeier model.

Figure 2.4 shows an example of using the Cauchy dispersion for the determination of the refractive index of a $Ba_xSr_{1-x}TiO_3$ (BST) layer prepared by metal organic chemical vapor deposition (MOCVD). *k* was here zero for most deposited films, and it revealed a non-zero value only for lower quality layers.

The fact that the dispersion of the refractive index can be described by simple polynomials in certain wavelength ranges can be used for the determination of the layer thickness and to use this value as a fixed parameter later for more complex optical models in the wavelength range that shows absorption and a more complex dispersion. In this case, the accurate knowledge of the layer thickness is very helpful for a successful determination of the optical properties using B-Spline models or model-independent point-by-point inversions (Agocs et al., 2014; Collins & Ferlauto, 2005).

ZrO_2 was also modeled by the Cauchy dispersion in both for sputtered layers and for oxidized Zr surfaces for nuclear applications. The latter was also an environmental ellipsometry study (Figure 2.5). The comparison shown here demonstrates very well the difference between the optical properties of the same material but using different preparation methods. The anodic oxide and the cubic zirconium oxide had the highest refractive index, the thermal oxidation being the next, and the sputtered layers the lowest. This behavior also depends on the

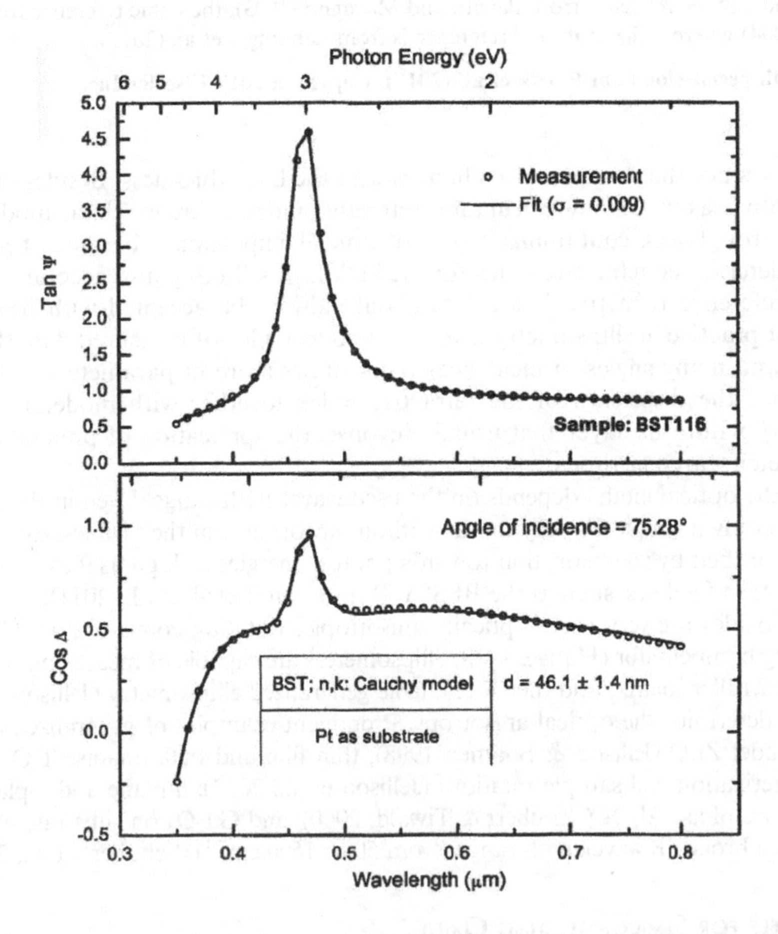

FIGURE 2.4 Fitted tan Ψ and cos Δ spectra using the Cauchy model shown in the inset for the dispersion of a BST layer.

Reproduced with permission from Petrik, et al. (2002). Copyright 2002 Elsevier Inc.

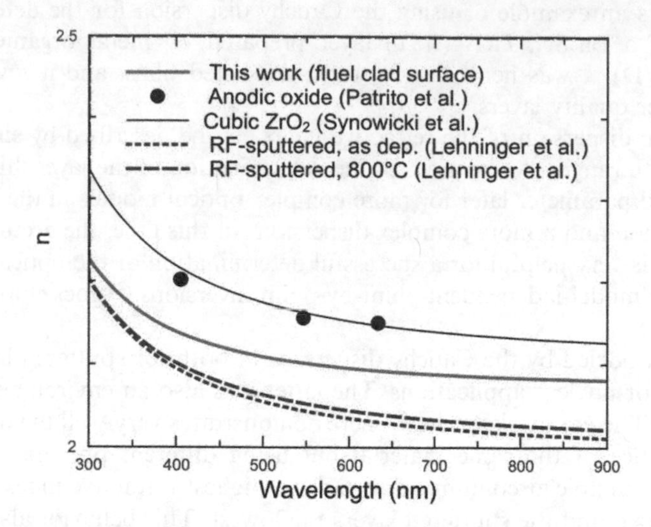

FIGURE 2.5 ZrO_2 references from different kinds of layers. The thicker solid line is from Petrik et al. (2017), the anodic oxide reference from Patrito and Macagno (1993), the cubic reference from Synowicki and Tiwald (2004) whereas the sputtered reference is from Lehninger et al. (2015).

layer thickness, since the structure also changes with the layer thickness. Besides other factors, this is the main reason that by calculating refractive index reference data, modeling of the surface (nano-roughness, contamination) is of crucial importance, because it substantially modifies the determined refractive index for the bulk layers. Taking into account all the above experiences, reference refractive index data should always be accepted with healthy skepticism. The best practice in ellipsometry is to try to collect a lot of measured data (broad wavelength range and many angles of incidence). This allows more fit parameters to be used and to parameterize the dispersion of the refractive index together with modeling the vertical non-uniformity within the layer that usually involves the application of preferably advanced global parameter search methods.

Naturally, the optical model depends on the used wavelength range. Even in the case of SiO_2 which requires only a simple Cauchy model without absorption in the visible-near-UV range, it needs to be described by an absorption towards photon energies as high as 9 eV, usually accessible at synchrotron facilities, such as the BESSY II in Berlin (Petrik et al., 2011).

Crystalline oxides are very often optically anisotropic. Rotating compensator (Collins, 1990) or dual rotating compensator (Hauge, 1978) ellipsometers are capable of measuring multiple components of the Muller matrix, and therefore enable generalized ellipsometry (Jellison, 2004; Schubert, 1998) to determine the optical anisotropy. Prominent examples of generalized ellipsometry on oxides include: ZnO (Jellison & Boatner, 1998); thin film and bulk anatase TiO_2 (using Sellmeier parameterization and sample rotation) (Jellison et al., 2003); *a*-plane and *c*-plane one-side polished sapphire phase Al_2O_3 (Schubert & Tiwald, 2000); and Ga_2O_3 on substrate with different orientations in a broad IR wavelength range (from 50 to 1500 cm^{-1}) (Schubert et al., 2016).

2.3.2 Models for Semiconducting Oxides

Once the band gap becomes lower, reaching the UV-Vis photon energy region accessible by regular ellipsometry, the above model for the description of dielectric oxides gets more complicated. If only the onset region of absorption close to the gap energy is in the focus of the study, applications of the Tauc or Cody approaches dominate to describe the photon energy dependence of the absorption (typically applied for the imaginary part of the dielectric function). In higher energy regions, where interband transitions take place, the descriptions become even more complicated (see section 2.4.2). In many cases, the Tauc gap parameterization is only extended by a couple of Lorentz oscillators in the high photon energy region.

One of the most widely used approaches to parameterize the dielectric function of the gap region is a combination of the Tauc expression for the imaginary part of the dielectric function for photon energies above the band edge

$$\varepsilon_{2T}(E) = A_T + \frac{(E - E_g)^2}{E^2} \tag{6}$$

with A_T and E_g being the amplitude and the optical band gap, and a Lorentz oscillator of

$$\varepsilon_{2L}(E) = 2nk = \frac{A_L E_0 C E}{\left(E^2 - E_0^2\right)^2 + C^2 E^2} \tag{7}$$

which, after combining Eqs (5) and (6), results in the expression of

$$\varepsilon_{2TL}(E) = \frac{A_L E_0 C (E - E_g)^2}{\left(E^2 - E_0^2\right)^2 + C^2 E^2} \frac{1}{E} \tag{8}$$

if $E_g > 0$, otherwise $\varepsilon_2 = 0$, with E_0 and C being the peak transition energy and the broadening, respectively, each parameter having the dimension of eV. This approach is called the Tauc-Lorentz (TL) oscillator model (Jellison & Modine, 1996), which among others has two great features. The first one is that to be Kramers-Kronig consistent, there is an analytical formula for ε_1, instead of using the numerical integration through Eq. (3). The use of this formula (although it is almost an entire page in the article) makes the calculation very fast, much faster than the numerical integration. The other important feature is that it provides a numerical value of E_g right from the fit, so the Tauc plot doesn't have to be applied. The TL model is mainly used for amorphous semiconductors, and it can be best used at photon energies around the optical gap.

The TL model has later been improved by Ferlauto and coworkers (2002) to add an Urbach tail to the Tauc gap expression. Here, the photon energy region near the optical gap is already divided in three parts, each having its own photon energy-dependent function for the imaginary part of the dielectric function. Very importantly, the analytical formula for the real part of the dielectric function is provided also in this case (similar to that for the TL model), enabling again the fast calculation.

One of the most investigated semiconducting oxides is ZnO, the properties of which also depend very much on the deposition conditions. The interband transitions can be described by oscillators, including excitonic effects (Major, Juhász, et al., 2009). This way, the method is sensitive to the electrical properties (conducting, semiconducting) and therefore suitable for a quick and reliable qualification of the materials. The Drude tail analysis towards the infrared wavelength region is a method to detect free carriers in the layer, and therefore correlates very well with the conductance (Németh et al., 2008).

In general, since the imaginary part of the dielectric function (which can be measured with an accuracy of $\approx 10^{-4}$) is proportional to the joint density of states that is closely related to the crystal structure, ellipsometry provides sensitive indirect information on the crystallinity and related characteristics, such as the conductance (Németh et al., 2008). To build optical models for semiconductors, in general, it is a useful approach to fit the spectra first in the photon energy region below the band gap using a few-parameter dispersion model (see, for example, the Cauchy model of Eq. (1)) and determine the layer thickness and the below-gap properties. Because now the thickness is known, the B-Spline, point-by-point, or other parametric models can be applied with a higher reliability (Figure 2.6), especially when checking the "semiconductor criterion," i.e. choosing a thickness for which the imaginary part of the dielectric function smoothly approaches zero towards the low photon energies around and below the band gap (Agocs et al., 2014; Arwin & Aspnes, 1984; Aspnes et al., 1984).

It is important to point out that the result very often depends on the photon energy range used for the fit (see Figure 2.7). The obvious reason for the increasing deviations is that lowering the upper limit of the photon energy range, the part of the exciton peak involved in the fit decreases, and the sensitivity of the Tauc-Lorentz parameters responsible for this oscillator decreases. However, usually there is the second, less obvious reason. At the scale of ellipsometric sensitivity, most thin films are vertically non-uniform. Most of the grain growth mechanisms result in evolving and vertically changing structures with nucleation layers, V-shaped grains, surface nanoroughness, etc. On the other hand, in the above-gap region of the spectrum, the light penetration depth in semiconductors can be very small (even in the range of 10 nm or below) due to the large absorption. Therefore, by decreasing the upper limit of the photon energy, the penetration depth increases and results in a kind of "depth scan," i.e. variation of the photon energy range changes the depth region of the layer used in the fit. In perfectly uniform films, this wouldn't cause a deviation; however, most thin films reveal this effect due to the vertical gradient in the optical properties. This "penetration depth effect" is not only a disadvantage. In case of proper modeling, it can be used to gain more information from the films, and to achieve a model-independent depth scan by scanning the wavelength range (Petrik et al., 2014).

2.3.3 Models for Conducting Oxides

Transparent conducting oxides have been key materials in microelectronics with indium-tin-oxide as the most important one, and with further efforts to replace it, for example, by ZnO

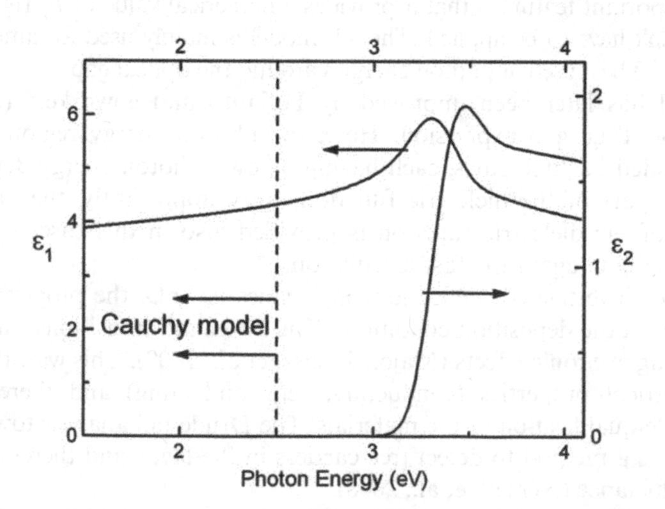

FIGURE 2.6 Dielectric function of a 50 nm thick ZnO layer prepared by sputtering and annealing. The range below 1.5 eV was described by the Cauchy model, whereas that above 1.5 eV by the B-Spline parameterization.

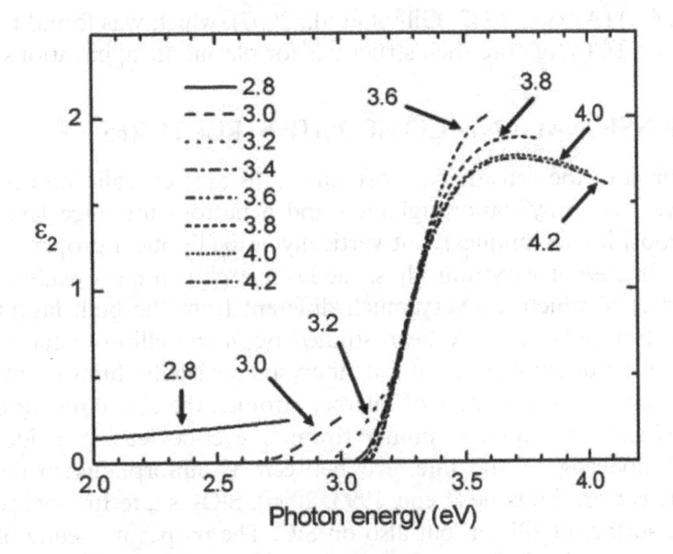

FIGURE 2.7 Imaginary part of the dielectric function determined by fitting the Tauc–Lorentz parametric model to the measured data using different wavelength ranges. The legend shows the upper limits of the range in eV. The lower limit was 0.7 eV in each case.

with optimized deposition and doping. The conductance can indirectly be measured by the analysis of the Drude tail in combination with Lorentz oscillators:

$$\varepsilon(E) = 1 - \frac{A_0^2}{E(E - i\nu_0)} + \sum_{j=1}^{N} \frac{A_j^2}{E_j^2 - E(E - i\nu_j)} \tag{9}$$

where A_0 and ν_0 are the amplitude and broadening parameter of the Drude term and A_j, ν_0, and E_j are the amplitude, broadening, and center energy of the Lorentz oscillators, respectively (Roeder et al., 2008). The combination of a Cauchy model and a Lorentz oscillator model is sufficient to properly describe both the visible and the infrared range (Synowicki, 1998). It has been shown that the near gap absorption features correlate well with the electrical properties using either the Cauchy model or Lorentzian line shapes (Major, Nemeth, et al., 2009; Németh et al., 2008). More specifically, there was a good correlation between the specific resistance and the gap energy, as well as between the specific resistance and the exciton strength (related to the absorption features in the UV photon energy range) using the parameterization published in Yoshikawa and Adachi (1997).

In case of indium tin oxide (ITO) not only the transparency can be utilized in the visible wavelength range, but also the infrared range for plasmonic effects (Rhodes et al., 2008). Utilizing the Drude dispersion and measurements in the Kretschmann geometry, it was shown that the optimum thickness for surface plasmon polaritons was 160 nm in ITO, and the thickness of the ITO layer can be changed in a much broader range than in case of gold plasmonics (Kalas et al., 2017), modifying the resonant photon energy position with the layer thickness significantly.

Plasmonics has also been demonstrated for low-dimensional metal–metal oxide structures utilizing a transparent conductive oxide (TCO) layer (Khosroabadi et al., 2014). Here, the Drude-Lorentz model was combined with the generalized Maxwell-Garnett effective medium approximation (EMA) (Aspnes, 1982; Gilliot et al., 2007), which was found to be most suitable for the description of TCO/Ag core shell structures for plasmonic applications.

2.4 LOW-DIMENSIONAL AND COMPOSITE STRUCTURES

On the scale of ellipsometric sensitivity, most thin films are vertically non-uniform, having at least a surface layer (typically nanoroughness) and a bottom interface layer (e.g. nucleation layer in case of deposition techniques), but vertically graded optical properties within the films aren't rare as well. In case of very thin films, the layer itself can be considered as an interface, the optical properties of which are very much different from the bulk layers. This effect and the formation of interface layers have been studied by *in situ* ellipsometry already in the late 1980s for the deposition of amorphous and nanocrystalline silicon films (Nguyen et al., 1994).

Also, in case of one of key materials of microelectronics, the above mentioned SiO_2, the direction of technological development points towards ever-decreasing oxide thicknesses, for which the proper modeling of the interface between an amorphous oxide and the single-crystalline substrate is a crucial issue (Irene, 1993, 2006). SiO_2 is a technologically critical material not only on the surface of silicon, but also on SiC. The proper modeling of ultra-thin oxide is of primary importance for SiC as well (Petrik, Szilágyi, et al., 2009). Ellipsometry is also actively used for the characterization of two-dimensional oxide layers or layer stacks (Zheng et al., 2017). However, the sample preparation, the interface, and interpretation issues largely influence the reliability of the data, as shown in the example of graphene (Pápa et al., 2017).

The second subtopic of this section will be the case of a 3D matrix with low-dimensional inclusions, in which the oxides are either the host materials (e.g. embedded low-dimensional semiconductor nanoparticles) or the inclusions (e.g. for changing the refractive index of the layers). The applications of semiconductor nanocrystals in dielectric matrices cover a broad range starting from silicon in silicon oxide (Agocs et al., 2011) or silicon nitride (Basa et al., 2007) for memory application to germanium in oxides such as HfO_2 and ZrO_2 (Agocs et al., 2017).

2.4.1 Low-Dimensional Oxide Structures

Oxide nanodots can be embedded in thin films in order to modulate the refractive index. In case the size of inclusions is much smaller than the wavelength of measuring light, the EMA can be used. If both components are dielectrics, such as in the case of TiO_2 nanoparticles in polyimide (Petrik et al., 2012), simple polynomial models (e.g. Cauchy dispersion) can be applied, only fitting the imperfections at the interfaces, or the graded refractive index within the layer.

In the TiO_2/polyimide system (Petrik et al., 2012), the refractive index of the composite (at the wavelength of 633 nm) can be varied from 1.65 to 1.80 when increasing the volume fraction of TiO_2 nanoparticles from 0 to 60%. The intrinsic refractive indices of the components are 1.65 and 2.5. Even at the wavelength of 1300 nm (important for waveguide applications), the variation with the composition is close to 0.1. However, as expected, although the composite layer created by doctor blading was very homogeneous on electron microscopy images with uniformly distributed nanoparticles (average size of approximately 10 nm), the fit result of ellipsometry was much better when assuming a thin layer at the interface between the composite layer and the silicon substrate. Nevertheless, the layer quality was very good, taking into account that doctor blading is a roll-to-roll compatible, no-vacuum, low-cost thin film preparation technique being widely used in the industry. In this approach, the solution of the film material is placed in front of the blade, and the blade is moved over the sample to produce a layer behind the blade. The parameters of the film can be controlled among others by the concentration, the speed of the blade, the gap between the blade and the substrate as well as the temperature.

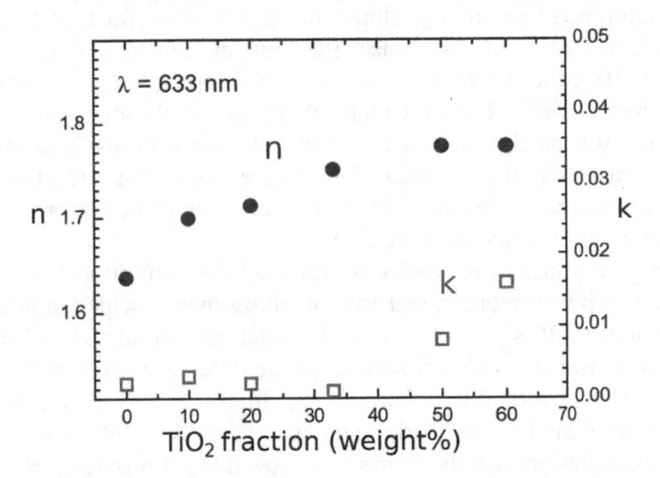

FIGURE 2.8 Refractive indices and extinction coefficients measured by ellipsometry on approximately 300 nm thick polyimide layers with embedded 10-nm size TiO_2 nanoparticles up to 60 wt% TiO_2, at the wavelength of 633 nm.

Besides polyimide, industrial prototype epoxy polymer can also be used as host material for TiO_2 nanoparticles of sizes below 10 nm (Landwehr et al., 2014) targeting applications such as the high throughput production of micro-optical elements (Figure 2.9). Due to the small size of the components, the effective refractive index of the layer can also be described by a simple Cauchy dispersion with the imaginary part of the refractive index set to k=0, meaning a low absorption of the composite. Since the refractive index of the starting material is lower here, and the volume fraction of TiO_2 was only increased to the value of 23 wt%, the refractive index could be varied in the range from 1.54 to 1.63. In case of this material, an important feature was the stability and low shrinkage at high temperatures (shown up to 220°C).

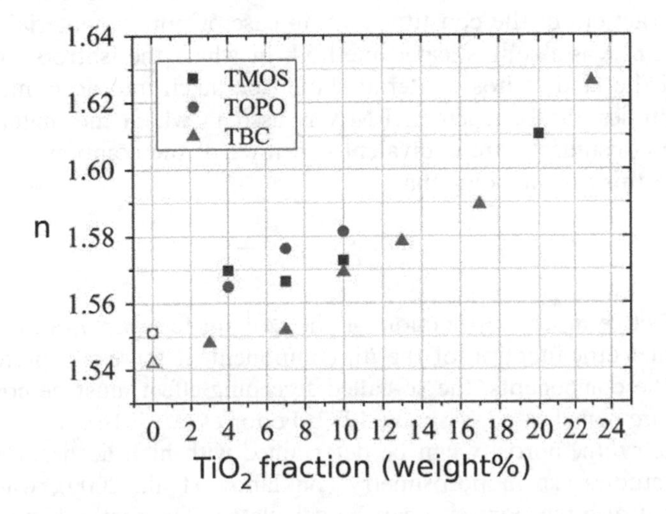

FIGURE 2.9 Refractive index at the wavelength of 635 nm of KATIOBOND OM VE 1107/07 composite containing with trimethoxy(7-octen-1-yl)silane (TMOS), trioctylphosphine oxide (TOPO) or 4-tert-butylcatechol (TBC) stabilized TiO_2 nanoparticles in weight fractions ranging from 0% (pure organic resist) to 23%. Layer thickness ranges from 100 to 300 nm.

Due to its sub-nanometer sensitivity, ellipsometry is a powerful tool for the measurement of low-dimensional, ultrathin layers. Since the optical constants can be measured with a precision down to 10^{-4}, the deviation of the optical properties from that of the bulk value can sensitively be characterized. The first high-quality and high-importance oxide investigated by ellipsometry was SiO_2 on Si surfaces for electronic applications. The models in this case (and later on SiC) can be based on tabulated references, since the refractive index of thermal and native oxide is dense and reproducible (this is an important feature that makes silicon-based microelectronics technology so powerful).

The thickness of gate oxides in microelectronics has been continuously decreasing, requiring more and more the sub-nanometer precision of ellipsometry, which holds even in case of embedded layers. Figure 2.10 shows the optical model and simulation with and without the oxide layer. The high sensitivity is clearly seen from the difference between the simulated curves, especially in the region of larger wavelength, in which the polysilicon layer on the top is transparent. In this application, the oxide layer was used as a diffusion barrier that doesn't completely hinder the diffusion; it only allows a shallow dopant profile after a high temperature diffusion process for bipolar transistors. Consequently, the thickness of this barrier oxide was only a few nanometers. However, in spite of this challenge, its thickness could be measured with high accuracy, depending on the crystallinity of the silicon layer on top, as it influences the absorption, and the range of wavelengths that can reach the embedded oxide.

The tabulated oxide reference for low-dimensional films, however, requires density correction in many cases. A prominent example is the ultrathin oxide on the surface of native and oxidized SiC, which also depends on the termination of the hexagonal 6H-SiC surfaces. Model 1 of Figure 2.11 from Petrik, et al. (2009) applies a Bruggeman EMA model (Aspnes, 1982; Bruggeman, 1935) using "voids" as a second component and also a transition layer (Model 2) that significantly improves the fit quality when dealing with ultra-thin layers. For most oxides, the Cauchy or Sellmeier dispersion has to be combined with a transition layer model, in which the transition layer is usually an effective medium with components using the optical functions of the layers above and below the transition layer.

Density correction is a major issue for porous films (Fodor et al., 2016; Fried et al., 1996; Petrik, Fried, et al., 2009). In these cases, the effective dielectric function is in almost all cases calculated using EMA. The reason is that EMA provides a quantitative result in terms of volume fractions of the constituents – in case of porous materials it is the volume fraction of void. EMA is a sophisticated method, in which the isotropy of the boundaries of the phases and the kind of host material should be taken into account. However, in the majority of the studies, the Bruggeman EMA is used in which the materials are spatially isotropic, and the constituents are equivalent – neither of the components is considered as a host material. In this case, it holds that

$$0 = \sum_{i}^{N} f_i \frac{\varepsilon_i - \varepsilon}{\varepsilon_i + 2_\varepsilon} \tag{10}$$

where ε is the effective dielectric function of the composite layer; f_i and ε_i denote volume fraction and the dielectric function of the ith component. If there is a preferential direction of the phases of the components, the so-called screening effect must be considered and the evaluations get more complicated (Aspnes, 1982; Fodor et al., 2016).

Using ellipsometry, the porosity can be determined with high accuracy that can be used for sophisticated studies like in porosimetry (Baklanov, et al., 2000), where not only the volume fraction but also the pore size can be calculated. The method can also be used for the study of the saturation of pores with water or other gases and, for example, for the degradation of glasses (Ngo et al., 2018).

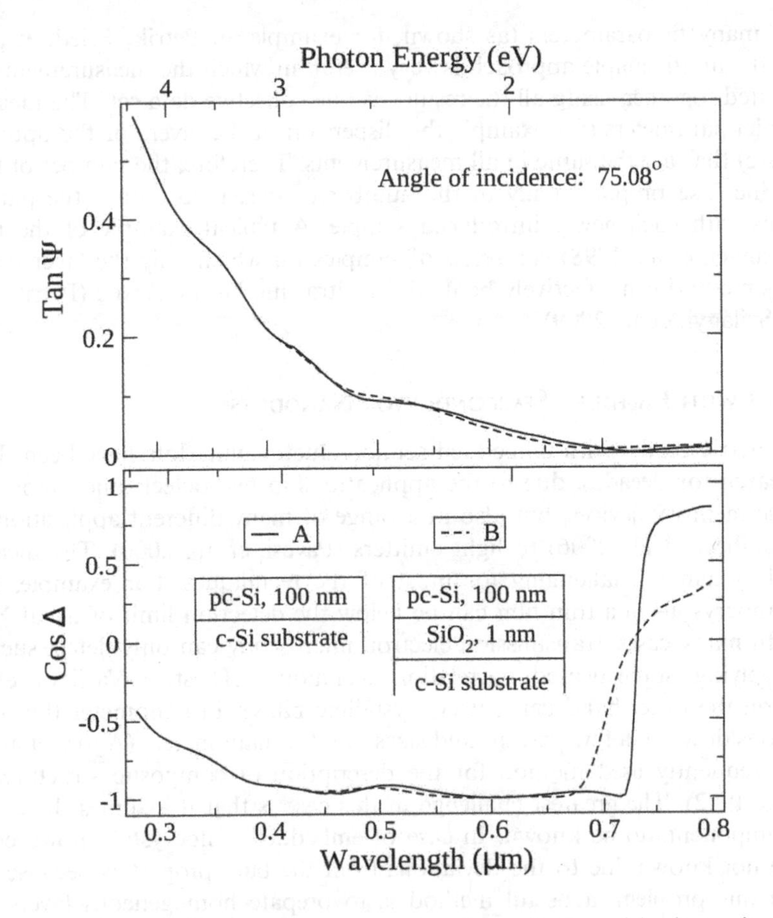

FIGURE 2.10 Simulation of ellipsometry spectra using the optical models shown in the inset. The dielectric function of pc-Si was taken from Jellison, et al., (1993).

Reproduced with permission from Petrik, et al. (2002). Copyright 2002 American Institute of Physics.

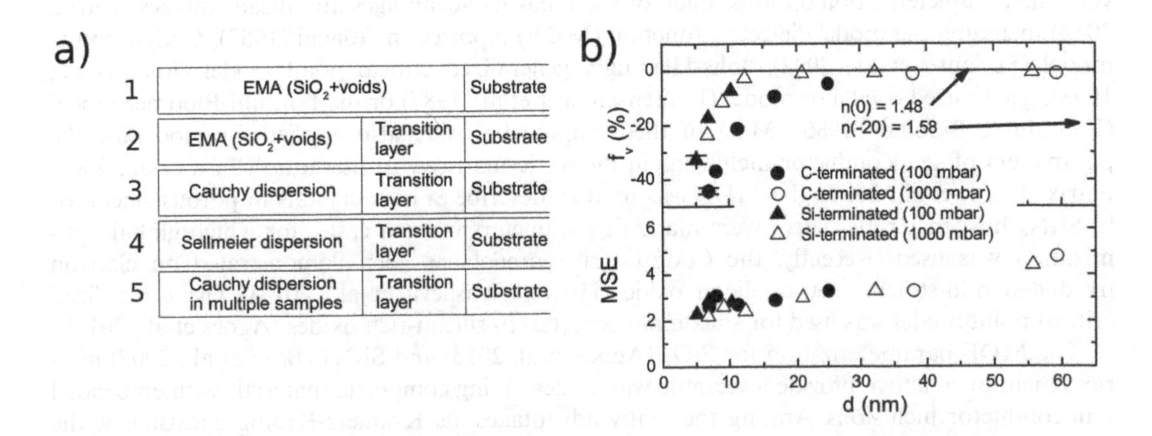

FIGURE 2.11 (a) Optical models to describe ultra-thin SiO_2 layers on SiC. (b) Volume fraction of void when using Model 1. MSE denotes the mean squared error, i.e. the quality of the fit.

Reproduced with permission from Petrik, Szilágyi, et al. (2009). Copyright 2009 American Institute of Physics.

In case of many fit parameters (as shown, for example, in Petrik, Fried, et al. (2009) for Figure 2.11), the multi sample approach is very useful, in which the measurements on different samples are fitted together, using all the results as one collective data set. The idea is that there are many model parameters (for example the dispersion of the layers or the optical properties of the substrate) that are the same in all measurements. Therefore, the number of fitted parameters doesn't increase proportionally to the number of measured data – the parameters/data ratio decreases with each newly introduced sample. A typical example of the multi sample method (Herzinger, et al., 1998) is a series of samples for which only the layer thicknesses are different. This method can effectively be used for ultra-thin layers as well (Petrik, Fried, et al., 2009, Petrik, Szilágyi, et al., 2009).

2.4.2 Oxides with Embedded Semiconductor Nanodots

Non-stoichiometric oxides with embedded semiconductor nanodots have been the subject of intensive research for decades, due to the applications in microelectronics, most importantly in non-volatile memory devices, but also in a range of many different applications from solar absorbers (Conibeer et al., 2006) to light emitters (Pavesi, et al., 2000). The measurement of such material system is challenging for most of the techniques. For example, the size and amount of nanocrystals in a thin film can be below the detection limit of usual X-ray diffraction devices. In many cases, transmission electron microscopy can only detect such nanoparticles when applying sophisticated correlation techniques (Hristova-Vasileva et al., 2018). Ellipsometry on the other hand can detect crystalline phases in nanometer thin layers and in volume fractions down to a few percent and sizes of a few nanometers (Agocs et al., 2017).

The most frequently used method for the description of composite structures is also the EMA (Aspnes, 1982). The greatest challenge in this case is that it assumes the dielectric function of the components to be known. In case of embedded nanocrystals, however, the optical properties are not known due to the deviations from the bulk properties because of the small size. To solve this problem, a useful method is to prepare homogeneous layers with optical properties possibly identical to those found in the composite material, to determine the reference dielectric function with a higher accuracy. This has been demonstrated for Ge nanocrystals in a ZrO_2 matrix (Agocs et al., 2017). Figure 2.12 shows the reference dispersion curves for amorphous and crystalline Ge measured in homogeneous thin films and from the literature.

The dielectric function of the embedded semiconductor phase can be modeled using many versatile parameterization options. Each of them has its advantages and disadvantages (Petrik, 2014) including the model dielectric function (MDF) reported in Adachi (1987), Cody-Lorentz model (Ferlauto et al., 2002), Johs-Herzinger generalized critical point model (Johs, et al., 1998), generalized oscillator model (Lautenschlager et al., 1987) or the Forouhi-Bloomer model (Forouhi & Bloomer, 1986). Most of these approaches were also applied for modelling the parameters of semiconductor inclusions in dielectric matrices. In literature (Basa et al., 2007; Petrik, Fried, et al., 2009), the MDF was used to describe Si nanocrystals in porous silicon or in Si_3N_4 host materials. There were many fit parameters in both cases, for which global optimization was used. Recently, the Cody-Lorentz model has been demonstrated on electron irradiated non-stoichiometric silicon oxide (Hristova-Vasileva et al., 2018). The generalized critical point model was used for silicon nanocrystals in silicon-rich oxides (Agocs et al., 2011).

The MDF parameterization for SiO_2 (Agocs et al.,2013) and Si_3N_4 (Basa et al., 2007) matrices mentioned above provide a versatile way of describing composite materials with embedded semiconductor inclusions. Among the many advantages, its Kramers-Kronig consistency, the physical relevancy and the analytical description has to be emphasized at first. The MDF model can be programmed easily and the Kramers-Kronig consistent computation is quick and flexible. Due to the large number of fit parameters, limitations of parameters must carefully be considered, for example, by coupling or fixing some of them (Petrik, Fried, et al., 2009).

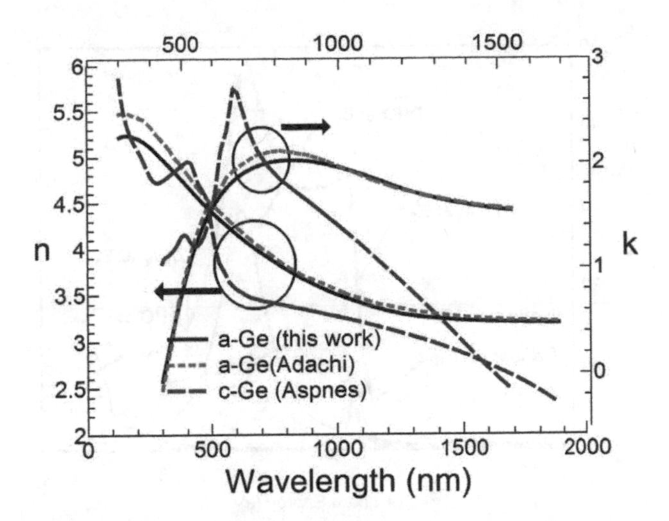

FIGURE 2.12 Optical properties of amorphous (a-Ge) and single-crystalline (c-Ge) Ge used for the modelling of Ge clusters embedded in ZrO_2. The reference shown by black lines was determined from a single-layer sample, and compared to other references from the literature (a-Ge from Ref. (Adachi, 1999)) and single-crystalline (c-Ge from Aspnes and Studna (1983)) references.

Reproduced with permission from Agocs et al. (2017). Copyright 2017 Elsevier Inc.

The structure of MDF oscillators is demonstrated for silicon in Si_3N_4 in Figure 2.13. The (2D)M_0 critical point and the photon energy of 3.4 eV is described by

$$\varepsilon(E) = -B_1\chi_1^{-2} ln(1-\chi_1^2) \tag{11}$$

with

$$\chi_1 = \frac{E + i\Gamma_1}{E_1} \tag{12}$$

where B_1 and Γ_1 denote the strength and broadening parameters, respectively. For the (2D)M_2 contribution, we use the equation

$$\varepsilon(E) = -F\chi_{2m}^{-2} ln\frac{1-\chi_{cl}^2}{1-\chi_{2m}^2} \tag{13}$$

with

$$\chi_{2m} = \frac{E + i\Gamma_2}{E_2} \tag{14}$$

$$\chi_{cl} = \frac{E + i\Gamma_2}{E_{cl}} \tag{15}$$

where F and Γ are the strength and broadening parameters. E_{cl} is a low energy cut-off. The damped harmonic oscillators (DHO) are described by

$$\varepsilon(E) = \frac{c}{(1-\chi_2^2) - i\chi_2\gamma} \tag{16}$$

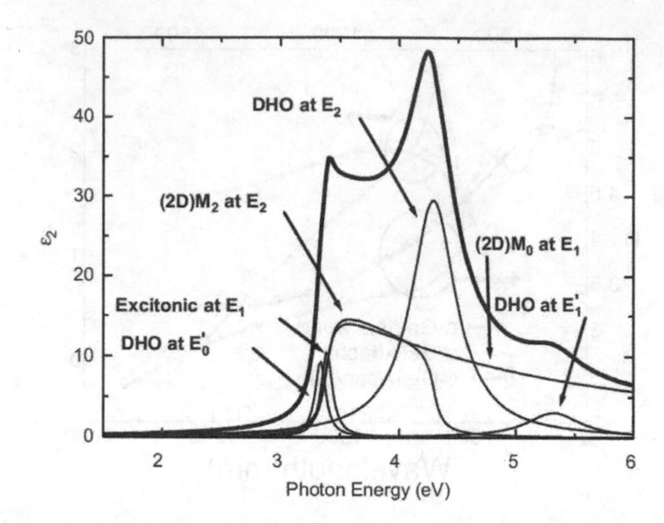

FIGURE 2.13 Oscillators of the MDF parameterization used for Si nanocrystals embedded in Si_3N_4. Two 2D critical points are denoted by $(2D)M_0$ and $(2D)M_2$, described by Eqs. (11–15). DHO denotes damped harmonic oscillators from Eqs. (16) and (17). The excitonic lineshape is closely related to the DHOs.

Reproduced with permission from Ref. (Basa et al., 2007). Copyright 2007 Elsevier Inc.

with

$$\chi_2 = \frac{E}{E_2} \tag{17}$$

where C and γ are the non-dimensional strength and broadening parameters.

2.5 APPLICATIONS

From the ellipsometric evaluation and optical modeling point of view of complex oxide materials, the main issue in applications is to handle a large number of fit parameters. It can be solved by a good (i) assumption for the starting parameters, (ii) a reliable approach to step by step move to a larger number of fit parameters by introducing new ones using strict rules or (iii) using parameter search approaches. In step (ii), a new parameter can be introduced if it is physically meaningful, if it improves the fit quality and if it does not correlate with other parameters. For step (iii), there are many global search approaches including grid search, random global search, simulated annealing, hill climbing, and Bayesian optimization (Polgár et al., 2006; Schneider et al., 2017). Some of them require a gradient fit algorithm such as the Levenberg-Marquardt method as the final step.

Limiting the number of fit parameters is a great challenge in most applications. For example, the models that describe semiconductor components in dielectric films are usually even more complex than those described in the previous section, because the films are typically non-uniform in the vertical direction. The reason is that on the scale of sensitivity of ellipsometry, most films have boundaries and interface layers both for the substrate and the surface (as, for example, in the case of porous silicon (Petrik et al., 2006) or polycrystalline silicon (Petrik et al., 2000)). Thus, the models include not only oscillator and volume fraction parameters, but these are usually distributed in many layers, the thicknesses of which are also unknown (Petrik, 2008). In semiconductors, the vertical uniformity can be determined using a special wavelength range scan utilizing the fact that the penetration depth is largely dependent on the photon energy around the band gap (Petrik et al. 2014).

The mapping of vertical non-uniformity in non-absorbing oxide is more challenging because of the typically low changes in the refractive index, and the lack of the imaginary part of *n*. The number of fit parameters can be limited for these materials if assuming that the dispersion parameters do not change in the Cauchy model (parameters "*B*" and "*C*" in Eq. (1)), and only the non-dispersive "*A*" parameter is varied as a function of the depth. This approach has been demonstrated to work for oxides prepared by laser deposition (Chandrappan, Murray, Kakkar, et al., 2015, Chandrappan, Murray, Petrik, et al., 2015), ion implantation (Bányász et al., 2012), or thermal oxidation (Petrik et al., 2017). The refractive index variations measured by ellipsometry were in the order of magnitude of 10^{-2}. Figure 2.14 shows that completely different profiles were revealed between Er and Yb co-doping or sequential doping using femtosecond laser ablation into a tellurite glass (Chandrappan, Murray, Kakkar, et al., 2015; Chandrappan, Murray, Petrik, et al., 2015). Using this technique, higher Er concentrations can be achieved without clustering, which is of key importance in integrated photonics platforms.

If the layers undergo processing steps (e.g. annealing), the vertical uniformity gets even worse due to diffusion, oxidation, and non-uniform phase changes (Agocs et al., 2011, 2017; Petrik et al., 2005). As expected, an even larger vertical inhomogeneity was found for oxidized zirconium tube for nuclear cladding (Petrik et al., 2017). As shown in Figure 2.15, both *n* and *k* varied largely as a function of the depth using the same modeling approach, i.e. fixing the dispersion parameters in the Cauchy model, and fitting only the non-dispersive part ("A_C" in Eq. (1)) for *n* and the amplitude for *k* as a function of depth. This study showed that the optical characterization is a powerful tool for metal surface in nuclear power plant technology. In general, the same technique is applicable not only for the Zr tubes, used as cladding materials for the nuclear fuel, but also for other metal surfaces. It allows monitoring of their oxidation and degradation as well as even *in situ* monitoring capabilities or surface mapping – due to the high speed of the measurement (a spectrum from the wavelength of 200 nm to 1700 nm in 708 points can be recorded within 1 s).

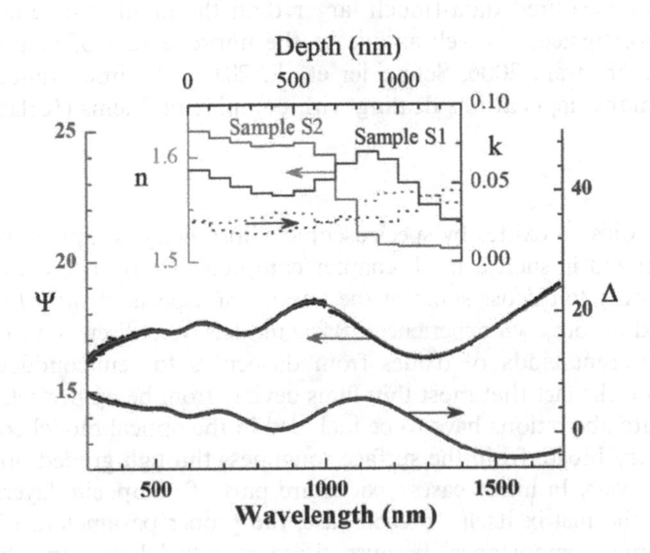

FIGURE 2.14 Refractive index profiles measured by ellipsometry on rare-earth doped tellurite modified silica prepared in a single-step (Sample 1) or two-step (Sample 2) process.

Reproduced with permission from. Chandrappan et al. (2015). Copyright 2015 Optical Society of America.

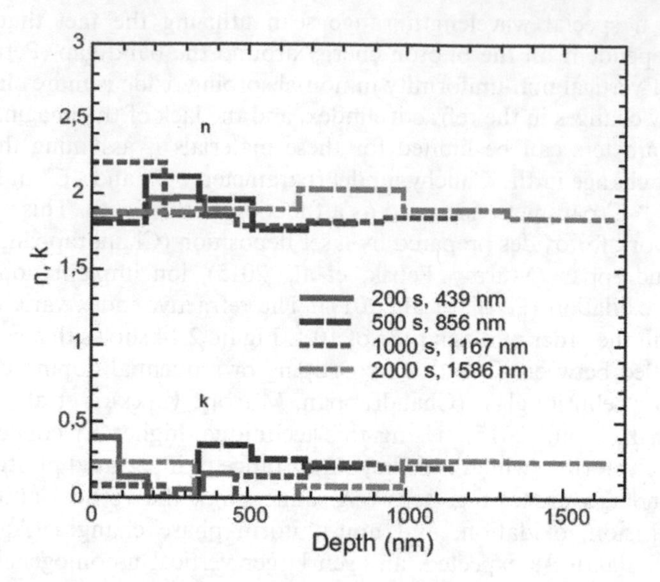

FIGURE 2.15 Depth profiles of n and k for different oxide thicknesses (at the wavelength of 633 nm) measured on Zr nuclear cladding material made of alloy E110G, oxidized for 200–2000 s in steam atmosphere at 600°C and 800°C.

Reproduced with permission from Petrik et al. (2017). Copyright 2017 Elsevier Inc.

Oxides are also used as high and low index layer sequences for Bragg multilayers to adjust wavelength-dependent absorption and reflection features, the application of which was demonstrated recently for high-sensitivity internal reflection sensors (Sinibaldi et al., 2015). In ellipsometry, the determination of high-low index sequences is possible because of the large number of measured data (much larger than the number of unknown parameters of the structure investigated), as well as due to the improvement of methods used for parameter search (Polgár et al., 2006; Schneider et al., 2017), the importance of which has to be emphasized in many applications dealing with complex problems (Kalas et al., 2019).

CONCLUSION

The literature of studies on oxides by spectroscopic ellipsometry is rapidly increasing. It is too large to be summarized in such a book chapter comprehensively. Therefore, in this work an attempt was made only to discuss some of the important aspects of measurements and evaluations, mainly based on our own experiences. Basic models were discussed to describe homogeneous layers of different kinds of oxides from dielectrics to semiconductors. The greatest challenge is, however, the fact that most thin films deviate from being perfect, and in ellipsometry even those minute aberrations have to be included in the optical model correctly. The range of these effects is very broad, from the surface roughness through graded optical properties in depth to interface layers. In many cases, oxides are part of composite layers being either the nano inclusions or the matrix itself. In each case, the proper parameterization of the optical functions is of primary importance, because those modified layers (partly because of size effects) cannot be usually described by literature references. In at least nine out of ten cases, the dielectric functions of those components cannot be found in databases, and cannot be prepared in form of bulk materials to be measured as references. This means that in most cases the dielectric functions have to be parameterized and fitted together with the composition and geometry parameters such as the volume fractions, layer thicknesses, and boundary layers. Therefore, also looking back historically, the direction of development in ellipsometry in general goes from simple structures such as a thermally oxidized silicon wafer, toward complex structures such as porous layers, superlattices, or nanodot layers. The instrumentation of ellipsometry is very mature, but the evaluation approaches have to be improved and developed significantly in the future.

REFERENCES

Adachi, S. (1987). Model dielectric constants of GaP, GaAs, GaSb, InP, InAs, and InSb. *Physical Review B*, *35*, 7454–7463. doi:10.1103/PhysRevB.35.7454.

Adachi, S. (1999). *Optical constants of crystalline and amorphous semiconductors.* Boston: Springer US. doi:10.1007/978-1-4615-5247-5.

Agocs, E., Fodor, B., Pollakowski, B., Beckhoff, B., Nutsch, A., Jank, M., & Petrik, P. (2014). Approaches to calculate the dielectric function of ZnO around the band gap. *Thin Solid Films*, *571*, 684–688. doi:10.1016/j.tsf.2014.03.028.

Agocs, E., Nassiopoulou, A. G., Milita, S., & Petrik, P. (2013). Model dielectric function analysis of the critical point features of silicon nanocrystal films in a broad parameter range. *Thin Solid Films*, *541*, 83–86. doi:10.1016/j.tsf.2012.10.126.

Agocs, E., Petrik, P., Milita, S., Vanzetti, L., Gardelis, S., Nassiopoulou, A. G., … Fried, M. (2011). Optical characterization of nanocrystals in silicon rich oxide superlattices and porous silicon. *Thin Solid Films*, *519*, 3002–3005. doi:10.1016/j.tsf.2010.11.072.

Agocs, E., Zolnai, Z., Rossall, A. K., van den Berg, J.A., Fodor, B., Lehninger, D., … Petrik, P. (2017). Optical and structural characterization of Ge clusters embedded in ZrO_2. *Applied Surface Science*, *421*, 283–288. doi:10.1016/j.apsusc.2017.03.153.

Arwin, H. (2005). Ellipsometry in life sciences. In E. G. Irene & H. G. Tompkins (Eds.), *In handbook of ellipsometry* (p. 42). Norwich, NY: William Andrew.

Arwin, H., & Aspnes, D. E. (1984). Unambiguous determination of thickness and dielectric function of thin films by spectroscopic ellipsometry. *Thin Solid Films*, *113*, 101–113. doi:10.1016/0040-6090(84)90019-1.

Aspnes, D. E. (1982). Optical properties of thin films. *Thin Solid Films*, *89*, 249–262.

Aspnes, D. E., & Studna, A. A. (1975). High precision scanning ellipsometer. *Applied Optics*, *14*, 220–228. doi:10.1364/ao.14.000220.

Aspnes, D. E., & Studna, A. A. (1983). Dielectric functions and optical parameters of Si, Ge, GaP, GaAs, GaSb, InP, InAs, and InSb from 1.5 to 6.0 eV. *Physical Review B*, *27*, 985–1009. doi:10.1103/PhysRevB.27.985.

Aspnes, D. E., Studna, A. A., & Kinsbron, E. (1984). Dielectric properties of heavily doped crystalline and amorphous silicon from 1.5 to 6.0 eV. *Physical Review B*, *29*, 768–779. doi:10.1103/PhysRevB.29.768.

Azzam, R. M. A., & Bashara, N. M. (1977). *Ellipsometry and polarized light. North-Holland personal library.* Amsterdam: North-Holland Pub. Co.

Baklanov, M. R., Mogilnikov, K. P., Polovinkin, V. G., & Dultsev, F. N. (2000). Determination of pore size distribution in thin films by ellipsometric porosimetry. *Journal of Vacuum Science & Technology B*, *18*, 1385–1391. doi:10.1116/1.591390.

Bányász, I., Berneschi, S., Bettinelli, M., Brenci, M., Fried, M., Khanh, N. Q., … Zolnai, Z., (2012). MeV energy N^+-implanted planar optical waveguides in Er-doped tungsten-tellurite glass operating at 1.55 μm. *IEEE Photonics Journal*, *4*, 721–727. doi:10.1109/JPHOT.2012.2194997.

Basa, P., Petrik, P., Fried, M., Dobos, L., Pécz, B., & Tóth, L. (2007). Si nanocrystals in silicon nitride: An ellipsometric study using parametric semiconductor models. *Physica E*, *38*, 76–79. doi:10.1016/j.physe.2006.12.021.

Bruggeman, D. A. G. (1935). Berechnung Verschiedener Physikalischer Konstanten von Heterogenen Substanzen. I. Dielektrizitätskonstanten Und Leitfähigkeiten Der Mischkörper Aus Isotropen Substanzen. *Annalen Der Physik*, *416*, 636–664. doi:10.1002/andp.19354160705.

Chandrappan, J., Murray, M., Kakkar, T., Petrik, P., Agocs, E., Zolnai, Z., … Jose, G. (2015). Target dependent femtosecond laser plasma implantation dynamics in enabling silica for high density erbium doping. *Scientific Reports*, *5*, 14037. doi:10.1038/srep14037.

Chandrappan, J., Murray, M., Petrik, P., Agocs, E., Zolnai, Z., Tempez, A., … Jose, G. (2015). Doping silica beyond limits with laser plasma for active photonic materials. *Optical Materials Express*, *5*, 2849–2861. doi:10.1364/OME.5.002849.

Cobet, C., Oppelt, K., Hingerl, K., Neugebauer, H., Knör, G., Sariciftci, N. S., Gasiorowski, J. (2018). Ellipsometric spectroelectrochemistry: An in situ insight in the doping of conjugated polymers. *The Journal of Physical Chemistry C*, *2018*(122), 24309–24320. doi:10.1021/acs.jpcc.8b08602.

Collins, R. W. (1990). Automatic rotating element ellipsometers: Calibration, operation, and real-time applications. *Review of Scientific Instruments*, *61*, 2029–2062. doi:10.1063/1.1141417.

Collins, R. W., & Ferlauto, A. S. (2005). Optical physics of materials. In E. G. Irene & H. G. Tomkins (Eds.), *In handbook of ellipsometry* (p. 93). Norwich, NY: William Andrew.

Collins, R. W., An, I., Fujiwara, H., Lee, J., Lu, Y., Koh, J., Rovira, P. I. (1998). Advances in multichannel spectroscopic ellipsometry. *Thin Solid Films*, *313–314*, 18–32. doi:10.1016/S0040-6090(97)00764-5.

Collins, R. W., Koh, J., Ferlauto, A. S., Rovira, P. I., Lee, Y., Koval, R. J., Wronski, C. R. (2000). Real time analysis of amorphous and microcrystalline silicon film growth by multichannel ellipsometry. *Thin Solid Films*, *364*, 129–137. doi:10.1016/S0040-6090(99)00925-6.

Conibeer, G., Green, M., Corkish, R., Cho, Y., Cho, E. C., Jiang, C. W., … Lin, K. L. (2006). Silicon nanostructures for third generation photovoltaic solar cells. *Thin Solid Films*, *511–512*, 654–662. doi:10.1016/j.tsf.2005.12.119.

Drude, P. (1889). Ueber Oberflächenschichten. I. Theil. *Annalen Der Physik*, *272*, 532–560. doi:10.1002/andp.18892720214.

Ferlauto, A. S., Ferreira, G. M., Pearce, J. M., Wronski, C. R., Collins, R. W., Deng, X., & Ganguly, G. (2002). Analytical model for the optical functions of amorphous semiconductors from the near-infrared to ultraviolet: Applications in thin film photovoltaics. *Journal of Applied Physics*, *92*, 2424–2436. doi:10.1063/1.1497462.

Fodor, B., Agocs, E., Bardet, B., Defforge, T., Cayrel, F., Alquier, D., … Petrik, P. (2016). Porosity and thickness characterization of porous Si and oxidized porous Si layers – An ultraviolet-visible-mid infrared ellipsometry study. *Microporous and Mesoporous Mater*, *227*, 112–120. doi:10.1016/j.micromeso.2016.02.039.

Forouhi, A. R., & Bloomer, I. (1986). Optical dispersion relations for amorphous semiconductors and amorphous dielectrics. *Physical Review B*, *34*, 7018–7026. doi:10.1103/PhysRevB.34.7018.

Fried, M., Juhász, G., Major, C., Petrik, P., Polgár, O., Horváth, Z., & Nutsch, A. (2011). Expanded beam (macro-imaging) ellipsometry. *Thin Solid Films*, *519*, 2730–2736. doi:10.1016/j.tsf.2010.12.067.

Fried, M., Lohner, T., & Petrik, P. (2001). Ellipsometric characterization of thin films. In H. S. Nalwa (Ed.), *Handbook of surfaces and interfaces of materials: Vol.4. Solid thin films and layers* (pp. 335–367). San Diego, CA: Academic Press.

Fried, M., Lohner, T., Polgár, O., Petrik, P., Vázsonyi, É., Bársony, I., … Stehle, J. L. (1996). Characterization of different porous silicon structures by spectroscopic ellipsometry. *Thin Solid Films*, *276*, 223–227. doi:10.1016/0040-6090(95)08058-9.

Fujiwara, H. (2007). *Spectroscopic ellipsometry: Principles and applications*. Chichester: John Wiley & Sons, Ltd.

Gilliot, M., En Naciri, A., Johann, L., Stoquert, J. P., Grob, J. J., & Muller, D. (2007). Optical anisotropy of shaped oriented cobalt nanoparticles by generalized spectroscopic ellipsometry. *Physical Review B*, *76*, 1–15. doi:10.1103/PhysRevB.76.045424.

Hauge, P. S. (1978). Mueller matrix ellipsometry with imperfect compensators. *Journal of the Optical Society of America A*, *68*, 1519–1528. doi:10.1364/JOSA.68.001519.

Herzinger, C. M., Johs, B., McGahan, W. A., Woollam, J. A., & Paulson, W. (1998). Ellipsometric determination of optical constants for silicon and thermally grown silicon dioxide via a multi-sample, multi-wavelength, multi-angle investigation. *Journal of Applied Physics*, *83*, 3323–3336. doi:10.1063/1.367101.

Hristova-Vasileva, T., Petrik, P., Nesheva, D., Fogarassy, Z., Lábár, J., Kaschieva, S., … Antonova, K. (2018). Influence of 20 MeV electron irradiation on the optical properties and phase composition of SiO_x thin films. *Journal of Applied Physics*, *123*, 195303. doi:10.1063/1.5022651.

Humlíček, J., & Šik, J. (2015). Optical functions of silicon from reflectance and ellipsometry on silicon-on-insulator and homoepitaxial samples. *Journal of Applied Physics*, *118*, 195706. doi:10.1063/1.4936126.

Irene, E. A. (1993). Applications of spectroscopic ellipsometry to microelectronics. *Thin Solid Films*, *233*, 96–111. doi:10.1016/0040-6090(93)90069-2.

Irene, E. A. (2006). SiO_2 films. In H. G. Tompkins and E. A. Irene (Eds.), *Handbook of ellipsometry* (pp. 569–636). Norwich: Springer Berlin Heidelberg. doi:10.1007/3-540-27488-x_8.

Jellison, G., & Boatner, L. (1998). Optical functions of uniaxial ZnO determined by generalized ellipsometry. *Physical Review B, 58*, 3586–3589. doi:10.1103/PhysRevB.58.3586.

Jellison, G. E. (1992). Optical functions of silicon determined by two-channel polarization modulation ellipsometry. *Optical Materials, 1*, 41–47. doi:10.1016/0925-3467(92)90015-F.

Jellison, G. E. (2004). Generalized ellipsometry for materials characterization. *Thin Solid Films, 450*(1), 42–50. doi:10.1016/j.tsf.2003.10.148.

Jellison, G. E., Boatner, L. A., Budai, J. D., Jeong, B. S., & Norton, D. P. (2003). Spectroscopic ellipsometry of thin film and bulk anatase (TiO_2). *Journal of Applied Physics, 93*, 9537–9541. doi:10.1063/1.1573737.

Jellison, G. E., Chisholm, M. F., & Gorbatkin, S. M. (1993). Optical functions of chemical vapor deposited thin-film silicon determined by spectroscopic ellipsometry. *Applied Physics Letters, 62*, 3348–3350. doi:10.1063/1.109067.

Jellison, G. E., & Modine, F. A. (1996). Parameterization of the optical functions of amorphous materials in the interband region. *Applied Physics Letters, 69*, 371–373. doi:10.1063/1.118064.

Johs, B., Herzinger, C. M., Dinan, J. H., Cornfeld, A., & Benson, J. D. (1998). Development of a parametric optical constant model for Hg 1-x Cd x Te for control of composition by spectroscopic ellipsometry during MBE growth. *Thin Solid Films, 313–314*, 137–142. doi:10.1016/S0040-6090(97)00800-6.

Kalas, B., Agocs, E., Romanenko, A., & Petrik, P. (2019). In situ characterization of biomaterials at solid-liquid interfaces using ellipsometry in the UV-visible-NIR wavelength range. *Physica Status Solidi A, 216*, 1800762. doi:10.1002/pssa.201800762.

Kalas, B., Nador, J., Agocs, E., Saftics, A., Kurunczi, S., Fried, M., & Petrik, P. (2017). Protein adsorption monitored by plasmon-enhanced semi-cylindrical Kretschmann ellipsometry. *Applied Surface Science, 421*, 585–592. doi:10.1016/j.apsusc.2017.04.064.

Khosroabadi, A. A., Gangopadhyay, P., Cocilovo, B., Makai, L., Basa, P., Duong, B., … Norwood, R. A. (2014). Spectroscopic ellipsometry on metal and metal-oxide multilayer hybrid plasmonic nanostructures: Erratum. *Optics Letters, 39*, 2810. doi:10.1364/ol.39.002810.

Kozma, P., Kozma, D., Nemeth, A., Jankovics, H., Kurunczi, S., Horvath, R., … Petrik, P. (2011). In-depth characterization and computational 3D reconstruction of flagellar filament protein layer structure based on in situ spectroscopic ellipsometry measurements. *Applied Surface Science, 257*, 7160–7166. doi:10.1016/j.apsusc.2011.03.081.

Landwehr, J., Fader, R., Rumler, M., Rommel, M., Bauer, A. J., Frey, L., … Spiecker, E. (2014). Optical polymers with tunable refractive index for nanoimprint technologies. *Nanotechnology, 25*, 505301. doi:10.1088/0957-4484/25/50/505301.

Lautenschlager, P., Garriga, M., Vina, L., & Cardona, M. (1987). Temp dependence of the dielectric function and interband critical points in Si. *Physical Review B, 36*, 4821–4830. http://prb.aps.org/abstract/PRB/v36/i9/p4821_1.

Lee, J., Koh, J., & Collins, R. W. (2001). Dual rotating-compensator multichannel ellipsometer: Instrument development for high-speed mueller matrix spectroscopy of surfaces and thin films. *Review of Scientific Instruments, 72*, 1742–1754. doi:10.1063/1.1347969.

Lee, J., Rovira, P. I., An, I., & Collins, R. W. (1998). Rotating-compensator multichannel ellipsometry: Applications for real time stokes vector spectroscopy of thin film growth. *Review of Scientific Instruments, 69*, 1800–1810. doi:10.1063/1.1148844.

Lehninger, D., Khomenkova, L., Röder, C., Gärtner, G., Abendroth, B., Beyer, J., … Heitmann, J. (2015). Ge nanostructures embedded in ZrO_2 dielectric films for nonvolatile memory applications. *ECS Trans, 66*, 203–212. doi:10.1149/06604.0203ecst.

Major, C., Juhász, G., Nemeth, A., Labadi, Z., Petrik, P., Horváth, Z., & Fried, M. (2009). Optical and electrical properties of Al doped ZnO layers measured by wide angle beam spectroscopic ellipsometry. *MRS Symp Proc, 1109*, 31–36.

Major, C., Nemeth, A., Radnoczi, G., Czigany, Z., Fried, M., Labadi, Z., & Barsony, I., (2009). Optical and electrical characterization of aluminium doped ZnO layers. *Applied Surface Science, 255*(21), 8907–8912. doi:10.1016/j.apsusc.2009.06.088.

Mikhaylova, Y., Ionov, L., Rappich, J., Gensch, M., Esser, N., Minko, S., … Hinrichs, K. (2007). In situ infrared ellipsometric study of stimuli-responsive mixed polyelectrolyte brushes. *Analytical Chemistry, 79*, 7676–7682. doi:10.1021/ac070853a.

Németh, Á., Major, C., Fried, M., Lábadi, Z., & Bársony, I. (2008). Spectroscopic ellipsometry study of transparent conductive ZnO layers for CIGS solar cell applications. *Thin Solid Films, 516*, 7016–7020. doi:10.1016/j.tsf.2007.12.012.

Ngo, D., Liu, H., Sheth, N., Lopez-Hallman, R., Podraza, N. J., Collin, M., … Kim, S. H. (2018). Spectroscopic ellipsometry study of thickness and porosity of the alteration layer formed on international simple glass surface in aqueous corrosion conditions. *Npj Materials Degradation, 2*, 20. doi:10.1038/s41529-018-0040-7.

Nguyen, H. V., An, I., Collins, R. W., Lu, Y., Wakagi, M., & Wronski, C. R. (1994). Preparation of ultrathin microcrystalline silicon layers by atomic hydrogen etching of amorphous silicon and end-point detection by real time spectroellipsometry. *Applied Physics Letters, 65*, 3335–3337. doi:10.1063/1.113024.

Pápa, Z., Csontos, J., Smausz, T., Toth, Z., & Budai, J. (2017). Spectroscopic ellipsometric investigation of graphene and thin carbon films from the point of view of depolarization effects. *Applied Surface Science, 421*, 714–721. doi:10.1016/j.apsusc.2016.11.231.

Patrito, E. M., & Macagno, V. A. (1993). Ellipsometric investigation of anodic zirconium oxide films. *Journal of the Electrochemical Society, 140*, 1576–1585. doi:10.1149/1.2221605.

Paulson, W., Johs, B., Herzinger, C. M., McGahan, W. A., & Woollam, J. A. (2002). Ellipsometric determination of optical constants for silicon and thermally grown silicon dioxide via a multi-sample, multi-wavelength, multi-angle investigation. *Journal of Applied Physics, 83*, 3323–3336. doi:10.1063/1.367101.

Pavesi, L., Dal Negro, L., Mazzoleni, C., Franzò, G., & Priolo, F. (2000). Optical gain in silicon nanocrystals. *Nature, 408*, 440–444. doi:10.1038/35044012.

Petrik, P. (2008). Ellipsometric models for vertically inhomogeneous composite structures. *Physica Status Solidi A, 205*, 732–738. doi:10.1002/pssa.200777847.

Petrik, P. (2014). Parameterization of the dielectric function of semiconductor nanocrystals. *Physica B, 453*, 2–7. doi:10.1016/j.physb.2014.03.065.

Petrik, P., Agocs, E., Volk, J., Lukacs, I., Fodor, B., Kozma, P., … Fried, M. (2014). Resolving lateral and vertical structures by ellipsometry using wavelength range scan. *Thin Solid Films, 571*, 579–583. doi:10.1016/j.tsf.2014.02.008.

Petrik, P., Egger, H., Eiden, S., Agocs, E., Fried, M., Pecz, B., … Giannone, D. (2012). Ellipsometric characterization of thin nanocomposite films with tunable refractive index for biochemical sensors. *Materials Research Society Symposium Proceedings, 1352*, 81–87. doi:10.1557/opl.2011.1342.

Petrik, P., Fried, M., Lohner, T., Polgár, O., Gyulai, J., Cayrel, F., & Alquier, D. (2005). Optical models for cavity profiles in high-dose helium-implanted and annealed silicon measured by ellipsometry. *Journal of Applied Physics, 97*, 123514. doi:10.1063/1.1937469.

Petrik, P., Fried, M., Vázsonyi, E., Lohner, T., Horváth, E., Polgár, O., … Gyulai, J. (2006). Ellipsometric characterization of nanocrystals in porous silicon. *Applied Surface Science, 253*, 200–203. doi:10.1016/j.apsusc.2006.05.087.

Petrik, P., Fried, M., Vazsonyi, E., Basa, P., Lohner, T., Kozma, P., & Makkai, Z. (2009). Nanocrystal characterization by ellipsometry in porous silicon using model dielectric function. *Journal of Applied Physics, 105*, 024908. doi:10.1063/1.3068479.

Petrik, P., Khánh, N. Q., Horváth, Z. E., Zolnai, Z., Bársony, I., Lohner, T., … Ryssel, H. (2002). Characterisation of $Ba_xSr_{1-x}TiO_3$ films using spectroscopic ellipsometry, rutherford backscattering spectrometry and X-ray diffraction. *Journal of Non-Crystalline Solids, 303*, 179–184. doi:10.1016/S0022-3093(02)00982-1.

Petrik, P., Lehnert, W., Schneider, C., Lohner, T., Fried, M., Gyulai, J., & Ryssel, H. (2001). In situ measurement of the crystallization of amorphous silicon in a vertical furnace using spectroscopic ellipsometry. *Thin Solid Films, 383*, 235–240. doi:10.1016/S0040-6090(00)01792-2.

Petrik, P., Lohner, T., Fried, M., Biró, L.P., Khánh, N.Q., Gyulai, J., … Ryssel, H. (2000). Ellipsometric study of polycrystalline silicon films prepared by low-pressure chemical vapor deposition. *Journal of Applied Physics, 87*, 1734–1742. doi:10.1063/1.372085.

Petrik, P., Lohner, T., Fried, M., Gyulai, J., Boell, U., Berger, R., & Lehnert, W. (2002). Ellipsometric study of the polysilicon/thin oxide/single-crystalline silicon structure and its change upon annealing. *Journal of Applied Physics, 92*, 2374–2377. doi:10.1063/1.1497694.

Petrik, P., Romanenko, A., Kalas, B., Péter, L., Novotny, T., Perez-Feró, E., … Hózer, Z. (2019). Optical properties of oxidized, hydrogenated, and native zirconium surfaces for wavelengths from 0.3 to 25 μm – A study by ex situ and in situ spectroscopic ellipsometry. *Physica Status Solidi A, 216*, 1800676. doi:10.1002/pssa.201800676.

Petrik, P., Sulyok, A., Novotny, T., Perez-Feró, E., Kalas, B., Agocs, E., … Hózer, Z. (2017). Optical properties of Zr and ZrO_2. *Applied Surface Science*, *421*, 744–747. doi:10.1016/j.apsusc.2016.11.072.

Petrik, P., Szilágyi, E., Lohner, T., Battistig, G., Fried, M., Dobrik, G., & Biró, L.P. (2009). Optical models for ultrathin oxides on Si- and C-terminated faces of thermally oxidized SiC. *Journal of Applied Physics*, *106*, 123506. doi:10.1063/1.3270424.

Petrik, P., Zolnai, Z., Polgar, O., Fried, M., Betyak, Z., Agocs, E., … Cobet, C. (2011). Characterization of damage structure in ion implanted SiC using high photon energy synchrotron ellipsometry. *Thin Solid Films*, *519*, 2791–2794. doi:10.1016/j.tsf.2010.12.070.

Poksinski, M., & Arwin, H. (2004). Protein monolayers monitored by internal reflection ellipsometry. *Thin Solid Films*, *455–456*, 716–721. doi:10.1016/j.tsf.2004.01.037.

Polgár, O., Petrik, P., Lohner, T., & Fried, M. (2006). Evaluation strategies for multi-layer, multi-material ellipsometric measurements. *Applied Surface Science*, *253*, 57–64. doi:10.1016/j.apsusc.2006.05.071.

Rhodes, C., Cerruti, M., Efremenko, A., Losego, M., Aspnes, D.E., Maria, J. P., & Franzen, S. (2008). Dependence of plasmon polaritons on the thickness of indium tin oxide thin films. *Journal of Applied Physics*, *103*, 093108. doi:10.1063/1.2908862.

Roeder, G., Manke, C., Baumann, P. K., Petersen, S., Yanev, V., Gschwandtner, A., … Ryssel, H. (2008). Characterization of Ru and RuO_2 thin films prepared by pulsed metal organic chemical vapor deposition. *Physica Status Solidi C*, *5*, 1231–1234. doi:10.1002/pssc.200777865.

Rothen, A. (1945). The ellipsometer, an apparatus to measure thicknesses of thin surface films. *Review of Scientific Instruments*, *16*, 26–30. doi:10.1063/1.1770315.

Schneider, P.-I., Garcia Santiago, X., Rockstuhl, C., & Burger, S. (2017). Global optimization of complex optical structures using Bayesian optimization based on Gaussian processes. *Proc. SPIE 10335, Digital Optical Technologies*, 103350O. doi:10.1117/12.2270609.

Schubert, M. (1998). Generalized ellipsometry and complex optical systems. *Thin Solid Films*, *313–314*, 323–332. doi:10.1016/S0040-6090(97)00841-9.

Schubert, M., Korlacki, R., Knight, S., Hofmann, T., Schöche, S., Darakchieva, V., … Higashiwaki, M (2016). Anisotropy, phonon modes, and free charge carrier parameters in monoclinic β-gallium oxide single crystals. *Physical Review B*, *93*, 125209. doi:10.1103/PhysRevB.93.125209.

Schubert, M., & Tiwald, T. (2000). Infrared dielectric anisotropy and phonon modes of sapphire. *Physical Review B*, *61*, 8187–8201. doi:10.1103/PhysRevB.61.8187.

Shan, A., Fried, M., Juhász, G., Major, C., Polgár, O., Németh, Á., … Collins, R. W. (2014). High-speed imaging/mapping spectroscopic ellipsometry for in-line analysis of roll-to-roll thin-film photovoltaics. *IEEE J Photovoltaics*, *4*, 355–361. doi:10.1109/JPHOTOV.2013.2284380.

Sinibaldi, A., Anopchenko, A., Rizzo, R., Danz, N., Munzert, P., Rivolo, P., … Michelotti, F. (2015). Angularly resolved ellipsometric optical biosensing by means of bloch surface waves. *Analyt Bioanalyt Chem*, *407*, 3965–3974. doi:10.1007/s00216-015-8591-8.

Synowicki, R. A. (1998). Spectroscopic ellipsometry characterization of indium tin oxide film microstructure and optical constants. *Thin Solid Films*, *313–314*, 394–397. doi:10.1016/S0040-6090(97)00853-5.

Synowicki, R. A., & Tiwald, T. E. (2004). Optical properties of bulk c-ZrO_2, c-MgO and a-As_2S_3 determined by variable angle spectroscopic ellipsometry. *Thin Solid Films*, *455–456*, 248–255. doi:10.1016/j.tsf.2004.02.028.

Tompkins, H., & Irene, E. A. (2005). *Handbook of ellipsometry*. Norwich: Elsevier Science.

Welch, V. L., Louette, P., Vukusic, P., Mouchet, S. R., Su, B.-L., Deparis, O., … Tabarrant, T. (2017). Assessment of environmental spectral ellipsometry for characterising fluid-induced colour changes in natural photonic structures. *Materials Today: Proceedings*, *4*, 4987–4997. doi:10.1016/j.matpr.2017.04.105.

Wurstbauer, U., Röling, C., Wurstbauer, U., Wegscheider, W., Vaupel, M., Thiesen, P. H., & Weiss, D. (2010). Imaging ellipsometry of graphene. *Applied Physics Letters*, *97*, 78–81. doi:10.1063/1.3524226.

Yoshikawa, H., & Adachi, S. (1997). Optical constants of ZnO. *Japanese Journal of Applied Physics*, *36*, 6237–6244. doi:10.1143/jjap.36.6237.

Zheng, S., Tu, Q., Urban, J. J., Li, S., & Mi, B. (2017). Swelling of graphene oxide membranes in aqueous solution: Characterization of interlayer spacing and insight into water transport mechanisms. *ACS Nano*, *11*, 6440–6450. doi:10.1021/acsnano.7b02999.

3 Electrical Characterization of Oxide-Based Materials Using Impedance Spectroscopy

Aleksej Smirnov

3.1 INTRODUCTION TO IMPEDANCE SPECTROSCOPY APPROACH

The concept of impedance spectroscopy (IS) as a coupling coefficient between two quantities that harmoniously change in time with frequency is universally accepted in science and technology. One of the changing quantities is conventionally called an input quantity or perturbation, while the other – an output quantity or response. In the framework of thermodynamics of irreversible processes, the input quantities are identified, as a rule, with generalized thermodynamic forces, whereas the output quantities are identified with thermodynamic flows and the reciprocal relations of Onsager flows are used. In electrochemical systems, the role of the generalized force is played by the potential of the electrode and the role of the flow is played by the electric current (this is the Faraday impedance that we will talk about). But there are other types of electrochemical impedance: photoelectrochemical (input value is light flux, output value is electric current); electro-reflection method (input value is electrode potential, output value is light flux modulation); laser pulse (input value is heat, output value is the amount of electricity or electrode potential), and so on. All these techniques are included in the general method of IS.

The principle of the Faraday impedance method (electrochemistry term) is that a sinusoidal voltage of a sufficiently small amplitude (less than 10 mV) is applied to the electrode in equilibrium, when there is a linear relationship between current and voltage. The alternating current (AC) method in experimental electrochemistry belongs to the group of relaxation methods. The relationship between relaxation and perturbation is described by linear equations, that is, the properties of the electrochemical system exhibit a linear relationship.

Consider the basic laws that describe electric current in linear circuits. Ohm's law: the current strength I for a section of the circuit is directly proportional to the applied voltage U between the ends of this section and inversely proportional to the resistance of the conductor of this section of the circuit for direct current (DC): $\boldsymbol{I = U/R}$ **or** $\boldsymbol{R = U/I}$. If in the DC circuit there is an emf (E) source with an internal resistance r, then $\boldsymbol{I = E/(R+r)}$ **or** $\boldsymbol{R + r = E/I}$. The strength of the AC is determined at a given voltage U not only by the resistance R that this circuit has at DC, but also by the presence of capacitors C or inductors L in this circuit. In other words, the same circuit will have different resistances for DC and AC. The resistance that this circuit provides to DC is called active, whereas the resistance that a capacitance or inductance exerts to an AC is called reactive (capacitive or inductive, respectively). The total resistance of the circuit to AC denoted by Z, in the case when the circuit contains reactive and active resistance, is composed of these values, but actually, is not equal to the simple sum of these resistances. A quantitative characteristic of linear AC circuits is the complex resistance (impedance), which determines the ratio of the electrical system to the disturbance. In this case, the electrode–electrical system interface in its electrochemical properties at a given time is equivalent to an electric circuit consisting of a number of resistances and capacitors connected in a certain way.

In general, when a section of a circuit contains not only active but also reactive (capacitive X_C, inductive X_L, or both) resistance, then the voltage between the ends of this section is shifted in phase with respect to the current, and the phase shift ranges from π/2 to −π/2 and is determined by the ratio between the active and reactive resistance of a given section of the circuit. Phase shift (or the difference between the phases of the current I and voltage U) is shown by $tg\varphi = [\omega L - (\omega C)^{-1}]R^{-1}$, where L is the inductance, C is the capacitance. For a sinusoidal AC, the ratio of harmonic signals can be written as follows (Barsoukov, & Macdonald, 2005):

$$\begin{aligned} U &= U_0 \sin(\omega t + \varphi) = U_0(\sin(\omega t)\cos(\phi) + \sin(\varphi)\cos(\omega t) \\ &= U_0 \cos(\varphi)\sin(\omega t) + U_0 \sin(\varphi)\cos(\omega t) \end{aligned} \tag{3.1}$$

$$\begin{aligned} \frac{U}{I} &= (U_0 \cos(\varphi)\sin(\omega t) + (U_0 \sin(\varphi)\cos(\omega t))\frac{1}{I_0}\sin(\omega t) \\ &= (\frac{U_0}{I_0}\cos(\varphi) + \frac{\frac{U_0}{I_0}\sin(\varphi)\cos(\omega t)}{\sin(\omega t)} = \frac{U_0}{I_0}\cos(\varphi) + \frac{U_0}{I_0}\sin(\varphi)\mathrm{ctg}(\omega t) \end{aligned} \tag{3.2}$$

when the ratio U_0/I_0 is denoted by Z_0 (this will be a constant for the given amplitude values) and U/I by Z (this value will depend on time and frequency), then total $Z = Z_0 \cos(\varphi) + Z_0\sin(\varphi)\mathrm{ctg}(\omega t)$. This value of Z is the total (or complex) resistance in the system through which AC flows, which is measured in Ohms and is also called impedance. The reciprocal of the impedance is the admittance $Y = 1/Z$. Since the impedance Z is equal to the ratio of voltage to current strength, the impedance can also be written in a complex form: $Z = Z' - jZ''$, where $Z' = Z_0\cos(\varphi)$ is the frequency-independent real part and $Z'' = Z_0\sin(\varphi)$ is the frequency-dependent imaginary part of the impedance. In exponential notation $Z = Z_0e^{j\omega t}$. Admittance Y has a value of $1/Z$ and the same (but opposite in sign) phase angle, that is, $Y = 1/Z_0e^{-j\omega t}$.

3.2 IMPEDANCE PRESENTATION METHODS

The studied electrosystem and processes in it can be characterized by the complex quantities such as: impedance $Z = Z' - jZ''$; – admittance $Y = 1/Z = Y' - jY''$; – electrical module $M = 1/\varepsilon = M' - jM''$; – dielectric constant $\varepsilon = M - 1 = Y/(j\omega C_c) = \varepsilon' + j\varepsilon''$. In these expressions $C_c = \varepsilon_0 A_c/l$ there is an empty cell (without sample) for measurements with electrodes of area A_c and distance between electrodes l. The dielectric constant is often denoted by ε^* to emphasize the complex nature of the quantity. These characteristics are used to describe dielectric materials and processes associated with the reorientation and relaxation of dipoles. All relations between the impedance and its derivatives are presented in the Figure 3.1 and in Table 3.1.

The information about the polarization of electrons, atoms, dipoles, and interfaces could be extracted from IS data measured from 10^{16} to 10^{-3} Hz, as shown schematically in Table 3.1 and Figure 3.1. Then it will be possible to calculate the capacitance and conductivity for each of these objects. In practice, however, the available frequency range with commercial impedance analysis instruments is usually below 1 MHz. This is sufficient to study interfacial processes, such as corrosion of materials in liquids and diffusion in solid materials and liquid solutions.

Here, it should be noted that IS is a general term for this technique, while the term electrochemical IS (EIS) is commonly used when referring to an aqueous electrolyte. In this work, the term IS is mainly used, since the main goal was to study the resistive and capacitive properties

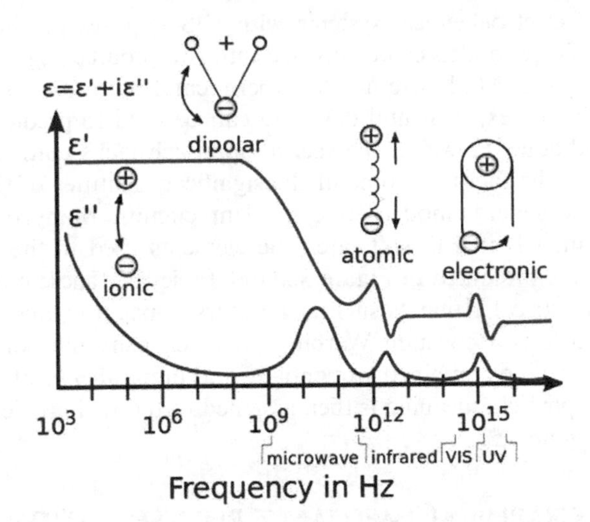

FIGURE 3.1 The dielectric properties of material, which can be at different frequencies of an applied perturbation signal.

TABLE 3.1
Relations between the impedance and its derivatives

parameter	Impedance, *Z*	Admittance, *Y*	Dielectric constant, *ε*	Electrical module *M*
Z		$Z' = Y'/\left(Y'^2 + Y''^2\right)$	$Z' = \varepsilon''/\omega C_0\left(\varepsilon'^2+\varepsilon'^2\right)$	$Z' = M''/\omega C_0$
		$Z' = Y'/\left(Y'^2 + Y''^2\right)$		
		$Z'' = -Y''/\left(Y'^2 + Y''^2\right)$	$Z' = -\varepsilon'/\omega C_0(\varepsilon'^2+\varepsilon''^2)$	$Z' = -M'/\omega C_0$
Y	$Y' = Z'/\left(Z'^2 + Z''^2\right)$		$Y' = \omega C_0\varepsilon''$	$Y' = \omega C_0 M''/\left(M'^2 + M''^2\right)$
	$Y'' = -Z'/\left(Z'^2 + Z'\right)$		$Y'' = \omega C_0\varepsilon'$	$Y'' = \omega C_0 M'/\left(M'^2 + M''^2\right)$
ε	$\varepsilon' = -Z'/\omega C_0\left(Z'^2 + Z''^2\right)$	$\varepsilon' = Y''/\omega C_0$		$\varepsilon' = -M'/\left(M'^2 + M''^2\right)$
	$\varepsilon'' = Z'/\omega C_0\left(Z'^2 + Z''^2\right)$	$\varepsilon'' = Y'/\omega C_0$		$\varepsilon'' = -M''/\left(M'^2 + M''^2\right)$
M	$M' = -Z''\omega C_0$	$\varepsilon' = \omega C_0 Y''/\left(Y'^2 + Y''^2\right)$	$M' = \varepsilon'/\left(\varepsilon'^2 + \varepsilon''^2\right)$	
	$M'' = -Z'\omega C_0$		$M'' = -\varepsilon'/\left(\varepsilon'^2 + \varepsilon''^2\right)$	

of solid oxide systems. AC IS is a valuable tool for studying both the bulk transport properties of materials and the electrochemical reactions at interfaces. Typically, AC impedance experiments are carried out over a wide range of frequencies (several mHz to several MHz), and the interpretation of the resulting spectra is aided by analogy to equivalent circuits involving simple components such as resistors and capacitors. In general, such equivalent circuits are not unique, and indeed there exists an infinite set of circuits that can represent any given impedance. It is common to select a physically plausible circuit containing a minimal number of components and, in a somewhat ad hoc way, assign physical significance to the derived parameters. Often, meaningful insight into material behavior can be gained from such analyses.

Characterization of electrochemical systems with EIS requires the interpretation of data using suitable models. These models can be divided into two broad categories: equivalent circuit models and process models. Models return to experimental data to estimate parameters that can adequately describe the experimental data and can be used to predict the behavior of the system in various conditions. However, the special approach lacks mathematical and physical rigor, and it can lead to the fact that some of the significant features of the data can be overlooked. We show some general models of equivalent circuits. Many of these models were included as standard models in EIS software. The elements used in the following equivalent circuits and equations are presented in Figure 3.2 and Table 3.2. These models are constructed using known passive electrical elements such as resistors, capacitors, and inductors, as well as distributed elements such as a constant Warburg phase element and constant phase element (CPE). These elements can be combined sequentially and in parallel to obtain complex equivalent circuits. A specific physical meaning is then assigned to the various elements of the equivalent circuit shown in Figure 3.2.

3.3 METHOD OF GRAPHICAL IMPEDANCE REPRESENTATION IN BODE AND NYQUIST COORDINATES

Let us have some system to which a harmonic voltage signal $U = U_0 \sin(\omega t)$ is applied and some signal $I = I_0 \sin(\omega t + \varphi)$ is taken. Our task is to graphically represent how the complex impedance $Z(j\omega)$ of the system changes when applying different frequencies ω. For the graphical representation of these complex functions, both polar and Cartesian coordinates can be used. Specifically, for the complex impedance $Z(j\omega)$, the plot of modulus $|Z(j\omega)|$ and phase angle $\theta(\omega)$ vs. frequency ω or $f = \omega/2\pi$ is called the Bode plot or Bode diagram. Like other spectroscopic plots, the frequency is explicit. The plot of imaginary part of the impedance, $\mathrm{Im}Z(j\omega)$ or $-\mathrm{Im}Z(j\omega)$, vs. real part of the impedance, $\mathrm{Re}Z(j\omega)$, is called the Nyquist plot. The

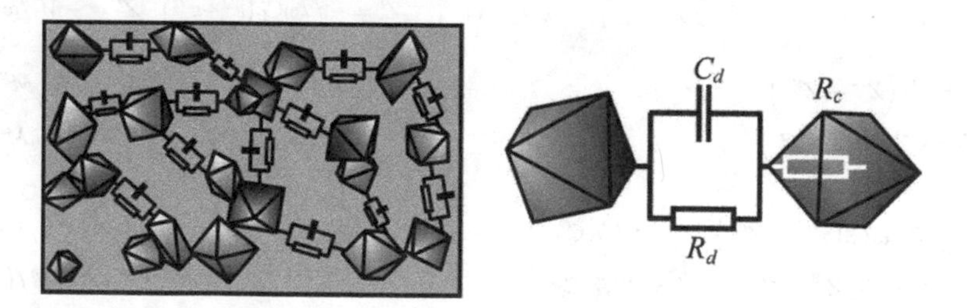

FIGURE 3.2 Example of material structure and equivalent circuit.

TABLE 3.2
Element of equivalent impedance circuits

Symbol	Element	Impedance
R1	R – resistor	$Z = R, Z' = R, Z'' = 0$
C	C – capacitor	$Z = (i\omega C)^{-1}, Z' = 0, Z'' = -(\omega C)^{-1}$
L	L – inductor	$Z = \omega L, Z' = 0, Z'' = \omega L$
CPE	CPE – constant phase element	$Z = (iT\omega)^{-\alpha}$
GE	GE – Gerischer element	$Z = (Y_0(R_0 + i\omega)^{0.5})^{-1}$
W_s	Warburg Short Circuit Terminus	$Z = R \tanh[(IT\omega)^{\alpha}] (IT\omega)^{-\alpha}$
W_o	Warburg Open Circuit Terminus	$Z = R \coth[(IT\omega)^{\alpha}] (IT\omega)^{-\alpha}$

Nyquist coordinates are the complex ohmic plane, the real component Z' is plotted along the x axis, and the imaginary resistance component Z'' along the y axis. As mentioned above, when representing the impedance as a complex number $Z = Z' - jZ''$, $Z' = Z_0 \cos(\varphi)$, and $Z'' = Z_0 \sin(\varphi)$. In this case, the frequency is implicit. It is worth mentioning that the plot of Imε $(j\omega)$ vs. Reε $(j\omega)$ is called the Cole–Cole plot and sometimes in the literature the name is exchangeable with the Nyquist plot. Unless specified, both the Nyquist plot and Bode plot are called the impedance spectrum. Figure 3.3 gives an example of the Nyquist plot and Bode plot for the same impedance function. The phase curve expresses the dependence of the phase angle φ in degrees on the frequency logarithm: φ (also denoted by θ) = $f(\log\omega)$. Such coordinates are used for mathematical analysis of functions, finding asymptotic dependences, and so on.

AC electric field is an informative working tool in various fields of electrochemistry, physics, and material science. The frequency ranges of an external alternating electric field superimposed on a sample usually varies from 10^{-6} to 10^{12} Hz (Table 3.3). In this frequency range, it is possible to study the dielectric properties of the sample bulk and electrode processes, namely:

- the interface between electrolyte and metal, metal oxide, or semiconductor electrode;
- transport properties of materials; establishing the mechanism of electrochemical reactions (from studying corrosion processes to reactions in biological and living objects);
- adsorption processes on the surface of the electrodes;
- the properties of porous electrodes.

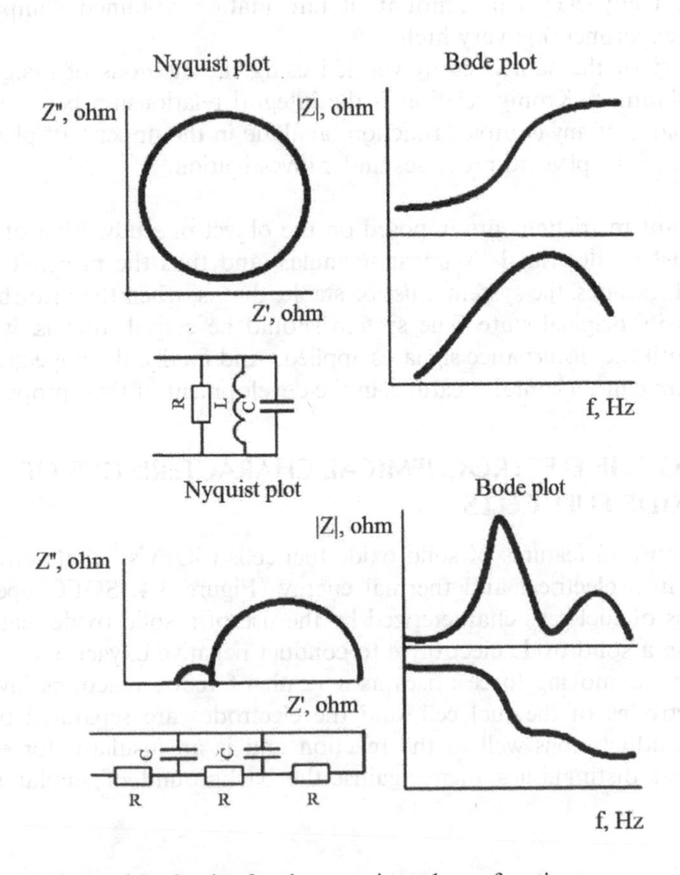

FIGURE 3.3 Nyquist plot and Bode plot for the same impedance function.

TABLE 3.3
Value of capacity theoretically expected for certain processes

Capacitance, F	Phenomenon responsible
10^{-12}	Bulk
10^{-11}	Minor, second phase
10^{-11}–10^{-8}	Grain boundary
10^{-10}–10^{-9}	Bulk ferroelectric
10^{-9}–10^{-7}	Surface layer
10^{-7}–10^{-5}	Sample-electrode-interface
10^{-4}	Electrochemical reaction

3.4 ADVANTAGES OF THE IS METHOD

(1) The linearity of the method involves the interpretation of the results in terms of the theory of linear systems – the output signal is a linear function of the input signal: $U(\text{output}) = I\,(\text{input})Z$.
(2) The impedance measured in a fairly wide frequency domain contains all the information that can be obtained using various constant current methods.
(3) Experimental efficiency (the amount of information obtained compared with the cost of the experiment) is very high.
(4) The reliability of the data is easily verified using the methods of integral transformation (the Kramers–Kronig relation is the integral relationship between the real and imaginary parts of any complex function, analytic in the upper half-plane), which are independent of the physical processes under investigation.

However, important restrictions are imposed on the object of study. First of all, the response of the system must be described by linear formulas, and thus the principle of superposition must be respected. Besides, the system must be stable, that is, when the disturbance is removed, it must return to its original state. The system should be causal, that is, it should not give a click-through until the disturbance signal is applied. And finally, the impedance must be finite and physical system cannot contain features in the development of their properties.

3.5 STUDYING THE ELECTROCHEMICAL CHARACTERISTICS OF SOLID OXIDE FUEL CELLS

Consider the design and features of solid oxide fuel cells (SOFCs)[1] that convert the chemical energy of a fuel into electrical and thermal energy (Figure 3.4, SOFC operation concept). SOFCs are a class of fuel cells characterized by the use of a solid oxide material as the electrolyte. SOFCs use a solid oxide electrolyte to conduct negative oxygen ions from the cathode to the anode. Electro moving force arises as a result of redox reactions involving free electrons on the electrodes of the fuel cell, and the electrodes are separated by an electrolyte, a material that conducts ions well in the reaction and is an insulator for electrons. SOFCs have a feature that distinguishes them against the background of similar systems. Oxygen

1 https://mypages.iit.edu/~smart/garrear/fuelcells.htm.

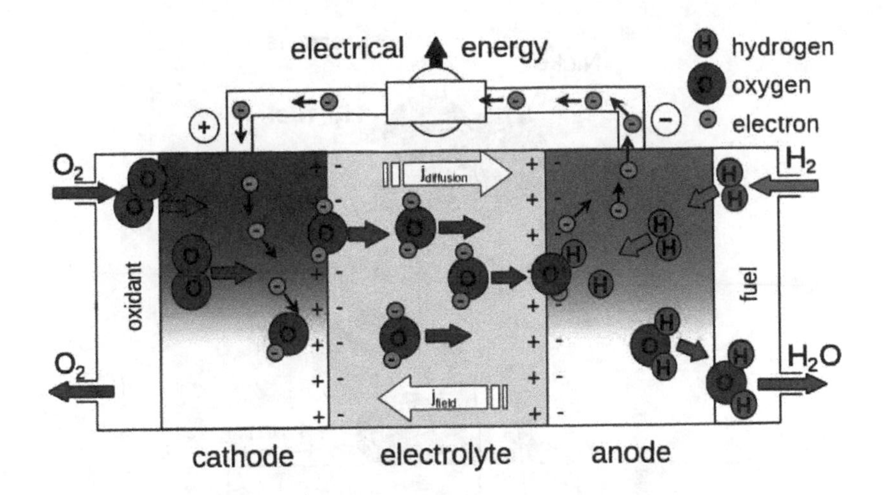

FIGURE 3.4 SOFC operation concepts.

Reproduced from Xia, Zou, Yan, and Li (2018) under the terms of CCA License.

ions are not rare or toxic. All functional parts of SOFC, including electrolyte, are solid oxides that are highly stable and chemically inert compounds. The flip side of the chemical stability of oxygen compounds is the low mobility of its ions in oxides. Ionic conductivity has an activation character that grows with temperature. SOFC operating temperatures are in the range of 700–900°C. In laboratory samples, satisfactory current densities can be achieved even at 500°C; however, as a rule, such samples have a critically small resource.

A fundamental component of SOFC is an electrode. From the generalization of the literature on various types of yttria-stabilized zirconia (YSZ) electrodes, it is clear that the resistance is highly dependent on the nature of the electrode. Different processes can limit depending on the composition of the electrode, processing parameters, and measurement conditions. Some trends may be revealed for high-, medium-, and low-frequency modes. At high frequency mode, the contribution presumably associated with the structure of the electrode may appear for porous and composite electrodes.

Wilson and Barnett (2008) propose 8YSZ systems (8 mol% of Y_2O_3-stabilized ZrO_2) with alloy content of yttrium oxide of 8 mol% with a characteristic composite structure size of 0.1 – 5 μm. Fuel (for example, hydrogen) is oxidized by oxygen ions coming from the electrolyte membrane at the anode according to the following reaction: $H_2(g) + O^{2-} = H_2O(g) + 2e$ (Figure 3.5). The electrons obtained in the reaction pass through an external circuit to the cathode, where they reduce oxygen: $\frac{1}{2}O_2(g) + 2e^- = O_2$. The reaction includes reagents from the gas phase and occurs on the surface of the electrodes, and therefore, to reduce their resistance, they are made porous. Typical power density obtained from a unit area of an electrolyte membrane is 200–2000 mW/cm^2 (Gewies, & Bessler, 2008).

The EIS method (Fedotov, & Bredikhin, 2013; Lasia, 1999) was adapted to studying SOFC. IS provides information on the characteristic times and resistances of various transfer processes that occur during operation of the fuel element. Each process is modeled, to a first approximation, by a chain of parallel connected resistances R and capacitance C, where the resistance corresponds to the active resistance introduced by this process into the total resistance of the element. The characteristic time ($t = RC$) corresponds to the relaxation time of the process. Consider how the resistance of gas flows appears. The reaction requires access of hydrogen and water vapor to the anode. The difficulty in transporting hydrogen and water vapor leads to a change in their chemical potentials in the reaction zone, which leads to a decrease in the cell voltage linear

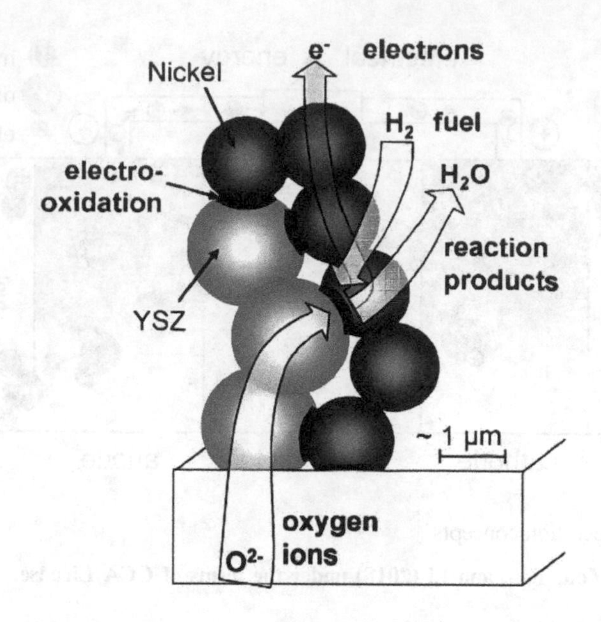

FIGURE 3.5 Scheme of the reaction at the three-phase boundary in the porous anode.

Adapted with permission from Waag et al., 2013, Copyright 2013, American Chemical Society.

in current at relatively low currents. Thus, this effect looks like the appearance of additional active resistance in the circuit of the element. When measuring impedance, in addition to resistance, a capacitance is also associated with this process, which reflects the ability of gas channels to accumulate matter.

In the Nyquist, coordinates of such a model corresponds a semicircle. Real processes are displayed by curves close in shape to a semicircle. In Figure 3.6, authors have presented the impedance spectrum calculated in the framework of this model. The impedance of the system Ni-8YSZ composite consists of two main parts: the low-frequency and the high-frequency semicircles, as well as the third, mid-frequency part. The anode was impregnated with nickel nanograins to test this model (see Figure 3.5). Nickel nanograins significantly increase the length of the three-phase

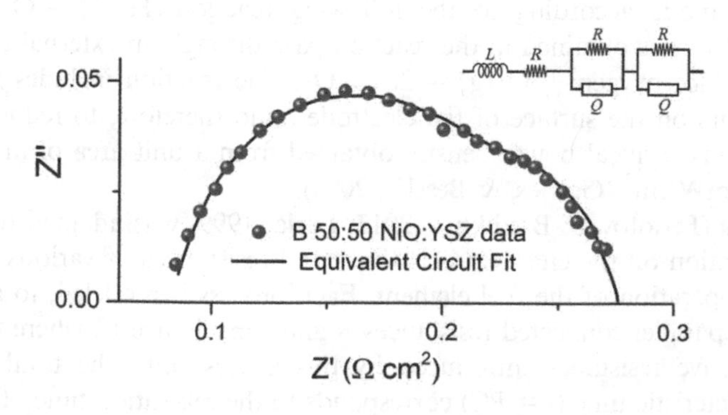

FIGURE 3.6 Nyquist plots of EIS data measured and equivalent circuit fit of the EIS data SOFC sample ($x_{NiO} = 50$).

Reproduced with permission from Wilson and Barnett (2008). Copyright 2008, Electrochemical Society.

boundary, slightly affecting the volume ratios of the conducting phases, which leads to the required decrease in the reaction resistivity with virtually no change in other system parameters.

The equivalent electric circuit is shown in Figure 3.6. From left to right: the first RC circuit with a characteristic frequency of the order of megahertz corresponds to the resistance and capacity of the electrolyte. At high frequencies, the cell begins to behave like a capacitor due to the inertness of oxygen ions compared to electrons. The electrolyte is sandwiched between two electron-conducting anodes, like an insulator between the plates of a capacitor. The last RC circuit describes the gas resistance, or low-frequency part of the impedance.

In Nakajima, Kitahara, and Konomi (2010) an impedance separation analysis of the anode and cathode of a practical SOFC is conducted. EIS with a two-electrode setup is applied to an anode-supported intermediate temperature microtubular SOFC composed of a $Ni/ZrO_{20.9}Y_2O_3$ 0.1 cermet anode, a $La_{0.8}Sr_{0.2}Ga_{0.8}Mg_{0.2}O_{2.8}$ electrolyte, and a $La_{0.6}Sr_{0.4}Co_{0.2}Fe_{0.8}O_3$ cathode (Figure 3.7). Measurements are carried out for the cell operated at 700°C with varying flow rates and compositions of the H_2/N_2 mixture gas fed into the anode and the O_2/N_2 mixture gas fed into the cathode. The anode and cathode impedances are thereby separately assigned to low- and high-frequency impedance spectra, respectively. An equivalent circuit model is applied to the spectra to acquire the polarization resistances and associated capacitances for the charge and mass transfer processes at the anode and cathode and the cell ohmic resistance. The variations in these circuit parameters are then obtained in accordance with current densities and anode gas-feed conditions. In addition, the hydrogen diffusion length correlated with the Nernst loss in the axial direction of the anode substrate tube is estimated. These parameters obtained separately and simultaneously for each part of the cell are informative for a detailed analysis and diagnosis of practical SOFCs under operation.

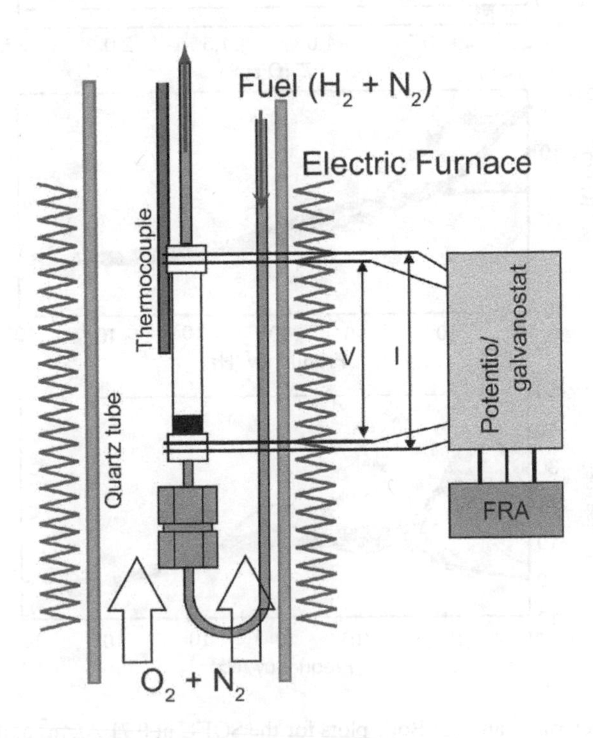

FIGURE 3.7 Experimental setup of the microtubular SOFC.

Reproduced with permission from Nakajima et al. (2010). Copyright 2010, Electrochemical Society.

SOFC has advantages including high-efficiency power generation by operation at 500–1000°C, which results in a low environmental load. Moreover, SOFCs provide high-quality waste heat, and the use of hydrocarbon fuels such as city gas, liquefied petroleum gas, and alcohol is relatively easy. However, for practical use, optimization of the electrode and electrolyte materials and the structure of the cell are required to improve the performance. In addition, optimization of the operation conditions and improvement of cell durability need to be addressed. Thus, fundamental data for the diagnosis of cell status under operation are of great importance. Monitoring the status is also necessary to prevent a serious accident due to the breakdown of the cell. EIS has been widely employed for such analysis. In particular, methods are needed for diagnosis of the operating status of the polymer electrolyte fuel cell by analyzing the variation of resistances and capacitances of equivalent circuit models of the polymer electrolyte fuel cell (Nakajima, Konomi, Kitahara, & Tachibana, 2008). Many EIS analyses for SOFCs have been reported in Ishihara, Shibayama, Nishiguchi, and Takita (2000).

An EIS analysis of a microtubular intermediate temperature SOFC with a two-electrode setup that can be operated in the intermediate temperature range of 500–800°C was carried out under operation with varying anode and cathode gas flow rates and compositions (Figure 3.8a, b). All

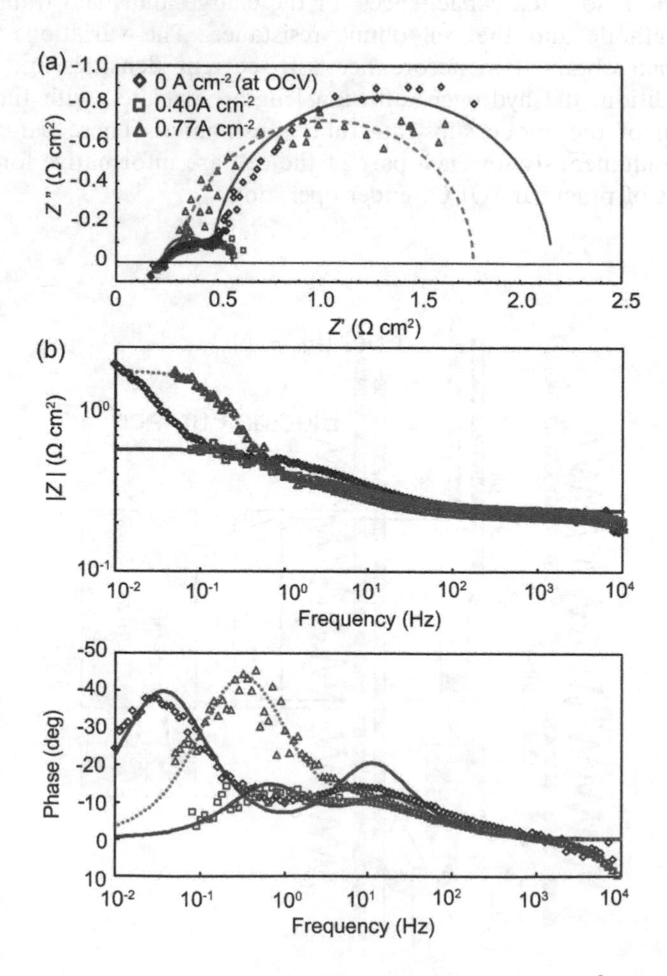

FIGURE 3.8 (a) Complex plane and (b) Bode plots for the SOFC at 0.71 A/cm^2 and 700°C under different anode gas flow rates.

Reproduced with permission from Nakajima et al. (2010). Copyright 2010, Electrochemical Society.

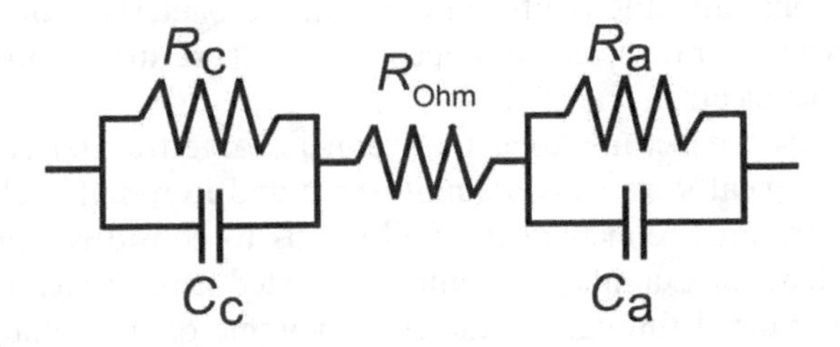

FIGURE 3.9 Equivalent circuit of SOFC.

Reproduced with permission from Nakajima et al. (2010). Copyright 2010, Electrochemical Society.

measurements were performed using an anode-supported microtubular SOFC (Kawakami et al., 2006) with outer and inner diameters of 5 and 3 mm, respectively. The thickness of the electrolyte was 30 mm (Saito et al., 2005). The anode substrate tube was made of $NiO/ZrO_20.9Y_2O_30.1$. An anode interlayer of $NiO/Ce_{0.9}Gd_{0.1}O_{1.95}$ for low-temperature operation was coated onto the anode substrate. $La_{0.8}Sr_{0.2}Ga_{0.8}Mg_{0.2}O_{2.8}$ was employed as the electrolyte. A layer of $Ce_{0.6}La_{0.4}O_{1.8}$ was inserted between the anode interlayer and electrolyte to prevent undesirable nickel diffusion during cell fabrication at high temperature.

The behaviors of the impedances and the equivalent circuit parameters of the cell were clarified. In a cell with distinctively different anode and cathode time constants, the impedances associated with the anode and cathode reactions and ohmic resistance appeared in different frequency regions in the impedance spectra, depending on the anode and cathode gas-feed conditions. Hence, those impedances can be analyzed simultaneously and separately without the reference electrode. This is informative for a detailed analysis and diagnosis of practical SOFCs. Thereby, the variations in the equivalent circuit parameters (Figure 3.9) for the charge and mass transfers at the anode and cathode and ohmic resistance under operation condition can be obtained separately. In addition, average diffusion length associated with the Nernst loss in the anode can be estimated from the parameters. Monitoring these parameters allows diagnosing the status of the electrodes, gas feed, and electrolyte/current collector individually. This enables the detection of the degradation of each part of a cell and abnormal operating state before serious accident.

3.6 STUDY OF ELECTRONIC TRANSPORT MECHANISMS IN BIOLOGICAL NANOOBJECTS

Let us see the adaptation of methods for the study of charge transfer for biosystems and a demonstration of their application on the specific example of one nanoobject. The problem of studying the mechanisms of electron transport in biological systems was first formulated in 1941 (Gralnick, & Newman, 2007). Until this moment, charge transfer was considered in living systems from the point of view of ion currents on which the work of excitable tissues – nerves, myocardium – is built. In the post-war period, a series of important works was carried out that determined the development of ideas about electronic transport in proteins. In 1956, R. Marcus (Richter et al., 2009) proposed his theory of external sphere electron transfer for simple ions, the coordination sphere of which did not undergo significant changes as a result of the redox reaction. It was quickly noticed that Markus ions are complexing agents of many prosthetic groups of proteins involved in electron transport in the respiratory chain of mitochondria, chloroplasts, and bacteria, such as heme iron and copper, iron–sulfur clusters, and so on. The experiments of Chance (Gorby, & Yanina, 2006), Nishimura, de Volta (Pirbadian et al., 2014; Subramanian, Pirbadian, El-Naggar, & Jensen, 2018), Vredenberg and Duisens (Reardon, & Mueller, 2013), performed in the late 50s–mid 60s, by estimating the electron transfer rate between *c*-type cytochrome and the photocenter in purple bacteria showed that starting from 100K and lower, the reaction rate ceases to depend on temperature. A possible explanation for this process is sub-barrier electron transfer.

In the following decades, it became clear that tunnel charge transfer is a very common phenomenon for many biological systems containing gems and iron–sulfur clusters. The theory of localized states for amorphous semiconductors allows us to consider a protein as a system in which there is a set of loci, for example, aromatic conjugated heme systems, in which part of the valence electrons are socialized throughout the entire volume of the conjugated system, that is, are quasi-free (Malvankar et al., 2011). The creation and study of doped aromatic semiconductors, such as polyaniline and polyacetylene, and subsequently organic superconductors, led to the understanding that large classes of organic compounds, including metabolites synthesized in biological systems (Mostert et al., 2012), theoretically can under certain conditions organize in conducting phases. Apparently, they did not deal with a directed search for such structures until the discovery of the phenomenon of long-range (by the standards of microbiology) charge transport in some bacterial colonies. Initially, a positive correlation was established between the number of filaments in cells, growing on the anode, and the amount of current generated by them. A wave of research on the conductivity of protein nanofilaments, hereinafter referred to as "pili" or "nanowires" (nanowires can be extended over long distances from cells to restore soluble iron oxide, turning it into a crystalline structure), began in 2000, after the discovery of facultative or obligate anaerobic species capable of restoring various organic and inorganic compounds to ensure the respiratory cycle in the body (Breuer, Rosso, Blumberger, & Butt, 2015).

Shewanella oneidensis MR-1 is a facultative anaerobe that respires using a variety of inorganic and organic compounds. MR-1 is also capable of utilizing extracellular solid materials, including anodes in microbial fuel cells, for electricity generation. Therefore MR-1 has been extensively studied to identify the molecular systems that are involved in electricity generation in microbial fuel cells. Kouzuma, Kasai, Hirose, and Watanabe (2015) have demonstrated the importance of extracellular electron-transfer (EET) pathways that electrically connect the quinone pool in the cytoplasmic membrane to extracellular electron acceptors. Figure 3.10 shows EET pathways (Mtr pathway) in *S. oneidensis* MR-1 for transferring electrons from intracellular electron carriers (e.g., NADH and quinones), across the inner membrane (IM) and outer membrane (OM), to extracellularly located insoluble electron acceptors.

Okamoto, Hashimoto, and Nakamura (2012) investigated the role of *c*-type cytochromes (*c*-Cyts) in electronic conductivity using *S. oneidensis* MR-1 biofilms and the importance of electronic conductivity in generating biological current. Single and multilayer biofilms were formed on indium tin oxide (ITO) electrodes. Current–voltage dependences were obtained from *c*-Cyts, which were electrically connected to the surface of the electrode. The thickness of the multilayer biofilm is 16 μm, which indicates a redox peak with an 8-fold larger pendant area than a single-layer biofilm (about 0.5 μm thick), which indicates an excess of *c*-Cyts that are capable of performing the redox process. To determine whether this *c*-Cyts electron channel is involved in the generation of biological current, authors performed voltammetry with slow scanning of multilayer biofilms. The high anode current of *c*-Cyts, caused by the microbial oxidation of lactate, was observed during scanning of the slow potential, demonstrating the transfer of respiratory electrons through the sequential redox cycle of c-Cyts. Therefore, it is most likely that the redox cycle of *c*-Cyts in multilayer biofilms is a consequence of sequential electronic exchange through the physical cycle between OM *c*-Cyts (the outer membrane (OM) of the cell c-type cytochromes (*c*-Cyts)) (Figure 3.11a). As shown in the proposed model, if electrons formed as a result of metabolic oxidation of lactate along the surface of the cell are delivered to the farther electrode, the surface filling OM *c*-Cyts should fill the percolation threshold, which is estimated as 50% in two dimensions according to the network model with resistance. The authors suggest that the mobility of cells present in the biofilm helps establish flow conditions by increasing the apparent surface filling of *c*-Cyts in the biofilm matrix,

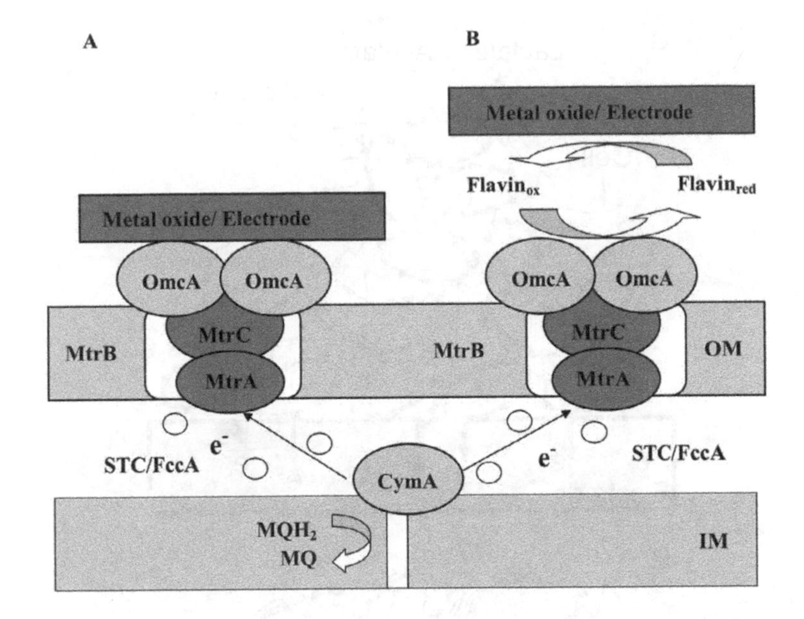

FIGURE 3.10 Proposed EET pathways (Mtr pathway) in *S. oneidensis* MR-1 involved in direct EET (a) and mediated EET (b). OM, outer membrane; IM, inner membrane; MQH_2, reduced form of menaquinone; MQ, oxidized form of menaquinone. Genetic and biochemical studies have identified five primary protein components, CymA, MtrA, MtrB, MtrC, and OmcA, comprising the EET pathway.

leading to electronic conductivity through the multilayer biofilm (Figure 3.11b). Experiments with deletion mutants deficient in *c*-Cyts (OM) *c*-Cyts (ΔmtrC/ΔomcA, ΔpilD) and the biosynthetic protein of the capsular polysaccharide (ΔSO3177) suggested that cell-surface-bound *c*-Cyts, those located on pili or extracellular polymeric substrates, play a predominant role in the long-range electron conduction in the biofilm of *S. oneidensis* MR-1.

Cytochrome *c* molecules with multiple homes transfer electrons from membrane-soluble quinones in the inner membrane (MQH_2) to the periplasm (CymA). Electrons then move through the outer membrane (MtrA, MtrB protein) to the extracellular surface (MtrC and OmcA), where they can contact insoluble metals directly (pathway to the right) or be shuttled by flavins to metals at a farther distance from the cell surface (pathway to the left) (Fredrickson et al., 2008). In Schmid et al. (2010), pilli was studied using the methods atomic force and transmission electron microscopy, surface-enhanced Raman scattering, chromatography-mass spectrometric analysis, nuclear magnetic resonance, magnetic susceptibility and electron paramagnetic resonance, scanning probe microscopy, and IS. Kinetics is carried out by special filaments (saws) that some bacteria use to transfer electrons to the environment. The chemical composition and structural organization of filaments can have many differences. The common properties of such filaments are large spatial anisotropy (length ≈ 10 μm at a cross section of ≈ 10 nm), the presence of a large number of aromatic amino acids, and/or aromatic prosthetic groups (gems) in the protein components, as well as the ability to form branched networks that combine bacteria of one or several species constituting the colony. Protein structures are sensitive to physical and chemical effects – pits. Owing to these classical methods, photo and electronic lithography are not applicable in this case, as a result of widespread use in technological organic solvent lines, for example, commercially available and widely used electronic resistive polymethylmethacrylate dissolved in toluene. After application of the irradiation of the necessary areas of electrons, the manifestation occurs by means of a mixture of methyl isobutyltones and isopropyl alcohol (1:3 vol.). The most common

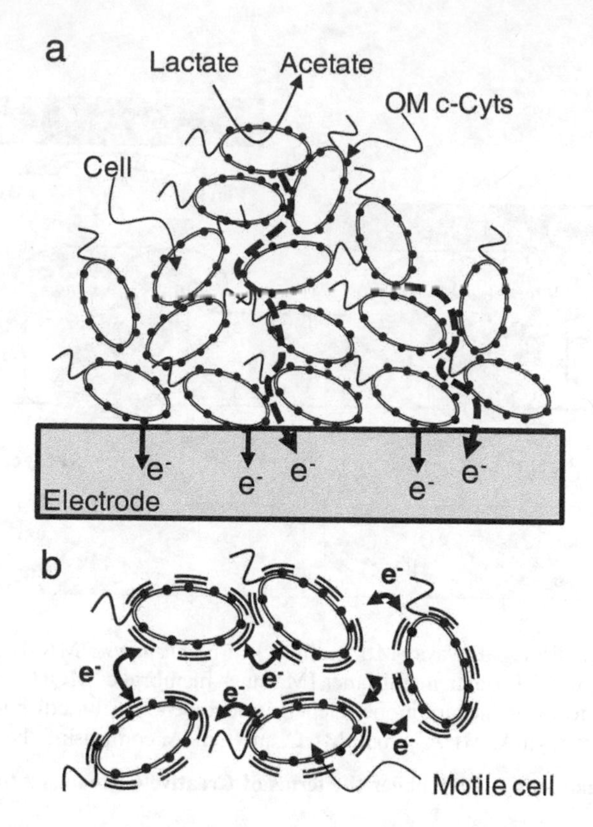

FIGURE 3.11 (a) Schematic illustrations for electron conduction in a multilayer biofilm of MR-1 cells mediated via OM *c*-Cyts. Metabolically generated electrons are transferred from the upper and middle parts of the biofilm to a distant electrode. (b) Speculated model for biofilm-mediated electron conduction. Electron exchange along the cell-surface *c*-Cyts developed a percolation pathway of electrons across multilayer biofilm by the assistance of cell motility to increase the apparent surface coverage of OM *c*-Cyts.

Reproduced with permission from Okamoto et al., 2012, Copyright 2012, Elsevier.

methodology is use of *Pt(acac)2* and *W(CO)6* and focused gallium ion beam. Under the conditions of the experiment, this method allows the formation of contact longer than 10–15 μm.

Grebenko et al. (2018) demonstrate that the apparent conductivity of nanofibers is critical to moisture and contains several components, including those that have unusual behaviors assigned to electron transport. It was demonstrated that in the case of *S. oneidensis* MR-1 charge transfer, water is strongly mediated within these objects. Based on current analysis of conductivity data, it was concluded that the studied filaments of *S. oneidensis* MR-1 are capable of hybrid (conjugated) electron and ionic conductivity.

The shape and location of contacts are maximally reproduced case sample "Short circuit" – shorting two photolithographic gold contacts platinum or tungsten deposited with a potassium beam. Figure 3.12 shows nanofilaments with organized platinum nanocontacts marked as IBID Pt on image. "Dummy" is a section of silicon between ePt + iPt contacts that does not contain an anode. In some cases, "dummies" were described by a pure capacitance semicircle (blue vertical line on saw sample lay within the high-frequency range from 2 MHz to ≈ 10 kHz, see Figure 3.13).

The test structure (see Figure 3.13a) can be described by the classical Warburg element (the linear part of the graph) at low frequencies (below 5 kHz), which usually corresponds to the diffusion of the reagent in the electrolyte solution. To determine the parameters of redox

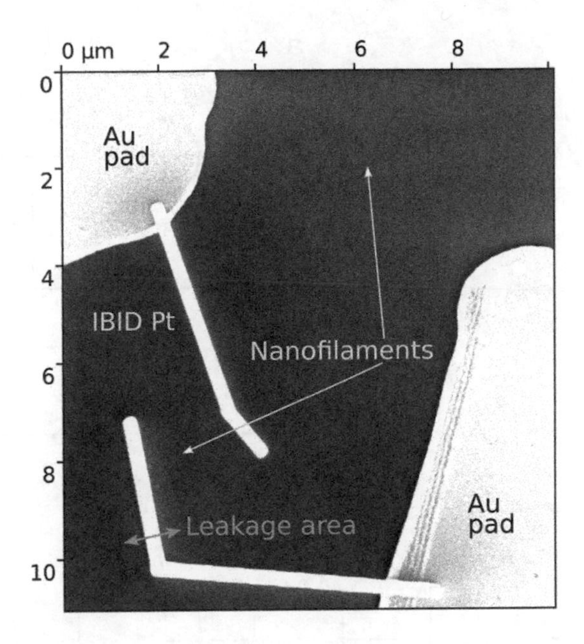

FIGURE 3.12 Several nanofilaments with organized platinum nanocontacts fabricated by ion beam (marked as IBID Pt on image). The leakage area can be seen as a brighter area around the contact and is indicated by green arrow.

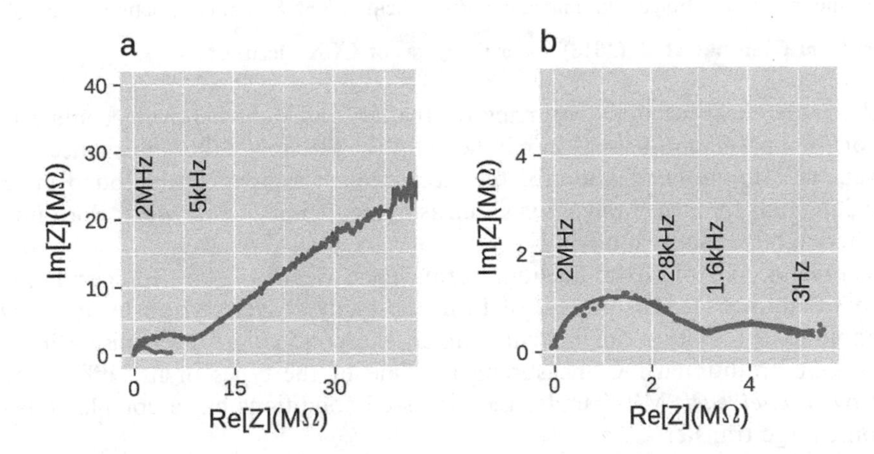

FIGURE 3.13 Impedance Nyquist plots. (a) Comparison of the filament (red line) and empty circuit with almost identical nanocontact geometry (blue line); half circle that continued by straight line with the slope of ca. 45° represents reactant diffusion (Warburg element) in test structure, while nanofilament hodograph consists only of several half-circles. (b) Characteristic nanofilament hodograph: experimental data (red dots) and calculated line for equivalent scheme (blue line).

reactions and diffusion of reagents in the surrounding medium, the authors selected the impedance of the test circuit structure according to the ES scheme (empty structure) (see Figure 3.14). This scheme describes the sample with nanofilament (three bottom *R*-*C* blocks,

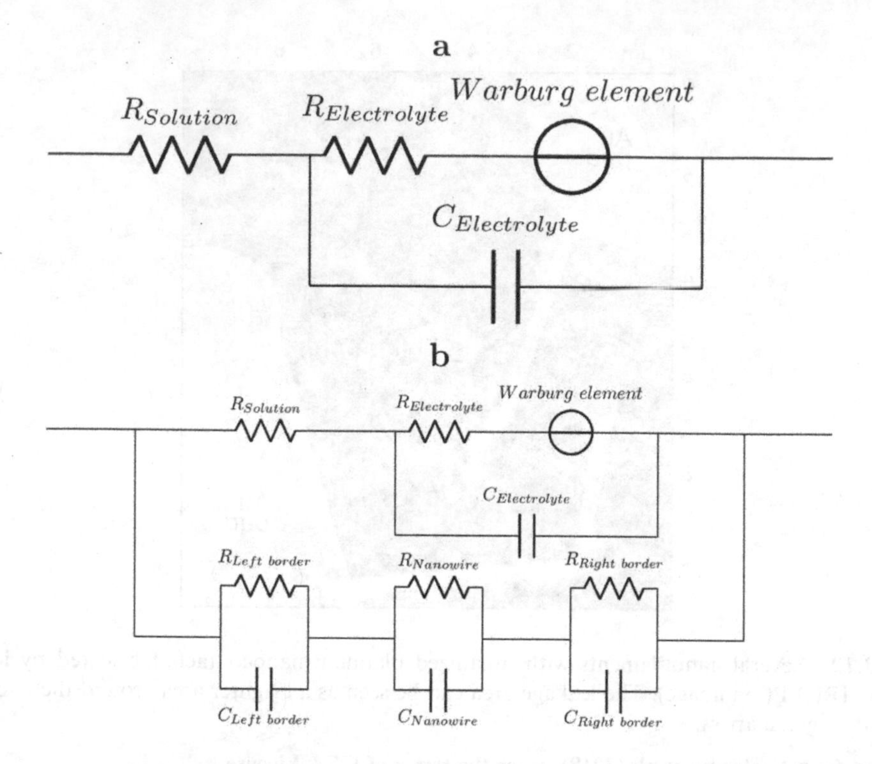

FIGURE 3.14 Equivalent schemes. (a) Scheme ES (empty structure), equivalent scheme for empty circuit test structure, which models redox reactions and diffusion of reactants in the surrounding medium. Technically, this scheme models electrolyte film impedance. (b) Scheme ES+NF (empty-structure + nanofilament).

two for electrochemical reactions occurring on the left and right nanowire–contact interfaces and one for the nanofilament itself) in parallel with the electrolyte film impedance.

However, the straight-line feature for the nanofilament sample can be found in lower frequency region. Full frequency range for all measurements is 1 Hz –2 MHz. Both impedances were acquired at 80% relative humidity.

Finally, precise control of the medium during measurements revealed specific properties. The authors demonstrate that the use of IS in the study of charge transfer in *S. oneidensis* MR-1 nanofibers provides more information on the conduction mechanisms in microbial nanofibers than traditional DC measurements. One of the types of nanofibers intensively produced by *S. oneidensis* MR-1 under the discussed conditions has a complex type of conjugated ion charge transfer.

3.7 STUDY OF IONIC TRANSFER IN SOLID ELECTROLYTES WITH NASICON STRUCTURE

Sodium (Na) Super (S) Ionic (I) Conductors (CON) (or NASICONs, for example $Na_{1+x}Zr_2P_{3-x}Si_xO_{12}$) are crystalline solids $A_1B_2(PO_4)_3$, where A is a monovalent cation and B is either a single or combination of tri, tetra, and penta valent ions. These compounds are a class of structurally isomorphous 3D framework compounds possessing high conductivity, often comparable to that of liquid electrolytes at higher temperatures. The high ionic conductivity of these materials is used in making devices such as membranes, fuel cells, and gas sensors. Advantages of IS method for such materials' investigation consist in the fact that it comes down to measuring the response of the system to very weak external influences (polarization by AC), deviating the system from equilibrium; therefore, in the process of measurements there is a noticeable change in solid electrolyte. Ionic conductivity of NASICONs even at room temperature is quite high (~10^{-4} –10^{-3} S/cm), which allows to uniquely interpret hodographs of impedance.

In Ahmadu, Tomas, Jonah, Musa, and Rabiu (2013), impedance spectra of mixed alkali $Na_{0.25}Li_{0.75}Zr_2(PO_4)_3$ were modeled based on an equivalent electric circuit consisting of resistor (R) and capacitance (C) elements. Polycrystalline materials are known to exhibit intragrain and grain boundary impedances and thus they can be represented by the equivalent circuit shown in Figure 3.15. The circuit has two RC elements joined together in series representing the grain (R_g, C_g) and grain boundary (R_{gb}, C_{gb}) (Figure 3.16).

Figure 3.17 shows the frequency dependence of the imaginary part of impedance (Z'') at different temperatures. Only one prominent peak is observable at each temperature and it systematically shifts toward higher frequencies with increase in temperature after an initial decrease. This implies a thermal activation of the conductivity relaxations that are within a decade of frequency (1.55 × 10^7–2.32 × 10^6 Hz) in the temperature range (470– 600K). The electrical conductivity and dielectric behavior of $Na_{0.25}Li_{0.75}Zr_2(PO_4)_3$ was investigated in the GHz range of frequency using a model RC circuit (Figure 3.15). The values that authors received for activation energy for the grain boundary and bulk were ~0.40 and 0.36 eV, respectively. Similarly, the maximum conductivity obtained for the bulk was 0.3 S/m, which compares favorably with typical NASICON materials. The dielectric permittivity ε' showed

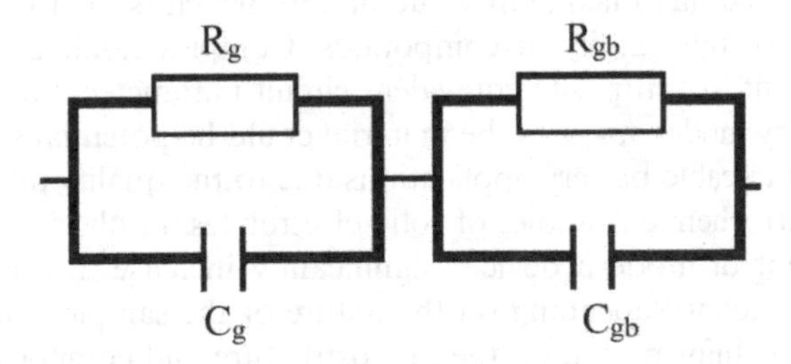

FIGURE 3.15 Model RC circuit for polycrystalline materials.

Reproduced from Ahmadu et al., 2013, Diamond Open Access Publishing Policy

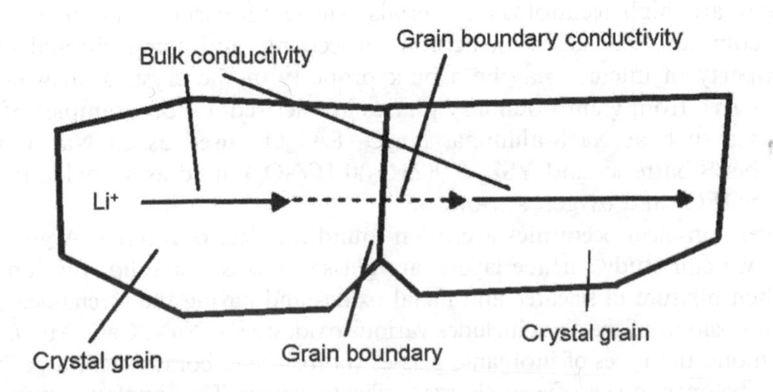

FIGURE 3.16 Presenting the grain (R_g, C_g) and grain boundary (R_{gb}, C_{gb}).

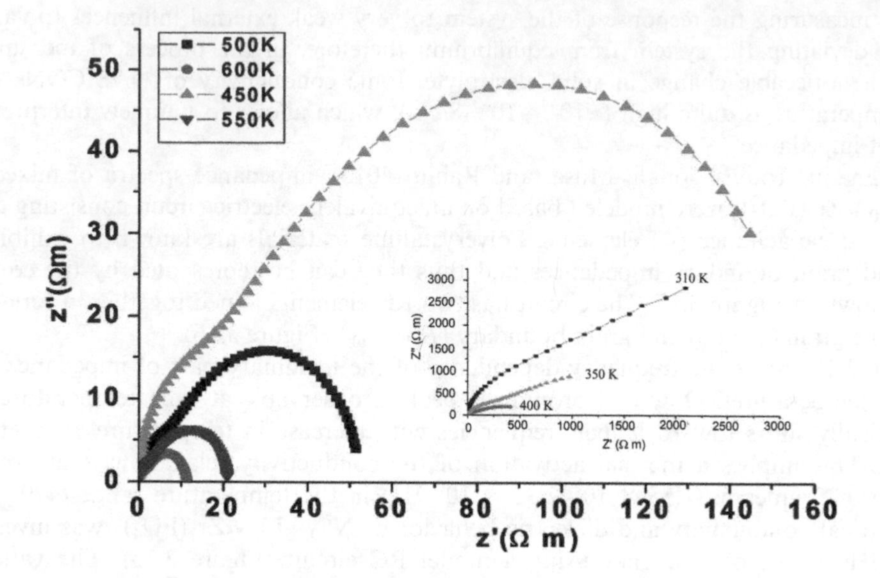

FIGURE 3.17 Complex plane plots (Z'' vs. Z') at different temperatures.
Reproduced from Ahmadu et al., 2013, Diamond Open Access Publishing Policy.

a non-Debye behavior and has maximum value of 136, which is high in comparison with the most reported values for this family of compounds. Complex nonlinear least squares fitting results indicate a good fit to composite equivalent circuit parameters. Moreover, further fitting using the generic battery model suggests the material could be potential solid electrolyte material for lithium ion rechargeable battery applications due to the quality of the fitting parameters obtained. Thus, in electrochemical studies of solid electrolytes, IS plays a huge role, and its capabilities with the advent of modern devices significantly increase. But a variety of responses should be taken into account depending on the nature of the sample being studied and set of experimental conditions. Information on the microstructure and composition of the solid electrolyte gives the opportunity to make a choice in favor of one or another equivalent circuit for the interpretation of the impedance spectrum.

3.8 CHARACTERIZATION ELECTROCERAMICS BY IS

Electroceramics are high technology materials whose properties and, hence, applications depend on a complex interplay of structural, processing, and compositional variables. The particular property of interest may be a bulk property of the crystals, in which case, fully dense ceramics free from grain boundary phases are desired. Good examples of this are ceramic electrolytes such as Na 8-alumina, Na_2O- $8A1_2O_3$, used as an Na^+ ion conducting membrane in Na/S batteries and YSZ, $0.9(ZrO_2)0.1(Y_2O_3)$, used as an oxide ion conducting membrane in SOFCs and oxygen sensors.

Surface layers on electroceramics are often found on electroceramics. A good example of objects where we can study surface layers, are glasses. Glass is a solid solution obtained by cooling a molten mixture of silicates and metal oxides and having the mechanical properties of solids. The composition of the glass includes various oxides: $SiO_{21}Na_{20}$, CaO, MgO, B_2O_3, Al_2O_3, and others. Among the types of inorganic glasses (borosilicate, borate, etc.), a particularly large role in practice belongs to silica-fused glasses – silicate glasses. The simplest composition is glass obtained by melting pure silica to form a vitreous mass.

A good example is provided by lithium silicate (Irvine, Sinclair, & West, 1990). Although this transparent glass is apparently atmosphere-stable and water-insoluble, it does in fact form a hydrated/carbonated surface layer very quickly during cooling of the melt. This shows up in the impedance spectrum (see Figure 3a in Irvine et al., 1990) as an additional poorly resolved semicircle. A glass without this layer (see Figure 3b in Irvine et al., 1990) shows only one semicircle, corresponding to lithium ion conduction through the bulk of the glass and a nearly vertical, low frequency "spike" representing charge build-up at the blocking metal electrodes.

Using IS, it is possible to study the conditions for formation/removal of the surface layer and the associated kinetics. For example, by analyzing the impedance spectra of electrically conductive ceramics based on two-component oxide ionic conductors of the $CaO–Al_2O_3$ system by the quantities related to intragranular, it is possible to distinguish conductivity and electrical conductivity through grain boundaries. As an example, in Irvine, Sinclair, and West (1990), Figure 1a presents impedance diagrams for oxygen-ion conductor $Ca_{12}Al_{14}O_{33}$. Each semicircle in the diagram corresponds to a parallel combination capacitance and resistance. The numerical values of the resistances can be obtained from the points of intersection of the circle with the Z axis, and the capacitance values can be obtained from the equation for the maximum frequency value $f_{max} = RC\omega$. For correlation between experimental values of capacity and theoretically expected for certain processes, we can use Table 3.3. Thus, in the given example, the left semicircle, for which the experimental value of C is 1×10^{-12} F corresponds to the volumetric resistance of individual grains of the sample. Right semicircle with value $C = 4 \times 10^{-9}$ F characterizes the resistance of grain boundaries. Using these quantities, it is possible to construct the temperature dependences of the volumetric and grain-boundary conductivity of the sample. Thus, using IS, conditions and kinetic patterns of formation or removal of the surface layer can be studied.

Perovskite family lead-free ceramics such as for example $Ba(Fe_{0.5}Nb_{0.5})O_3$_$BaTiO_3$ (Meera et al., 2012) has been widely studied for their high dielectric and ferroelectric properties in different temperature and frequency ranges, up to about 10 MHz and ~700 K. Similar studies have been carried out on other compositions from this family by some researchers $(1\text{-}x)Ba(Fe_{0.5}Nb_{0.5})O_3\text{-}xBaTiO_3$ (Singh, Kumar, & Rai, 2011), $Bi(Mg_{0.5}Ti_{0.5})O_3\text{-}PbTiO_3$ (Sharma, Rai, Hall, & Shackleton, 2012), and $(K_{0.5}Bi_{0.5})(Fe_{0.5}Nb_{0.5})O_3$ (Sahoo, Pradhan, Choudhary, & Mathur, 2012). Both modulus spectroscopy and dielectric conductivity formalism were employed to study dielectric relaxation phenomena. The temperature dependence of dielectric and loss spectra was investigated. These compounds show a typical negative temperature coefficient behavior like that of semiconductors. The frequency dependence of AC conductivity is well fitted to Jonscher's single power law.

3.9 THE IMPEDANCE OF THIN DENSE OXIDE CATHODES

Mixed conductivity oxides find important applications in SOFCs, both as cathodes and anodes, and in semi-permeable membranes for (partial) oxidation reactions. A very important aspect of these applications is the transfer and reduction of the amount of oxygen in the environment at the gas/solid interface or vice versa. The general transfer reaction can be represented as $O_{2g} + 4e' + 2V_o^{\infty} \longleftrightarrow 2O_o^x$. The article (Boukamp, Hildenbrand, Nammensma, & Blank, 2011) evaluates the effects of layer thickness, oxygen diffusion, and surface exchange rate on the general finite length diffusion. A simple model was obtained for the impedance of a dense $La_{0.6}Sr_{0.4}Co_{0.2}Fe_{0.8}O_{3\text{-}\delta}$ (LSCF) cathode with different thickness deposited by pulsed laser deposition (PLD) on 2.5 mm thick $Ce_{0.9}Gd_{0.1}O_{1.95}$ pellets in a three-electrode arrangement. The PLD was performed with a KrF excimer laser, using a fluency of 2.6 J/cm^2 and a frequency of 20 Hz. The LSCF target was an isostatically pressed pellet on a rotating holder.

The laser ablation occurred in a vacuum chamber in 0.02 mbar oxygen ambient. The CGO substrates were heated to 750°C during deposition. The reference electrode was provided by a Pt-wire bonded with a little Pt-ink into a groove at half height at the cylindrical side of the pellet. For the counter electrode, a similar PLD layer with different thickness was applied. The proper derivation of the impedance of such an arrangement of electrodes is not yet available in the literature. The impedance is obtained for an electrode with a dense layer of mixed conductive oxide, provided that the electronic resistance can be ignored. The effect of layer thickness on electrode properties and some preliminary results on the addition of chromium are also presented. Analysis of the surface shows that the PLD process easily leads to significant contamination of the Cr surface of the LSCF. An analysis of EIS shows that the effect on the exchange rate of this Cr contamination is still negligible. The use of a porous Pt-counter electrode is not advisable, as the electrode properties are inferior to the LSCF electrodes and will result in a strong pseudo-inductive artifact in the impedance (see Figure 3.18).

3.10 IS STUDY OF POROUS INDIUM TIN OXIDE-BASED GAS SENSOR

It was reported that many transparent conductive oxides, such as ZnO, In_2O_3, and SnO_2 can be used in a gas sensor (Hsueh, & Hsu, 2008). Their popularity as gas-sensitive materials is due to their suitable physicochemical properties, such as natural nonstoichiometric structures (Kaur, Kumar, & Bhatnagar, 2007), and free electrons arising from oxygen vacancies contribute to a change in electronic conductivity with a change in the composition of the surrounding atmosphere (Chu, Zeng, Jiang, & Masuda, 2009). In Saadoun, Boujmil, El Mir, and Bessais (2009), the gas-sensitive properties of ITO films were studied in DC measurements, which provide information on the global response of the sensor. The study of AC impedance is necessary to understand the nature of the conductivity processes and the gas/solid interaction mechanism).

To investigate the ITO sensing properties toward NO_2, Madhi, Saadoun, and Bessais (2012) have studied the nature of conduction processes and model-sensitive mechanisms in the impedance of AC of ITO films obtained by screen printing. All impedance spectra appear to be single semicircles without displacement along the Z' axis (Figures 3.19–3.21).

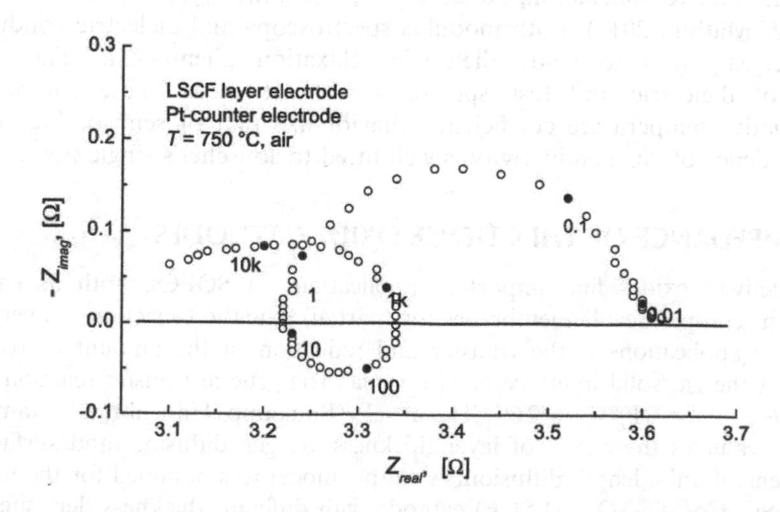

FIGURE 3.18 Inductive artifact in the electrode impedance in a three-electrode cell with a porous Pt electrode with a significant larger polarization resistance than the LSCF working electrode.

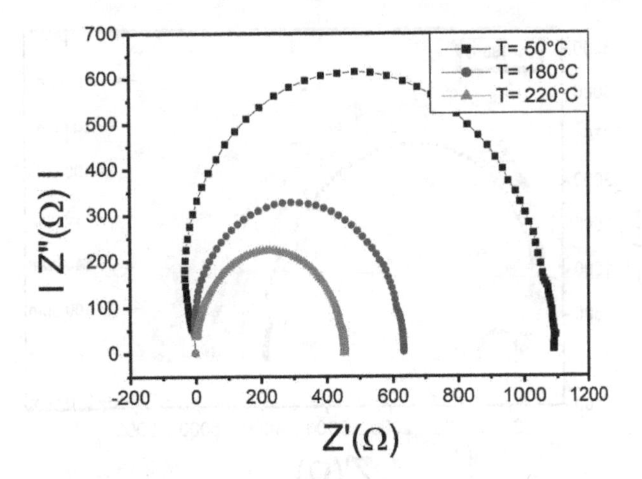

FIGURE 3.19 Nyquist diagram at various operating temperatures in ambient air.

Reproduced with permission from Madhi et al. (2012). Copyright 2012, Elsevier.

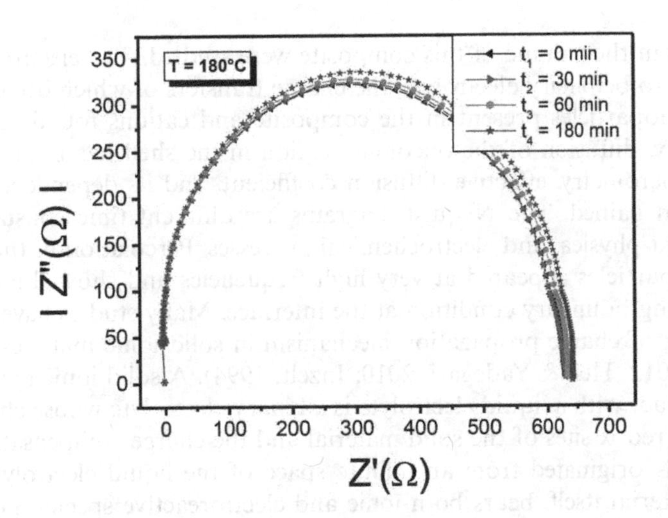

FIGURE 3.20 Impedance variation consecutives to repetitive excitation by 200 ppm of NO_2.

Reproduced with permission from Madhi et al. (2012). Copyright 2012, Elsevier.

The impedance spectra have a semicycle shape indicating the homogeneity of the grains (Figure 3.19). The spectrum remains quasi-stable and no memory effects were observed (Figure 3.20). Figure 3.21 illustrates the effect of NO_2 concentration on a Nyquist pattern. The impedance (Z') increases when exposed to NO_2 and begins to saturate when the concentration exceeds 160 ppm. The equivalent circuit can be decomposed in the simplest case in the form of a parallel *R–C* circuit, similar to other transparent conductive oxides.

3.11 STUDIES OF PROPAGATION MECHANISM IN SOLID IONIC MATRICES

Heli, Sattarahmady, and Majdi (2012) made a composite electrode of graphite, Nujol, and nanoparticles of the core of Fe_2O_3 – hexacyano-cobalt ferrate shell was prepared and charge

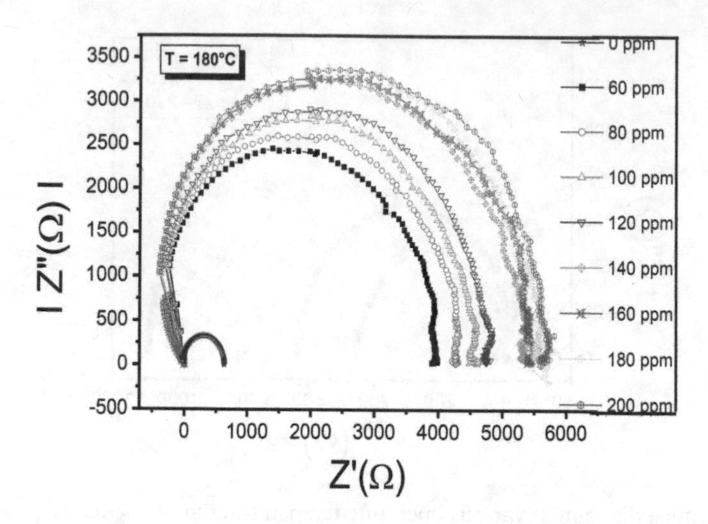

FIGURE 3.21 IS of the screen-printed ITO film vs. NO_2 concentration.

Reproduced with permission from Madhi et al. (2012). Copyright 2012, Elsevier.

transfer processes in the volume of this composite were studied. The electrode/solution interface was assumed to be binary electrolyte, the charge transfers of which occurred between the redox sites of nanoparticles present in the composite and cations found in solution. Using cyclic voltammetry, diffusion of the oncoming cation in the shell has been investigated. The use of chronoamperometry, effective diffusion coefficient, and its dependence on the applied potential has been gained. The Nyquist diagrams are different time constants appeared in relation to different physical and electrochemical processes. Percolation of the electrons in the shell of the nanoparticles appeared at very high frequencies and showed a diffusion feature process with passing boundary condition at the interface. Many studies have been devoted to the understanding of charge propagation mechanism in solid ionic matrices (Heli, Sattarahmady, & Majdi, 2012; Heli & Yadegari, 2010; Inzelt, 1994). A solid ionic material-containing redox sites in contact with a liquid electrolyte is a binary electrolyte whose charge transport is occurred between redox sites of the solid material and the charge compensation is performed by the counterions originated from an infinite space of the liquid electrolyte. On the other side, the solid material itself, bears both ionic and electroreactive species. Depending on the experimental conditions (e.g., type of solid matrix, nature and concentration of redox centers, and mobility and availability of counterions), the overall charge propagation process can be controlled by a variety of phenomena such as electron self-exchange rate, counterion migration, and ion pairing (Dalton et al., 1990; Surridge et al., 1989).

However, lower slope line appears at low frequencies in the Nyquist diagrams (see Figure 2 in Heli et al., 2012). This behavior can be characterized using CPE instead of the pure capacitance.

The roughness at the electrode/interface (local inhomogeneity presents at the electrode surface and nonuniform distribution of local capacities) and consequently, a distribution of activation energies of the processes occurring in the double layer causes the appearance of CPE behavior. The Nyquist diagrams in this DC potential range can be characterized with the electrical equivalent circuit as shown in Figures 2 and 6 in Heli et al. (2012). In this circuit, R_S, W, and CPE are the solution resistance, semi-infinite Warburg element related to the electron diffusion, and a CPE related to the double-layer capacitance, respectively.

3.12 MESOPOROUS α-Fe_2O_3 Films

Cummings, Marken, Peter, Upul Wijayantha, and Tahir (2011) studied thin mesoporous α-Fe_2O_3 films that were obtained on conductive glass substrates using self-layer assembly of ~4 nm aqueous oxide nanoparticles followed by calcination. As noted above, α-Fe_2O_3 (hematite) has been widely studied as a potential photoanode material for cells that destroys water (Dare-Edwards et al., 1983). Although its forbidden band (2.0 eV) is suitable for collecting visible light, and its valence band has sufficiently low energy for holes to oxidize water, an external bias is necessary in order to raise the energy of free electrons in the conduction band sufficiently for control reaction of hydrogen evolution at the counter electrode for water splitting The necessary external voltage bias can be provided by the solar cell in the configuration of the tandem cell (Brillet, Gratzel, & Sivula, 2010). Performance limitations imposed by high doping lengths and short diffusion holes in hematite films were damaged to increase the probability of holes that reach the interface between the oxide/electrolyte interfaces (Brillet et al., 2010). Electrodes were used to study the oxygen evolution reaction (OER) in the dark and when illuminated using *in situ* potentially modulated absorption spectroscopy (PMAS) and light modulated absorption spectroscopy (LMAS) in combination with IS. The formation of surface-bound valence types of iron (or "surface traps" holes) was derived from the PMAS spectra measured in the region of the beginning of the OER. Similar LMAS spectra were obtained at more negative potentials in the region of the onset of photoelectrochemical OER, which indicates attraction of the same intermediaries. The original solution in the work was to attract the impedance characteristics of the mesoporous electrodes α-Fe_2O_3. Frequency allowed measurements of PMAS and LMAS revealed slow relaxation, which may be related to the impedance response and that indicates that the lifetime of the intermediates (or trapped holes) involved in the OER is extremely long.

Electronic transport and electron transmission can be represented in a transmission line model as shown in Figure 3.22a (additional series resistance, R, arising mainly from substrate and fluorine-doped tin oxide contacts, not shown). The model describes charge transfer in the electrode material as well as transfer and storage of charge on the electrode/electrolyte interface: r_{trans} is a distributed transport resistance, R_{ct} is the charge transfer resistance, and C_{surf} is the capacitance of α-Fe_2O_3/electrolyte interface. The dominant contribution to potential is being made pseudo-capacitive associated with a surface redox reaction. The particle size is small enough (4 nm) to bend the strip, may be negligible, which means that the contribution space charge capacity can be neglected. Since the film α-Fe_2O_3 is mesoporous, electron (or hole) transport due to diffusion and charges in particles will be effectively protected by electrolyte, and the macroscopic electric field will be negligible. For low frequencies, the transmission line is reduced to parallel RC, the circuit shown in Figure 3.22b, which is in series with the resistance fluorine electrode substrate. Time constant $R_{ct}C$ corresponding to the discharge of pseudocapacitance through the Faraday resistance effectively represents the lifetime of surface-reduced redox species.

3.13 CPE BEHAVIOR CAUSED BY RESISTIVITY DISTRIBUTIONS

Further, our discussion will concern such an important element of equivalent impedance circuits as CPE. Generally, the analytic expression for the resulting impedance provides a useful relationship between system properties and CPE parameters. Application of such an approach to experimental data is reported in a paper by Hirschorn et al. (2010). Hirschorn et al. (2010) argue what behavior of CPE the normal distribution of resistivity will give for a system with a uniform dielectric constant. They show that under the assumption that the dielectric constant is position independent, the normal power distribution of local resistivity is consistent with

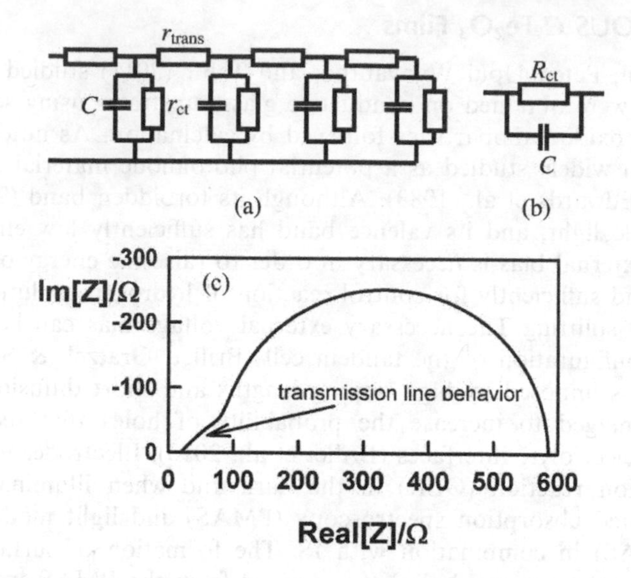

FIGURE 3.22 (a) Finite transmission line representing the impedance of the porous α-Fe_2O_3 electrodes. (b) Equivalent circuit in the low frequency limit. The additional series resistance is not shown in either circuit. (c) Typical impedance response for the transmission line circuit including a series resistance. Note the linear region at high frequencies, which is characteristic of transmission line behavior. The combination of spectroscopic and impedance techniques has given a deeper insight into the mechanism and kinetics of oxygen evolution at α-Fe_2O_3.

Reproduced with permission from Cummings et al. (2011). Copyright 2011, ACS Publications.

CPE. The original postulate is that does not take into account the formation of a space charge region for a material exhibiting semiconductor properties.

The power-law distribution of resistivity provides a physically reasonable interpretation of CPE for a wide class of systems where properties are expected to change in the direction perpendicular to the electrode. Excellent results have been described, for example, for applying a power-law distribution of resistivity to obtain the properties of a passive film on Fe-17Cr (Hirschorn et al., 2010). With comparison to the Young model, the impedance corresponding to a power-law distribution of resistivity was provided. The Young resistivity distribution can be expressed as $\rho(x) = \rho_0 \exp(-x\backslash\lambda)$, where ρ_0 is the resistivity at the surface and λ represents a characteristic length. Extraction of physical parameters: the frequency ranges, for which the impedance response is consistent with the CPE, are presented in Figure 3.23.

3.14 COPPER OXIDES IN NEUTRAL SOLUTIONS

Nakayama, Kaji, Notoya, and Osakai (2002), Nakayama, Kaji, Shibata, Notoya, and Osakai (2007), Nakayama, Kaji, Notoya, and Osakai (2008), Nakayama (2009) have used EIS to study the mechanism of reduction of copper oxides in alkaline and neutral solutions. Copper and copper alloys are widely used in industrial fields for the manufacturing of electrical wires, electronic components, electrical materials, and many other products. Although copper is a relatively corrosion-resistant material, oxide films of various thicknesses are formed on copper, which are exposed to air containing moisture and pollutants. The oxide films on copper consist of copper oxide (Cu_2O) and/or copper oxide (CuO). The selective determination of two oxides with different properties is important for the characterization of corrosion. A CuO/Cu sample and a CuO/Cu_2O/Cu oxide film of a Cu-duplex of Cu_2O/Cu oxide were prepared.

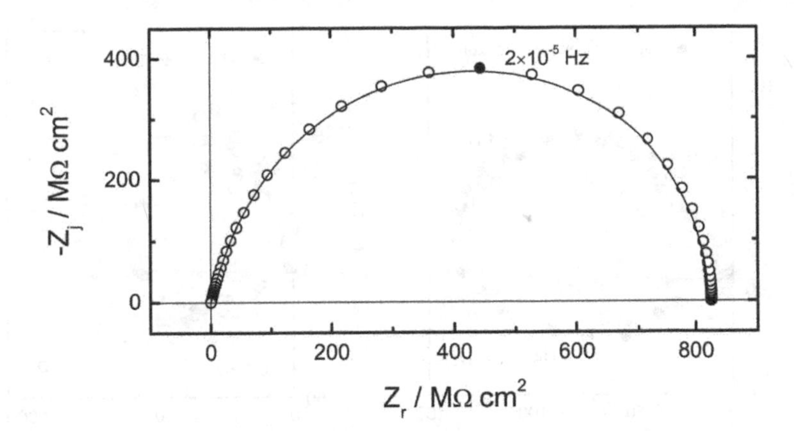

FIGURE 3.23 Nyquist representation of the impedance given in Figure 3.24 for CPE exponent = 6.67. The marked impedance at a frequency of 2×10^{-5} Hz is close to characteristic frequency $f_0 = 1.8 \times 10^{-5}$ Hz.

Reproduced with permission from Hirschorn et al. (2010). Copyright 2010, Electrochemical Society.

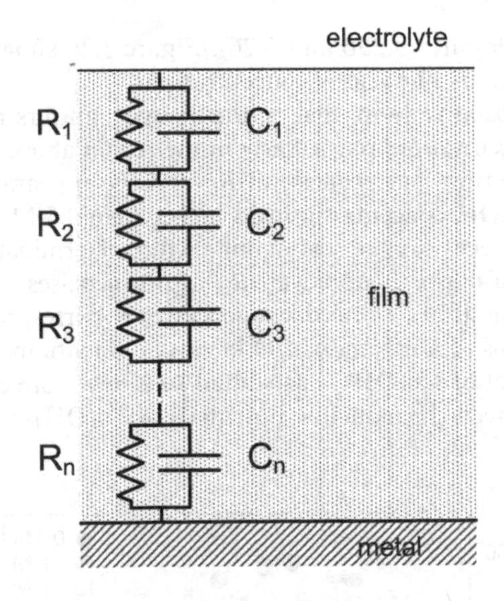

FIGURE 3.24 A distribution of *RC* elements that corresponds to the impedance response of a film.

Reproduced with permission from Hirschorn et al. (2010). Copyright 2010, Electrochemical Society.

Using X-ray diffraction, it was shown that Cu_2O and CuO coexist on the surface of the CuO/Cu_2O/Cu sample. The presence of a Cu_2O or CuO film on Cu_2O/Cu and CuO/Cu samples was also found by X-ray diffraction. In electrochemical measurements for standard samples, the test surface area was 1.0 cm^2. A Cu_2O, CuO, and $Cu(OH)_2$ powder samples were mixed with carbon paste (BAS Inc.) before each measurement (Nakayama et al., 2004). The mass ratio of the powder sample to carbon paste was 1:5. The amplitude of the superimposed current modulation was 0.1 mA, and the frequency range was from 50 MHz to 10 kHz.

The electrochemical impedance for the reduction of CuO was not strongly dependent on the type of alkaline solutions (see Figures 3.25a). Instead, the electrochemical impedance for the reduction of Cu_2O was significantly dependent on the type and concentration of

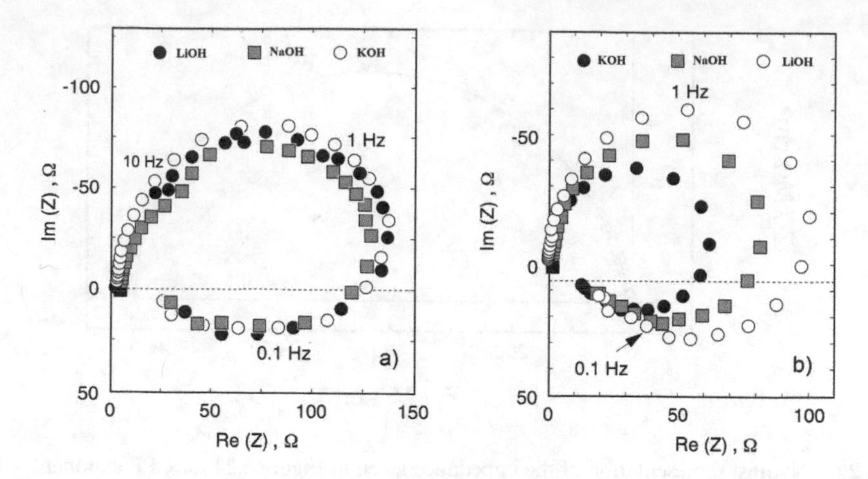

FIGURE 3.25 Electrochemical impedances of CuO/Cu (a) and Cu_2O/Cu (b) sample in 1 M LiOH, 1 M NaOH, and 1 M KOH (DC current: −1.0 mA; AC current: 0.1 mA).

Data adapted with permission from Nakayama et al., 2008. Copyright 2008, Elsevier.

alkali metal chloride (see Figures 3.25b and 3.26). Figure 3.26 shows Nyquist plot of Cu_2O/Cu samples in LiOH solutions of various concentrations.

The diameter of the capacitive loop, that is, the charge transfers resistance (R_{ct}), increased with increasing LiOH concentration, particularly in the region above 1 M. On the other hand, the specific behavior of the transient decrease in R_{ct} and the appearance of the inductive loop was confirmed when the LiOH concentration was higher than 0.5 M. A strongly alkaline solution containing Li ion is necessary for the simultaneous determination of Cu_2O, CuO, and $Cu(OH)_2$ on the surface of copper. Moreover, R_{ct} again increases with an increase in LiOH concentration of more than 1 M. These dependences may correspond to a good separation between reduction potentials of CuO and Cu_2O in chronopotentiometric measurements. However, the reduction potential of $Cu(OH)_2$ has shifted to a lower direction in neutral decisions. A sufficient separation between the reduction potentials of $Cu(OH)_2$ and Cu_2O was difficult.

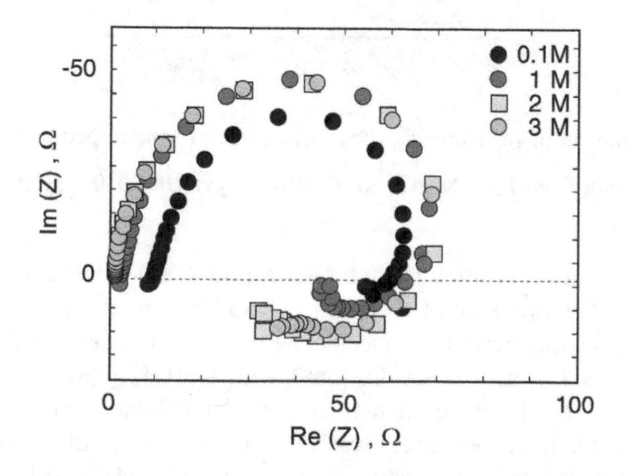

FIGURE 3.26 Nyquist plot of Cu_2O/Cu samples in KOH solutions of various concentrations (0.1 to 3 M). DC current: −1.0 mA, AC current: 0.1 mA.

Data adapted with permission from Nakayama et al., 2008. Copyright 2008, Elsevier.

3.15 LITHIUM NICKEL MANGANESE COBALT OXIDE POUCH CELLS

Lithium-ion batteries (LIBs) are becoming the main energy storage devices in the sectors of communication, transport, and renewable energy sources (Opitz, Badami, Shen, Vignarooban, & Kannan, 2017). However, the maximum possible energy supply for the LIB is insufficient for the long-term needs of society, for example, for long-range electric vehicles, hybrid electric vehicles and other portable devices (Nykvist, & Nilsson, 2015). Currently, the trend for electric vehicles is increasing sharply, especially in developed countries, as carbon dioxide emissions are becoming a serious problem in this decade. To meet the growing demand for energy, LIB scientists are still working to improve stability in a cyclic search, looking for more stable electrolytes with a voltage window, materials with a positive electrode (cathode) with a higher energy density, and negative cells with a larger capacity.

EIS has always been an effective, nondestructive technique that could analyze/characterize LIBs. The characteristic of the impedance is directly related to the capabilities of the battery and determines the voltage drop observed in the battery when current is applied (Waag, Fleischer, & Sauer, 2013). In Beelen, Raijmakers, Donkers, Notten, and Bergveld (2016), the authors demonstrated the use of IS to estimate battery temperature. It was proved that a simple equivalent circuit can imitate the behavior of the charge–discharge LIB (Liaw, Jungst, Nagasubramanian, Case, & Doughty, 2005). Many authors have shown that equivalent circuit models are the best choice for electric vehicle applications, compared to other physical models that require many parameters for model development (Franco, 2013). Deficiencies in this type of simulation exist, since this method is difficult to use in the design process of the cell, but these models are preferred in battery management applications (Fotouhi, Auger, Propp, Longo, & Wild, 2016). EIS method can be performed at different levels, either at the level of the electrodes (Andre et al., 2011) or at the level of the pouch cell. It can also be used for characteristics that go beyond lithium-ion technologies such as lithium–sulfur and lithium–air (Canas et al., 2013). Real-time impedance measurement methods have also been proposed for bag cells using inputted noise or signal; it also uses an algorithm that is used to estimate the impedance parameters from the model (Lohmann, Weßkamp, Haußmann, Melbert, & Musch, 2015). Fast Fourier transform is commonly used for signal detection (Christophersen, Motloch, Morrison, Donnellan, & Morrison, 2008). This is done to assess the state of battery health (Eddahech, Briat, Bertrand, Delétage, & Vinassa, 2012). Impedance studies are typically conducted to analyze the effect of parameters with respect to temperature, state of charge (SoC), and current velocity, as well as to quantify the health status and aging effect under various cycling/storage conditions. There have been several studies related to temperature and SoC studies in the LIB (Samadani et al., 2015). For aging studies, EIS was used as part of the verification procedure to quantify the change in impedance relative to the impedance of a new cell. This can give an idea of which cyclic conditions (SoC, current speed, or temperature) have a greater impact on battery life. Such a study was performed in European project on nickel manganese cobalt oxide cells [Batteries2020.eu… Towards Competitive European Batteries www.batteries2020.eu] and several other works, such as ISO 12405–3:2014 – Electrically propelled road vehicles *https://www.iso.org/standard/59224.html*. Based on these results, it is possible to select operating conditions that include security protocols, application, and the working environment.

In EIS study by Gopalakrishnan et al. (2019), a commercially available 20 Ah LIB (G1 – commercial cell) and a 28 Ah prototype (G2 – prototype) with the chemical composition of nickel–manganese cobalt oxide/graphite are used to determine the contribution of temperature and SoC to EIS. The SoC of the cell plays an important role in charge transfer processes in cells. The semicircle size in the Nyquist diagram is a criterion for the SoC of the cell. Its increase corresponds to a decrease in charge, which is also seen in Figure 4 in Gopalakrishnan et al. (2019).

Gopalakrishnan et al. (2019) have showed that advertising G1 cells work better in terms of less increase in charge transfer resistance compared to G2 cells, which means that G1 cells can protrude on larger SoC window as compared to G2 cells. The electrode structure, particle size, stacking of the electrodes, and other entities for both the cells are provided to compare the similarities and differences between both the cells. Equivalent circuit modeling is used to analyze and comprehend the variation in impedance spectrum obtained for both the cells. It is observed that the ohmic resistance varies with both temperature and SoC and the variation with temperature is more significant for the prototype cell. The prototype cell showed better charge-transfer characteristics at lower temperatures when compared to the commercial cell. Cells of generation 1 (cell G1 – commercial cell) and generation 2 (cell G2 – prototype) that were selected were different depending on various elements, including stoichiometry of the active material, particle size of the active material, electrode stacking, electrode architecture, separator, and electrolyte. The impedance spectrum of these two cells was studied as a function of SoC and temperature. Two equivalent circuits, one for low temperature and low temperature SoC, the second for all other conditions, were used to extract parameters from two different types of cells. CPEs were used to simulate the double layer capacity, as well as the diffusion process, since the semicircles obtained for both cells were depressed and the diffusion tail was not 45°. It was found that cells G1 have a lower ohmic resistance compared to cells G2 at all temperatures and SoC conditions, which can be associated with electrolyte solvents and, possibly, a separator (electrolyte absorption). The ohmic resistance varies with temperature and SoC, and the change with temperature is more significant for cell G2. A common reason for the reduced productivity of G2 cells may be due to the morphology of the electrodes and the separator used in the cell.

3.16 SPINEL-TYPE COBALT OXIDE THIN FILM ELECTRODES IN ALKALINE MEDIUM

For preparing spinel-type cobalt oxide thin film electrodes in alkaline medium, an important problem is the development of highly active, stable, and inexpensive electrocatalysts. Spinel-type cobalt oxide films have attracted much attention due to their excellent electrocatalytic properties with respect to the OER and good chemical stability in an alkaline solution. Due to their low cost, good conductivity, and electrocatalytic activity for OER, cobalt spinel oxides (e.g., Co_3O_4) have been the target of several studies (Hamdani et al., 2004). The electrochemical behavior of thin Co_3O_4 spinel films for the OER in 1 M KOH on stainless steel substrates using the thermal decomposition method at 400°C was studied by cyclic voltammetry and impedance methods (Laouini et al., 2008). The impedance measurements were carried out at various positive potentials, from the open circuit potential to the potential in the OER region. The value of the roughness coefficient, determined in the potential region where the charge transfer reaction is negligible, is similar to the value obtained by cyclic voltammetry. The importance of the oxygen evolution reaction stems from the fact that it is an anodic reaction in key processes such as aqueous electrolysis, electrochemical metal extraction, and organic electrosynthesis. However, due to the high overvoltage, electrolysis of alkaline water usually shows low energy efficiency. Efforts are being made to improve reaction kinetics and reduce overvoltage.

In recent years, the use of EIS has been expanded to characterize the electrochemical and electrocatalytic surface properties of oxide electrodes toward OER in order to understand solid state redox transitions. Palmas et al. (2007) have investigated the sol-gel behavior of prepared Co_3O_4 powder electrodes using cyclic voltammetry and EIS to gain access to solid-state surface redox transitions, which lead to the formation of active sites on the surface electrode for OER in alkaline solutions. These authors confirmed that the kinetic parameters – the exchange current density, and the electron transfer coefficient obtained from the EIS data are in good agreement with the data obtained in polarization n experiments. Castro, Real, and Pinheiro Dick (2004) studied porous nickel and cobalt oxides obtained by cathodic electrodeposition using approximation in terms of a finite conical pore transmission model connected in parallel.

3.17 GRAPHENE OXIDE SOLUTIONS

Graphene has been extensively studied for various applications due to its excellent electrical, mechanical, thermal, and optical properties (Neto, Guinea, Peres, Novoselov, & Geim, 2009). Alternatively, graphene oxide (GO) dispersed in a solvent could be a promising method for mass production of wafers in any size at low cost (Zhu et al., 2010). GO can be easily applied to the target substrate using centrifugation and/or spraying methods as a preliminary cursor, and then converted to graphene by thermal, chemical, or photocatalytic reduction (Stankovich et al., 2007). Meanwhile, GO performs its own functions with a hydroxyl/epoxy group on the basal plane and a carbonyl/carboxyl group on the edge, which is not like the final graphene (Geim, & Novoselov, 2007).

Thanks to such oxygen-containing functional groups, GO exhibits excellent chemical sensory properties and can be uniformly and stably dispersed in deionized (DI) water. In Yoon et al. (2017), the optical and electrical characteristics of a solution of GO (GS) with different concentrations of GO in DI water (see Table 3.4) were studied by EIS. GO has become very conductive in DI water. By oxidizing graphite and dispersing it in DI water, the authors formed GS for GO concentration ranges (from 0.25 to 7.00 g/L). The transmittance GS becomes completely opaque in the visible range from 300 to 700 nm at GO concentrations exceeding 2 g/L. The measurement results obtained by EIS for GS are presented by the Bode and Nyquist plots in the frequency range from 1 kHz to 10 MHz. In detail, EIS was performed, which was known to be a suitable method for studying the electrical properties of liquid materials. The observed experimental results were correlated with the simulation of an equivalent circuit for different GSs, which made it possible to extract physical parameters for explaining and interpreting the current flow mechanism of various GO concentrations in GS samples.

Based on the impedance analysis of Figure 3.27, authors proposed an equivalent GS circuit model, as shown in Figure 3.28. At first, the circuit model was developed for a stacked three-dimensional (3D) GO with an inductor (L_{GO}), resistor (R_{GO}), and two CPEs: (Q_{GO1} and Q_{GO2}), as shown in Figure 3.28. Generally, when circuits are not expressed as simple RC circuits, the CPE can be introduced with a frequency-independent Q-value, an imperfective resistive capacitance, and index α ($0 < \alpha < 1$; $\alpha = 0$ for a pure resistor and $\alpha = 1$ for an ideal capacitor). The stacked 3D graphite showed an inductive conductor property, while the fully oxidized GO stack showed highly resistive properties consisting of a large resistor and a CPE pair with a long phase delay (shown in Figure 3.28a). Once the oxidation was sufficiently completed, Q_{GO1} became similar to an ideal capacitor with an α of ~1. Although GOs were scattered in DI water in this work from the EIS measurements shown in Figure 3.28b, both circuit diagrams can be similar in principle. In circuit models of dielectric water, even a very small

TABLE 3.4
Various GO concentration in GS samples

Low-GO Samples	GS1	GS2	GS3	GS4	GS5	GS6
Concentration (g/L)	0.25	0.49	0.73	0.96	1.19	2.24
High-GO Samples	GS12	GS13	GS14	GS15	GS16	GS17
Concentration (g/L)	5.08	5.24	5.40	5.56	6.10	6.42

Reproduced from Yoon et al. (2017) under the Creative Commons Attribution License

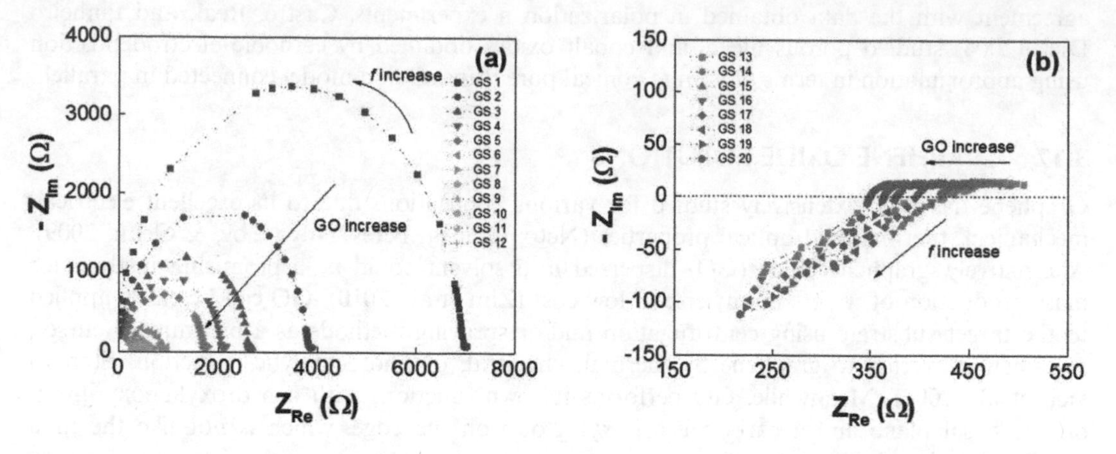

FIGURE 3.27 Nyquist plot of the GS samples: (a) for GS1 to GS11, which have low GO concentrations and (b) for GS12 to GS20, which have high GO concentrations.

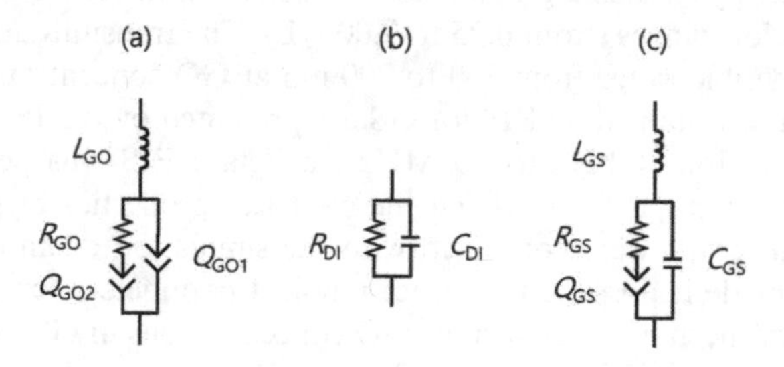

FIGURE 3.28 Equivalent circuit models: (a) stacked 3D GO10; (b) DI, and (c) GS inductor in series.

amount of water can contribute to a current conducting current.. Considering GO in GS as a component of a conducting network and DI water as a matrix material, it can be assumed that the model of the equivalent GS circuit may be similar to the GO model. However, one of the CPE GS elements (Figure 3.28a) was closer to the ideal capacitor due to the condenser component in the DI water, as shown in Figure 3.28c.

3.18 DENDRONIZED $CaSiO_3$-SiO_2-Si Nanoheterostructures

In recent years, calcium silicate ($CaSiO_3$) has been given an increasing attention for its promising applications based on its good bioactivity, biocompatibility, and biodegradability (Ni, Chang, Chou, & Zhai, 2007; Pan, Thierry, & Leygraf, 1996). At high-enough pressures, it is believed to have been crystallized with a perovskite structure and is, therefore, referred to as Ca-Si-perovskite. At lower pressures, Ca-Si-perovskite is not stable and converts to wollastonite. The tremendous improvements in high-quality film-formation techniques and compositional engineering of perovskite materials over the past 5 years have led to rapid improvements in the power conversion efficiency of perovskite solar cells (Yang et al., 2015). Although solar-to-electric conversion efficiencies of up to 18% have been reported for perovskite solar cells (Jeon et al., 2015), developing technologies further to achieve the efficiencies near theoretical values (>30%) continues to be an important challenge in making the solar cell industry economically competitive. The largest light loss mechanism for the perovskite/silicon solar cell is reflection. The majority of this reflection can be attributed to the sets of layer interfaces (Grant, Catchpole, Weber, & White, 2016), as the large index contrasts between the adjacent layers result in high Fresnel reflection. To increase light absorption within the silicon layer, a scattering surface can be introduced at the interlayer interface. A number of textured surfaces, such as random pyramids and random spherical caps, can be formed by a number of processes (Baker-Finch, McIntosh, & Terry, 2012).

A systematic study of charge carrier relaxation processes was carried out in sonochemically nanostructured silicon wafers, for which IS and transient photovoltage techniques are used (Savkina et al., 2019). Figure 3.29a shows the complex impedance plane plots of the samples (Savkina et al., 2019); they are semicircle shaped. The arrow shows the direction of the increase in frequency. Symbols are the experimental data and solid lines are the best-

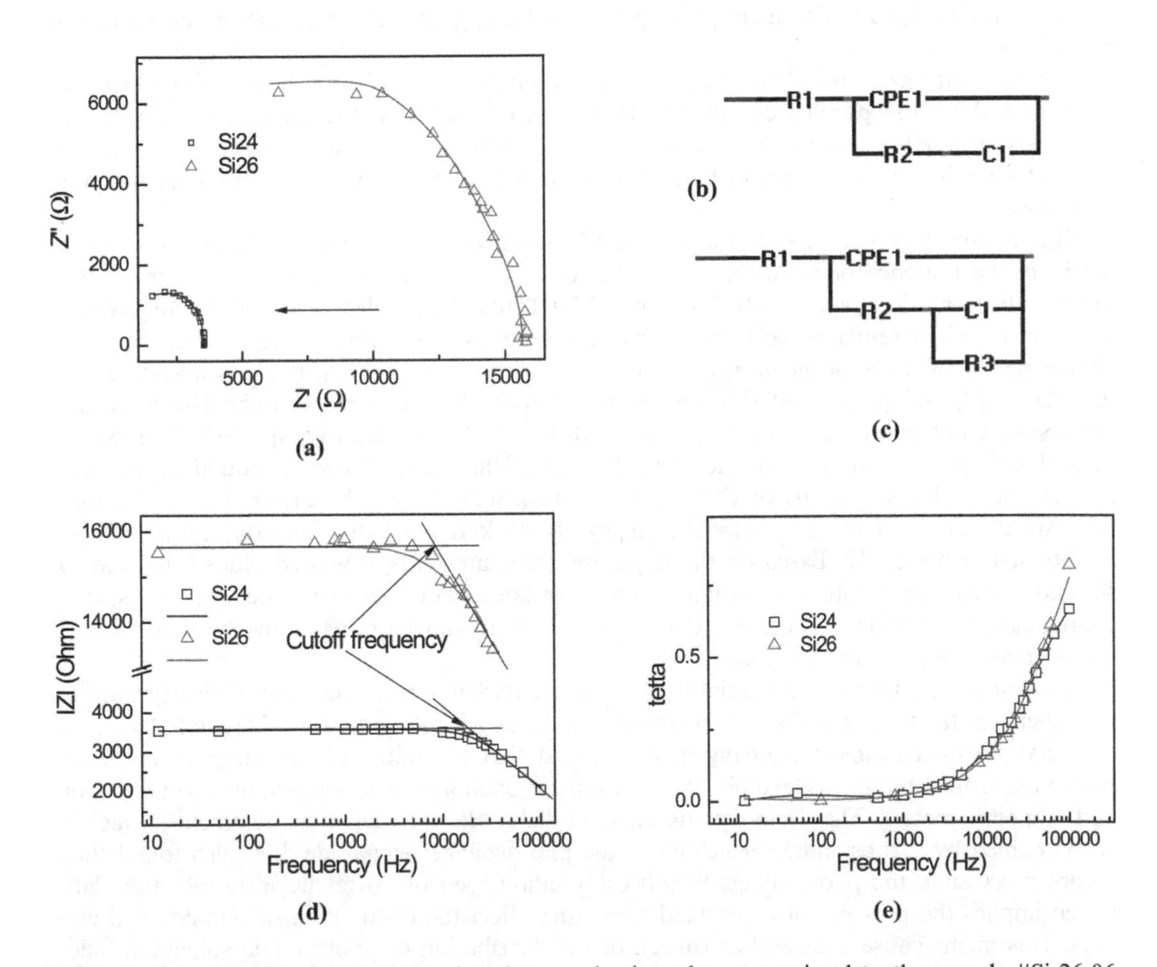

FIGURE 3.29 (a) Nyquist plots showing the complex impedance associated to the sample #Si-26-06 (triangles) and #Si-24-12 (circles) surfaces; (c) equivalent electrical circuits representing the interface behavior between the CaO-SiO_2 species and Si wafer for samples #Si-24-12 and #Si-26-06, respectively; (d) and (e) changes in the frequency-dependent impedance $|Z|$ and the phase shift θ, respectively, in samples #Si-24-12 and #Si-26-06.

TABLE 3.5
Parameters of the equivalent circuits

Samples	R1, Ohm	R2, kOhm	R3, kOhm	C1, F	CPE: $Z_{CPE} = A^{-1}(i\omega)^{-n}$ A	n	$\tau_1 = R2C1$	$\tau_2 = R3C1$
Si-26-06	200	2.1	1.2	2.4×10^{-9}	5.7×10^{-10}	0.98	~5 μs	~3μs
Si-24-12	1700	14	—	10^{-4}	4×10^{-10}	0.93	~1.4 s	—

fit curves to the measured spectra using the modified equivalent circuits of Figures 3.29c and 3.29d,e, which are obtained with the EIS spectrum analyzer (http://www.abc.chemistry.bsu.by/vi/analyser). The fitting was performed using the values of the circuit elements given in Table 3.5.

The conventional equivalent circuits used for samples investigated have series resistance *R*1 followed by the parallel circuit of CPE1, series *R*2-*C*1 and the parallel circuit *R*3-*C*1. The resistance *R*1 is usually associated with the contact resistance. However, since in this case $R1 \gg 1$ Ohm, it was believed that *R*1 combines both the contact and the bulk material resistances.

CPE is used to accommodate the nonideal behavior of the capacitance which may have its origin in the presence of more than one relaxation process with similar relaxation times (Bisquert & Fabregat-Santiago, 2010). The series *R*2-*C*1 and the parallel circuit *R*3-*C*1 (equivalent electrical circuit parameters see Figure 3.29c) corresponds to the charge transport in the space charge region of the sonochemically structured subsurface layer. Frequency dependence of the impedance $|Z|$ and phase shift θ is shown in Figure 3.29d and e, respectively. The frequency dispersion is not observed within the range of 10 Hz to 10 kHz (see Figure 3.29d). The experimental data give a cutoff frequency (the frequency that characterizes a boundary between a passband and a stopband) of about 8.5 kHz in #Si-26-06, which corresponds to the time constant of about 0.1 ms. The cutoff frequency of ≈18 kHz (with the time constant of ≈56 μs) is obtained in #Si-24-12. Based on the impedance measurements, it was concluded that the set #2 and #3 samples exhibit a capacitance-type impedance, which can be associated with several charge carrier relaxation processes. At least two of them can be revealed by the two different cutoff frequencies shown in Figure 3.29d.

It is found that interface potential in Si wafers remarkably increases upon their exposure to sonochemical treatments in Ca-rich environments. In contrast, the density of fast interface electron states remains almost unchanged. It is found that the initial photovoltage decay, taken before ultrasonic treatments, exhibits the involvement of shorter- and longer-time recombination and trapping centers. The decay speeds up remarkably due to cavitation treatments, which is accompanied by a substantial quenching of the photovoltage magnitude. It is also found that, before processing, the photovoltage is noticeably inhomogeneous over the surface of the plate, which implies the presence of distributed areas that affect the distribution of photoexcited carriers. Treatments cause a general expansion of the distribution of photovoltaic voltage. In addition, it was found that samples of sonochemically structured silicon are characterized by a capacitive-type impedance and demonstrate more than one process of relaxation of charge carriers after treatment in a cryoreactor and annealing after treatment with ultrasound. Sonochemical nanostructuring of silicon wafers with dendronized $CaSiO_3$ can provide a new promising way to create inexpensive multilayer solar cell structures with efficient use of solar energy.

3.19 *IN SITU* ANALYSIS OF OXIDE LAYER FORMATION IN LIQUID METAL

The EIS method must be a feasible technique to follow the formation of the functional oxide layer and the degradation of ceramic coating in liquid metals. An interesting application of IS is given in Kondo, Suzuki, Nakajima, Tanaka, and Muroga (2014). Some test materials (e.g., Fe, Cr, Y, and JLF-1 steel) were immersed in liquid metallic lead (Pb) mainly at 773 K as the working electrode of EIS. Some oxide layers formed on the electrodes in liquid Pb were analyzed by EIS. The impedance reaction was generalized as a semicircular Nyquist diagram, and the electrical properties and thicknesses of the oxide layers were evaluated in a nondestructive manner. EIS found a large impedance due to the formation of Y oxide formed in liquid Pb, although the impedance of Fe oxide and Cr oxide could not be detected due to their low electrical resistance. The time constant of the oxide layers was estimated from impedance information, and this value identified the types of oxides. A change in the time constant with the time of immersion indicated a change in the electrical properties determined by the chemical composition and crystal structure. The oxide layer thickness estimated by EIS is in good agreement with the evaluation of metallurgical analysis. The growth of the Y oxide layer in liquid Pb was successfully detected by EIS in a nondestructive mode. Liquid breeders, such as lithium (Li) and lead-lithium alloy (Pb–17Li), have the potential to produce high-performance blankets. Liquid metals, such as lead (Pb), tin (Sn), and gallium (Ga), are considered as a cooling fluid, a draining fluid. The use of functional layers, such as Y_2O_3 and Er_2O_3, for a loop pipeline is necessary to suppress tritium leaks and pressure drops of the magnetohydrodynamics flow in the general system (Muroga, & Pint, 2010). The fabrication of these functional layers was based on two methods. One is the fabrication of the ceramic coating on the tube wall by means of metal organic decomposition or metal organic chemical vapor deposition (Hishinuma et al., 2011) method before the exposure. The other is the *in situ* formation due to the oxidation of some metals plated or added to the steel. The plating of erbium metal to reduced activation ferritic/martensitic steel was carried out (Kondo, Takahashi, Tanaka, Tsisar, & Muroga, 2012) for the formation of Er_2O_3 layer by oxidation in liquid breeders (Jorcin et al., 2006). The metal such as aluminum added as the steel composition contributed to the formation of Al-rich compact oxide layer in liquid metal. These layers have a function of self-healing in the liquid metals, and enable a longer-term use in fusion blanket system. The corrosion reactions caused on the surface of these ceramic coatings and the oxides in the liquid metals were reported in the previous studies (Kondo, Tanaka, Muroga, Tsujimura, & Ito, 2012). Then, just the corrosion analysis of the specimen surface after the exposure to the liquid metals does not make clear the corrosion process in detail.

The electrical property of the oxide layers formed on some materials in the liquid Pb–Bi was analyzed by EIS (Chen et al., 2012). The EIS is a nondestructive way to figure out the change of the electrical properties and the thickness of the functional layer in liquid metals. However, the information on the performance of the EIS in the liquid metals is limited. In the present study, a series of EIS experiments was performed in the liquid metal system. Liquid lead (Pb) saturated with oxygen was used to form various oxides in the melt. The electrical properties of the oxide layers formed in the liquid metal were evaluated by EIS. The purpose is to investigate the potential of the EIS method in the liquid metal system and the effect of lead oxide formation on EIS in Fe test. Figure 3.30 shows that EIS results obtained at the exposure time of 983 h in the Fe test, and indicated a typical semicircular Nyquist plot of the Fe test. The semicircular Nyquist plot (or the capacitive semicircle) indicated the presence of oxide layer, which can be modeled as a parallel RC circuit with the elements of electrical resistance (R [Ω]) and capacitance (C [F]). These elements are expressed as $R = \frac{\rho d}{A}$; $C = \varepsilon\varepsilon_0\, A/d$. The electrical properties of the typical oxides are summarized in Table 3.6. Here, these values must be carefully referred since the data were not completely collected and the values must be different depending on the conditions. Though, the relative permittivity of Fe_2O_3 (hematite) and Fe_3O_4 (magnetite) was not available, these values must be close to that of FeO (wüstite). The impedance reaction indicated the presence of PbO on the surface of the sample, since it is determined through f_{max} and coincides with that for PbO (Table 3.6).

In these elements expressed here, ρ is the electrical resistivity (Ωm) of the oxide layer, A is the surface area of the specimen exposed to liquid Pb (m²), d is the thickness of the oxide layer (m), ε is the relative permittivity of the oxide layer, and ε_0 is the permittivity of vacuum (i.e., 8.854×10^{-12} F/m). The Nyquist plot can also give information about the time constant as $\tau = \frac{1}{2\pi f_{max}} = CR = \varepsilon\varepsilon_0\rho$, where f_{max} is the frequency at the maximum of the imaginary impedance measured as shown in Figure 3.30. The time constant is given by the product of the permittivity and the electro resistivity of the oxide layer. In other words, the time constant is the electrical properties of the oxide layer. The possible equivalent circuit of the oxide layers in liquid Pb is shown in Figure 3.31. Here, the electro resistance of liquid Pb and the wire was negligibly small. Then, the impedance measured at low frequencies (0.01 Hz in the present work) corresponds to the DC resistance of the oxide. The model can be simplified by omitting one parallel RC circuit by the oxide layer of liquid Pb or the specimens for the case of a single oxide layer. Here, the capacitor could be either a capacitor or CPE. It is reported that a CPE fit agreed with some results when the geometric distribution played significant role in the EIS (Jorcin et al., 2006).

TABLE 3.6
Electrical property of oxides

Materials	ρ, Ωm	ε	τ, sec	f_{max}, Hz
PbO	2.67×10^3	22	5.2×10^{-2}	3.1×10^5
Fe_3O_4 (magnetite)	11.32×10^{-1}	—	—	—
Fe_2O_3 (hematite)	6.24×10^1	—	—	—
FeO (wüstite)	1.56×10^{-4}	16	2.2×10^{-14}	7.2×10^{12}
Cr_2O_3	1.27×10^1	9.2	1.0×10^{-8}	1.5×10^4
Y_2O_3	1.18×10^6	14	1.5×10^{-1}	1.1

Adapted with permission from Kondo et al., 2014, Copyright 2014, Elsevier

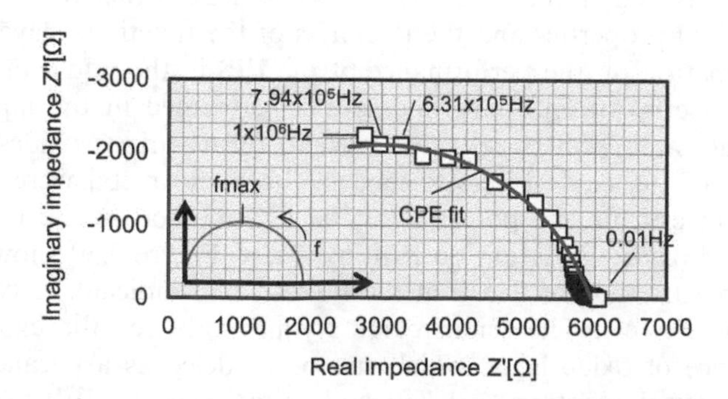

FIGURE 3.30 Nyquist plot obtained in Fe test at exposure time of 983 h.

Reproduced with permission from Kondo et al., 2014, Copyright 2014, Elsevier.

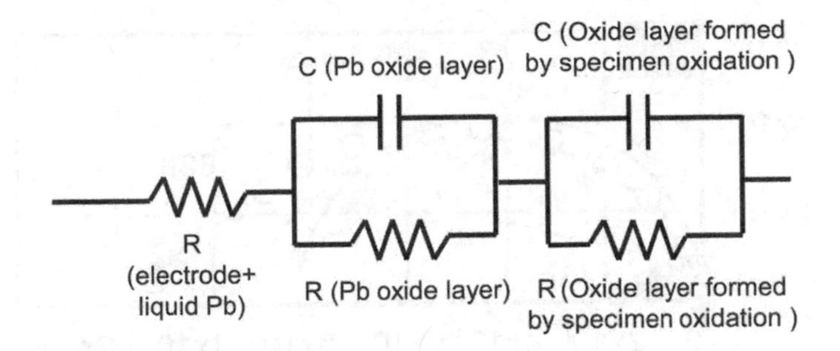

FIGURE 3.31 Equilibrium circuit for electrode surface – a capacitor can be replaced by a CPE.

Reproduced with permission from Kondo et al., 2014, Copyright 2014, Elsevier

The use of EIS for *in situ* analysis of the formation of an oxide layer in liquid lead was studied. The Nyquist graphs are shown in Figure 3.32a–c. A single capacitive semicircle was obtained after exposure at 19 h. The size of the capacitive semicircle has become larger with the function of time.

These results indicate that a single oxide layer forms on the surface and the layer grows with time. The curve was calculated on the capacitor and is indicated by the solid line in Figure 3.32a–c. Similarly, the fit curve calculated using the CPE was indicated by a dashed line in Figure 3.32a–c. At an early stage of exposure, the impedance response corresponded to the setup curve with CPE. Then, the capacitive semicircle gradually approached the curve of the installation with a capacitor, since the exposure time became longer. This trend showed that the geometric distribution became smaller when the thickness of the oxide layer became uniform and/or the surface roughness became smooth. Key findings:

(i) A semicircular Nyquist graph was obtained by the EIS method in the Fe test. The Nyquist site was rated as an equivalent circuit model.
(ii) The time constant of the Nyquist graph indicates the presence of a layer of lead oxide on the surface. The thickness of the oxide layer was theoretically estimated at about 500 microns. Metallurgical analysis of the sample surface showed the presence of a layer of lead oxide and a layer of iron oxide. The thickness of the lead oxide layer based on EIS results is consistent with the estimate obtained from metallurgical analysis.
(iii) A semicircular Nyquist graph with high electrical resistance was obtained in the Y test due to the formation of oxide Y. The temperature dependence of the electrical resistance of the layer was obtained, and the data dynamics corresponded to the previous literature.

3.20 METAL-OXIDE NANOCOMPOSITES

Admittance of oxide nanocomposites $0.97SnO_2 + 0.03CuO$; $0.9ZnO + 0.1CuO$; $0.9SnO_2 + 0.1CuO$, and $0.33SnO_2 + 0.33PbO + 0.33CuO$ were studied by IS (Tomaev et al., 2004). The nanocomposites were obtained by hydropyrolytic method. The thickness of the prepared films is estimated by the authors using ellipsometry and electron microscopy. The number of types of relaxers is shown, which is proportional to the number of phases in the composite. Due to several advantages, tin dioxide is one of the important materials used for the manufacturing of gas sensors (Davydov, Moshnikov, & Tomaev, 1998) based on multicomponent oxide systems of nanometer size. In particular (Gas'kov, & Rumyantseva, 2000),

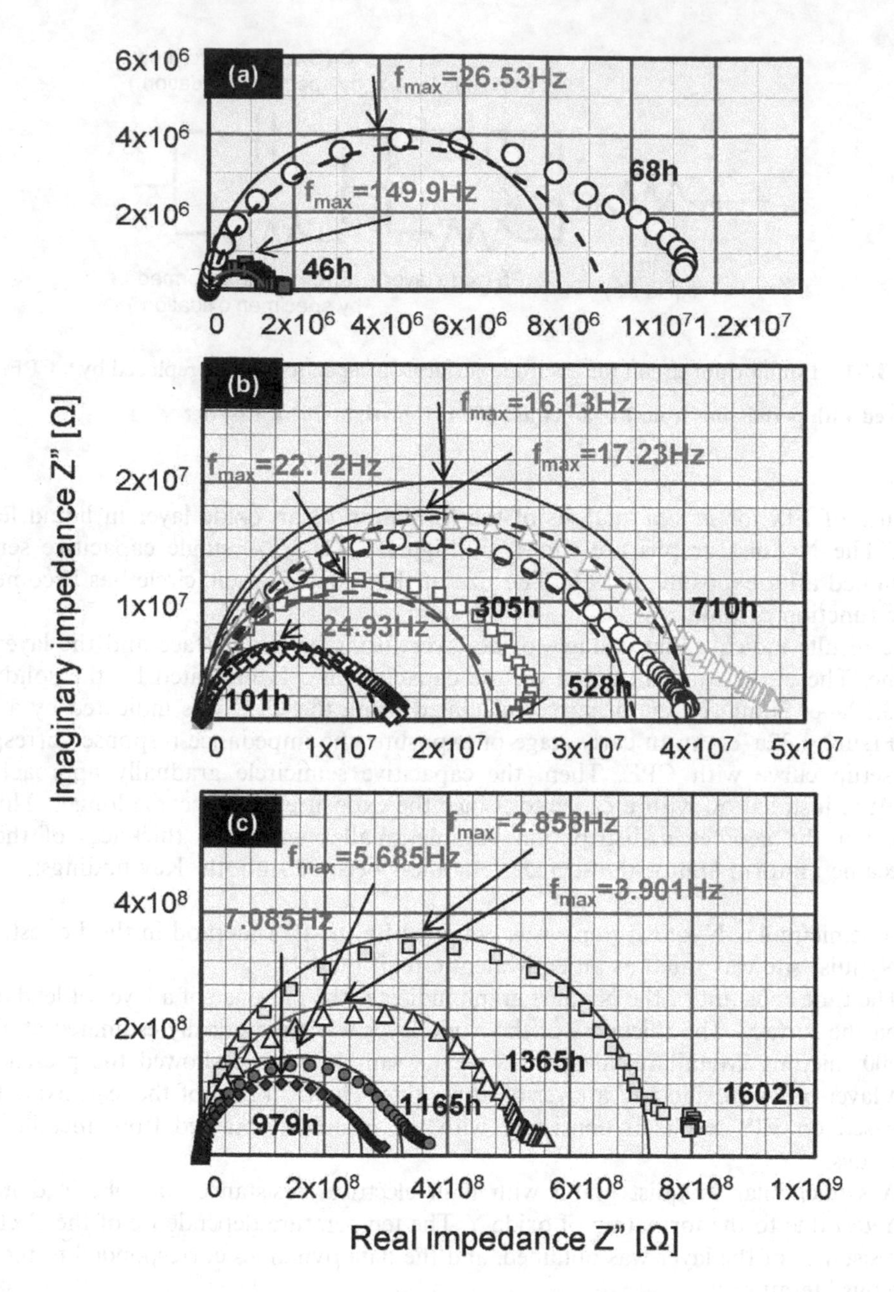

FIGURE 3.32 Nyquist plot obtained in Y test, (a) until 68 h, (b) until 710 h, and (c) until 1602 h.

Reproduced with permission from Kondo et al., 2014, Copyright 2014, Elsevier.

the high resistance of the SnO_2 + CuO nanocomposite in air is determined by the height of the potential barrier at the SnO_2 and CuO grain boundaries. In this case, in the presence of hydrogen sulfide, wide-gap CuO is converted to CuS with metallic conductivity. As a result, the resistance of the material becomes much lower and is determined by the electrical properties of the SnO_2 matrix and potential barriers that arise at grain boundaries, which, in turn, affect the gas sensitivity and selectivity of these materials.

Among the experimental methods used for studying the electrical properties of metal oxide semiconductors, admittance (impedance) spectroscopy plays a significant role. This is an effective method for studying the electrical characteristics of grain boundaries and boundaries, based on the analysis of the electrical response of the system when a disturbing electric effect is applied with a variable frequency (Ivanov-Shits, & Murin, 2000). Spectroscopy provides information on the densities of adsorption centers, carrier capture cross sections, relaxation times, and interface states.

The main results obtained by Tomaev et al. (2004) with the use of IS can be summarized as follows:

- (i) For the most part, the prepared metal-oxide films have a sufficiently homogeneous crystal structure characterized by at least two types of similar relocators. This can be judged by the presence of two arcs of semicircles with the centers located in the vicinity of the Z' axis in each impedance locus.
- (ii) The substrate (ceramic) material introduces minimum disturbances into the total impedance response of the substrate–film system.
- (iii) A comparison of the impedance loci of the metal-oxide films under investigation with similar data obtained by other authors (Göpel, 1994) allows to conclude that these films are promising to be used as gas sensors.
- (iv) An increase in the number of components in metal-oxide films complicates the graphical interpretation of the impedance loci.

3.21 OXIDE FILM ON TITANIUM FOR IMPLANT APPLICATION

Titanium is one of the most important materials for biomedical and dental implants. This is partly due to the excellent corrosion resistance of titanium and its alloys in many aqueous media, provided by the most protective passive film, which spontaneously forms on titanium. Attention is drawn to the surface oxide film on titanium and its long-term stability, which in biological environments play a decisive role for the biocompatibility of titanium implants. Observations may explain the unexpected growth of titanium oxide *in vivo* and the incorporation of ions on the surface of a titanium oxide implant. A passive film usually has a thickness of several nanometers and consists mainly of amorphous titanium dioxide. It is generally accepted that titanium surgical alloys have high corrosion resistance. However, there is a noticeable difference between the behavior of titanium *in vitro* and *in vivo*. It has been reported that a surface film formed on titanium implants in the human body after a few years can reach a thickness significantly exceeding the nanometer range. Under these conditions, the thickness and composition of the oxide layer changed with implantation time and some inclusion of mineral ions occurred. Despite the high corrosion resistance of titanium *in vitro*, there is increasing evidence that titanium is released and is accumulated in tissues adjacent to titanium implants.

In a study by Pan, Thierry, and Leygraf (1996), a passive oxide film formed on titanium and its natural growth in phosphate-buffered saline (PBS) with and without H and O were investigated using EIS for several weeks. In the absence of H_2O_2, the impedance response indicated a stable thin oxide film on titanium. However, the introduction of H_2O_2 into the solution led to significant changes in the EIS spectra (Figure 3.33), which changed with exposure time.

The interpretation of the results is based on a two-layer model of an oxide film consisting of a thin inner layer of the barrier type and a porous outer layer. Adding H_2O_2 to the solution led to a significant decrease in the corrosion resistance of titanium, as well as to a thickening of the porous outer layer. To explain the observed high *in vivo* oxidation/corrosion rates (equivalent circuits I and II), it has been suggested that H_2O_2 produced in biological systems plays an important role. An increase in the oxidation/corrosion rate of titanium was observed with an

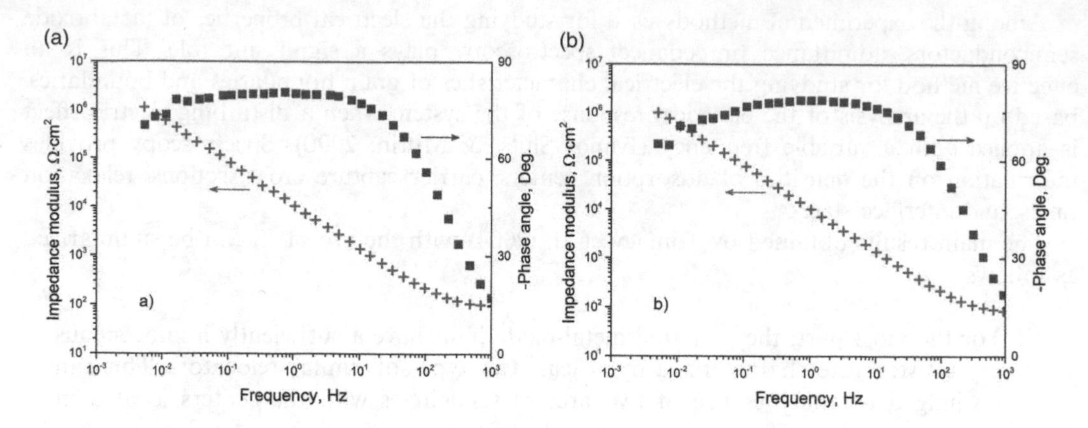

FIGURE 3.33 Bode plots for titanium exposed in the PBS solution without the addition of H_2O_2. (a) 1 day; (b) 30 days of exposure.

Reproduced with permission from Pan et al., 1996, Copyright 1996, Elsevier.

increase in the concentration of H_2O_2 in a solution with PBS (Tengvall, Elwing, Sjöqvist, Lundström, & Bjursten, 1989). Figure 3.34 provides a comparison of fitting quality for the two circuits when applied to a spectrum obtained after 30 days of exposure.

The presence of some hydrates/precipitates inside the oxide film has been verified by XPS showing that considerable number of ions had been incorporated into the outer part of the oxide film. The oxide model corresponding to different exposure conditions is schematically represented in Figure 3.35, together with the equivalent circuits I and II found suitable for representing the impedance characteristics. Notations: R_e is the solution resistance; C_b, R_b are the inner layer capacitance and resistance; C_p, R_p are the outer layer capacitance and resistance (or the electrolyte resistance inside pores); C_{h0}, R_{h0} represent the capacitance and resistance of hydrates/precipitates inside pores, respectively.

Although the passive film that forms on titanium in aqueous solutions is frequently described as a single TiO_2 layer, there is substantial evidence that this film in many exposure conditions exhibits a two-layer structure, that is, a dense inner layer and a porous outer layer.

By using surface-sensitive angular-resolved XPS, it was observed that the two-layer feature of the oxide film formed on titanium was exposed to PBS solution, and the porous layer had thickened due to the introduction of H_2O_2. The titanium oxide was found to be essentially TiO_2 with a transient region between the inner and outer layers. In addition, upon termination of the exposure with H_2O_2, ions from the PBS solution were found to be incorporated into the oxide film. It seems likely that the outer layer basically consists of the same oxide as the inner layer, but possesses microscopic pores that may be filled by either the solution or some hydrated/precipitated compounds depending on the exposure conditions. In summary, the two-layer model of the oxide film used for the impedance data fitting is supported by the results from independent techniques. From EIS study, the following conclusions may be drawn. First of all, EIS is a powerful method for *in situ* characterization of the passive oxide film on titanium. The evolution of the film due to the introduction of H_2O_2 can be monitored by EIS measurements as a function of exposure time. The EIS spectra can be interpreted in terms of a two-layer model of the oxide film, which consists of a thin barrier-type inner layer and a porous outer layer. The parameters obtained and their variation with exposure time indicate that the oxide film can be attacked by H_2O_2, which results in more defective inner layer and in a thicker and more porous outer layer. After prolonged exposure, the oxide film may reheal itself and exhibit a significantly higher corrosion resistance again due to some sealing processes inside the pores in the oxide film.

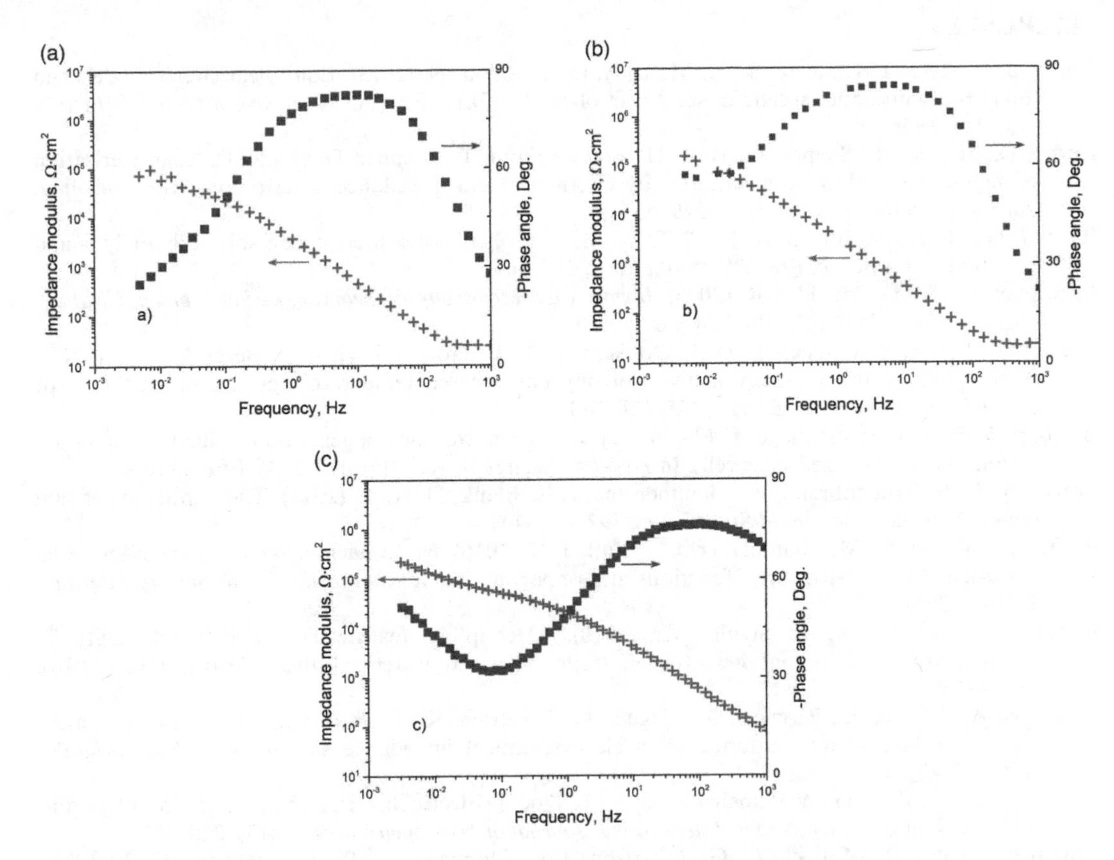

FIGURE 3.34 Bode plots for titanium exposed in the PBS solution with H_2O_2 addition having an initial concentration of 100 mM. (a): 1 day; (b): 15 days; (c): 30 days of exposure.

Reproduced with permission from Pan et al., 1996, Copyright 1996, Elsevier.

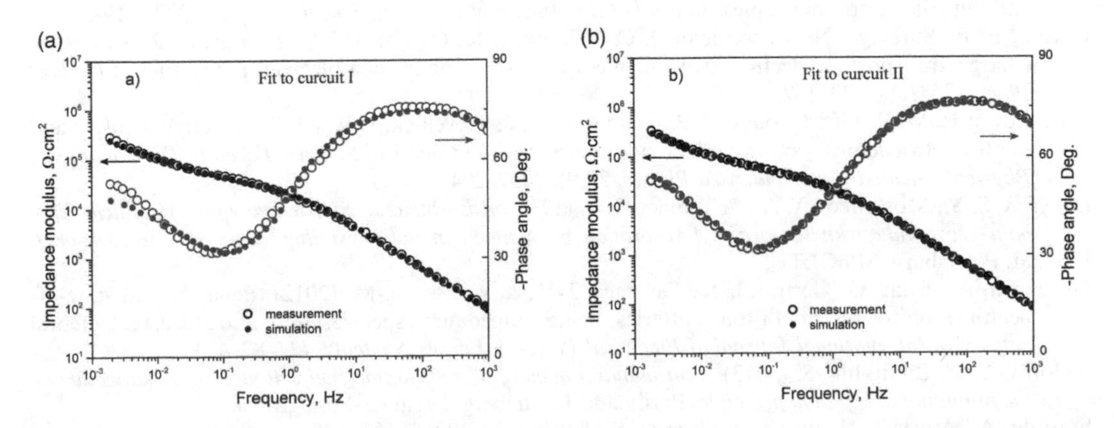

FIGURE 3.35 Comparison of fitting quality of the two equivalent circuits for a spectrum obtained after 30 days of exposure, when the titanium surface appears blue. (a) and (b): measured and simulated spectra vs. frequency comparing the measured and simulated data for circuits I and II, respectively.

Reproduced with permission from Pan et al., 1996, Copyright 1996, Elsevier

REFERENCES

Ahmadu, U., Tomas, S., Jonah, S. A., Musa, A. O., & Rabiu, N. (2013). Equivalent circuit models and analysis of impedance spectra of solid electrolyte $Na_{0.25}Li_{0.75}Zr_2 (PO_4)_3$. *Advanced Materials Letters, 4*(3), 185–195.

Andre, D., Meiler, M., Steiner, K., Walz, H., Soczka-Guth, T., & Sauer, D. U. (2011). Characterization of high-power lithium-ion batteries by electrochemical impedance spectroscopy. II: Modelling. *Journal of Power Sources, 196*, 5349–5356.

Baker-Finch, S. C., McIntosh, K. R., & Terry, M. L. (2012). Isotextured silicon solar cell analysis and modeling 1: Optics. *IEEE J Photovoltaics, 2*, 457–464.

Barsoukov, E., & Macdonald, J. R. (2005). *Impedance spectroscopy – Theory, experiment and applications* (2nd ed.). Hoboken, NJ: John Wiley & Sons. Inc.

Beelen, H. P. G. J., Raijmakers, L. H. J., Donkers, M. C. F., Notten, P. H. L., & Bergveld, H. J. (2016). A comparison and accuracy analysis of impedance-based temperature estimation methods for Li-ion batteries. *Applied Energy, 175*, 128–140.

Bisquert, J., & Fabregat-Santiago, F. (2010). Impedance spectroscopy: a general introduction and application to dye-sensitized solar cells. In *Dye-sensitized solar cells* (pp. 477–574). EPFL Press.

Boukamp, B. A., Hildenbrand, N., Nammensma, P., & Blank, D. H. A. (2011). The impedance of thin dense oxide cathodes. *Solid State Ionics, 192*, 404–408.

Breuer, M., Rosso, K. M., Blumberger, J., & Butt, J. N. (2015). Multi-haem cytochromes in *Shewanella oneidensis* MR-1: Structures, functions and opportunities. *Journal of the Royal Society Interface, 12*(102), 20141117.

Brillet, J., Gratzel, M., & Sivula, K. (2010). Decoupling feature size and functionality in solution-processed, porous hematite electrodes for solar water splitting. *Nano Letters, 10*(10), 4155–4160.

Canas, N. A., Hirose, K., Pascucci, B., Wagner, N., Friedrich, K. A., & Hiesgen, R. (2013). Investigations of lithium-sulfur batteries using electrochemical impedance spectroscopy. *Electrochimica Acta, 97*, 42–51.

Castro, E. B., Real, S. G., & Pinheiro Dick, L. F. (2004). Electrochemical characterization of porous nickel–cobalt oxide electrodes. *International Journal of Hydrogen Energy, 29*(3), 255–261.

Christophersen, J. P., Motloch, C. G., Morrison, J. L., Donnellan, I. B., & Morrison, W. H. (2008). Impedance noise identification for state-of-health prognostics. In *43rd Power Sources Conference*, Philadelphia, PA.

Chu, D., Zeng, Y.-P., Jiang, D., & Masuda, Y. (2009). In_2O_3-SnO_2 nanotoasts and nonorods: Precipitation preparation, formation mechanism, and gas sensing properties. *Sensors and Actuators B, 137*, 630–636.

Cummings, C. Y., Marken, F., Peter, L. M., Upul Wijayantha, K. G., & Tahir, A. A. (2011). New insights into water splitting at mesoporous α-Fe_2O_3 films: A study by modulated transmittance and impedance spectroscopies. *Journal of the American Chemical Society, 134*(2), 1228–1234.

Dalton, E. F., Surridge, N. A., Jernigan, J. C., Wilbourn, K. O., Facci, J. S., & Murray, R. W. (1990). Charge transport in electroactive polymers consisting of fixed molecular redox sites. *Chemical Physics, 141*(1), 143–157.

Dare-Edwards, M. P., Goodenough, J. B., Hamnett, A., & Trevellick, P. R. (1983). Electrochemistry and photoelectrochemistry of iron (III) oxide. *Journal of the Chemical Society, Faraday Transactions 1: Physical Chemistry in Condensed Phases, 79*(9), 2027–2041.

Davydov, S. Y., Moshnikov, V. A., & Tomaev, V. V. (1998). *Adsorbtsionnye yavleniya v polikristallicheskikh polu provodnikovykh sensorakh [Adsorption phenomena in polycrystalline semiconductor sensors].* St. Petersburg: SPbGETU.

Eddahech, A., Briat, O., Bertrand, N., Delétage, J.-Y., & Vinassa, J.-M. (2012). Behavior and state-of-health monitoring of Li-ion batteries using impedance spectroscopy and recurrent neural networks. *International Journal of Electrical Power & Energy Systems, 42*, 487–494.

Fedotov, Y., & Bredikhin, S. (2013). *Continuum modeling of solid oxide fuel cell electrodes: Introducing the minimum dissipation principle.* Berlin and Heidelberg: Springer-Verlag.

Fotouhi, A., Auger, D. J., Propp, K., Longo, S., & Wild, M. (2016). A review on electric vehicle battery modelling: From lithium-ion toward lithium-sulphur. *Renewable & Sustainable Energy Reviews, 56*, 1008–1021.

Franco, A. A. (2013). Multiscale modelling and numerical simulation of rechargeable lithium ion batteries: Concepts, methods and challenges. *RSC Advances, 3*, 13027.

Fredrickson, J. K., Romine, M. F., Beliaev, A. S., Auchtung, J. M., Driscoll, M. E., Gardner, T. S., … & Rodionov, D. A. (2008). Towards environmental systems biology of Shewanella. *Nature Reviews Microbiology, 6*(8), 592–603.

Gas'kov, A. M., & Rumyantseva, M. N. (2000). Choice of materials for solid-state gas sensors. *Inorganic Materials, 36*(3), 369–378.

Geim, A. K., & Novoselov, K. S. (2007). The rise of graphene. *Nature Materials, 6*, 183–191.

Gewies, S., & Bessler, W. G. (2008). Physically based impedance modeling of Ni/YSZ cermet anodes. *Journal of the Electrochemical Society, 155*(9), B937.

Gopalakrishnan, R., Li, Y., Smekens, J., Barhoum, A., Van Assche, G., Omar, N., & Van Mierlo, J. (2019). Electrochemical impedance spectroscopy characterization and parameterization of lithium nickel manganese cobalt oxide pouch cells: dependency analysis of temperature and state of charge. *Ionics, 25*(1), 111–123.

Göpel, W. (1994). New materials and transducers for chemical sensors. *Sensors and Actuators B: Chemical, 18–19*, 1–21.

Gorby, Y., & Yanina, S. (2006). Electrically conductive bacterial nanowires produced by *Shewanella oneidensis* strain MR-1 and other microorganisms. *Proceedings of the National Academy of Sciences of the United States of America, 106*(23), 9535.

Gralnick, J. A., & Newman, D. K. (2007). Extracellular respiration. *Molecular Microbiology, 65*, 1–11.

Grant, D. T., Catchpole, K. R., Weber, K. J., & White, T. P. (2016). Design guidelines for perovskite/silicon 2-terminal tandem solar cells: An optical study. *Optics Express, 24*, 1454–1470.

Grätzel, M. (2001). Photoelectrochemical cells. *Nature, 414*(6861), 338–344.

Grebenko, A., Dremov, V., Barzilovich, P., Bubis, A., Sidoruk, K., Voeikova, T., … Motovilov, K. (2018). Impedance spectroscopy of single bacterial nanofilament reveals water-mediated charge transfer. *PLoS One, 13*(1), e0191289. doi:10.1371/journal.pone.0191289.

Hamdani, M., Pereira, M. I. S., Douch, J., Addi, A. A., Berghoute, Y., & Mendonça, M. H. (2004). Physicochemical and electrocatalytic properties of Li-Co3O4 anodes prepared by chemical spray pyrolysis for application in alkaline water electrolysis. *Electrochimica Acta, 49*(9–10), 1555–1563.

Harbeke, Günther (ed.). (2012) *Polycrystalline Semiconductors: Physical Properties and Applications: Proceedings of the International School of Materials Science and Technology at the Ettore Majorana Centre* (Vol. 57). *July 1–15, 1984. Erice, Italy*: Springer Science & Business Media.

Heli, H., Sattarahmady, N., & Majdi, S. (2012). A study of the charge propagation in nanoparticles of Fe_2O_3 core-cobalt hexacyanoferrate shell by chronoamperometry and electrochemical impedance spectroscopy. *Journal of Solid State Electrochemistry, 16*(1), 53–64.

Heli, H., & Yadegari, H. (2010). Nanoflakes of the cobaltous oxide, CoO: Synthesis and characterization. *Electrochimica Acta, 55*(6), 2139–2148.

Hirschorn, B., Orazem, M. E., Tribollet, B., Vivier, V., Frateur, I., & Musiani, M. (2010). Constant-phase-element behavior caused by resistivity distributions in films I. Theory. *Journal of the Electrochemical Society, 157*(12), C452–C457.

Hishinuma, Y., Tanaka, T., Tanaka, T., Nagasaka, T., Tasaki, Y., Sagara, A., & Muroga, T. (2011). Er2O3 coating synthesized with MOCVD process on the large interior surface of the metal tube. *Fusion Engineering and Design, 86*(9–11), 2530–2533.

Horita, T., Yamaji, K., Sakai, N., Yokokawa, H., Weber, A., & Ivers-Tiffée, E. (2001). Oxygen reduction mechanism at porous La1− xSrxCoO3− d cathodes/La0. 8Sr0. 2Ga0. 8Mg0. 2O2. 8 electrolyte interface for solid oxide fuel cells. *Electrochimica Acta, 46*(12), 1837–1845.

Hsueh, T.-J., & Hsu, C.-L. (2008). Fabrication of gas sensing devices with ZnO nanostructure by the low-temperature oxidation of zinc particles. *Sensors and Actuators B, 131*, 572–576.

Inzelt, G. (1994). *Electroanalytical chemistry* (Vol. 18). A. J. Bard (Ed.). New York: Dekker.

Irvine, J. T., Sinclair, D. C., & West, A. R. (1990). Electroceramics: Characterization by impedance spectroscopy. *Advanced Materials, 2*(3), 132–138.

Ishihara, T., Shibayama, T., Nishiguchi, H., & Takita, Y. (2000). Nickel–Gd-doped CeO_2 cermet anode for intermediate temperature operating solid oxide fuel cells using $LaGaO_3$-based perovskite electrolyte. *Solid State Ionics, 132*(3–4), 209–216.

Ivanov-Shits, A. K., & Murin, I. V. (2000). *Ionika tverdogo tela [Ionics of solids]* (Vol. 1). St. Petersburg: St. Petersburg Gos.Univ.

Jeon, N. J., Noh, J. H., Yang, W. S., Kim, Y. C., Ryu, S., Seo, J., & Seok, S. I. (2015). Compositional engineering of perovskite materials for high-performance solar cells. *Nature, 517*, 476–480.

Jorcin, J. B., Orazem, M. E., Pébère, N., & Tribollet, B. (2006). CPE analysis by local electrochemical impedance spectroscopy. *Electrochimica Acta, 51*(8–9), 1473–1479.

Kaur, J., Kumar, R., & Bhatnagar, M. C. (2007). Effect of Indium nanoparticles on NO_2 gas sensing properties. *Sensors and Actuators B*, *126*, 478–484.

Kawakami, A., Matsuoka, S., Watanbe, N., Saito, T., Ueno, A., Ishihara, T., ... Yokokawa, H. (2006). Development of two types of tubular SOFCs at TOTO. In *Advances in Solid Oxide Fuel Cells II: Ceramic Engineering and Science Proceedings*, T. 27, C. 3–12.

Kondo, M., Suzuki, N., Nakajima, Y., Tanaka, T., & Muroga, T. (2014). Electrochemical impedance spectroscopy on in-situ analysis of oxide layer formation in liquid metal. *Fusion Engineering and Design*, *89*(7–8), 1201–1208.

Kondo, M., Takahashi, M., Tanaka, T., Tsisar, V., & Muroga, T. (2012). Compatibility of reduced activation ferritic martensitic steel JLF-1 with liquid metals Li and Pb–17Li. *Fusion Engineering and Design*, *87*(10), 1777–1787.

Kondo, M., Tanaka, T., Muroga, T., Tsujimura, H., & Ito, Y. (2012). Electroplating of erbium on steel surface in $ErCl_3$ doped LiCl-KCl. *Plasma and Fusion Research*, *7*, 2405069.

Kouzuma, A., Kasai, T., Hirose, A., & Watanabe, K. (2015). Catabolic and regulatory systems in *Shewanella oneidensis* MR-1 involved in electricity generation in microbial fuel cells. *Frontiers in Microbiology*, *6*, 609.

Laouini, E., Hamdani, M., Pereira, M. I. S., Douch, J., Mendonça, M. H., Berghoute, Y., & Singh, R. N. (2008). Electrochemical impedance spectroscopy investigation of spinel type cobalt oxide thin film electrodes in alkaline medium. *Journal of Applied Electrochemistry*, *38*(11), 1485.

Lasia, A. (1999). *Electrochemical impedance spectroscopy and its applications, modern aspects of electrochemistry* (Vol. 32, pp. 143–248). B. E. Conway, J. Bockris, & R. E. White (Eds.). New York: Kluwer Academic/Plenum Publishers.

Leung, K. M., Wanger, G., El-Naggar, M. Y., Gorby, Y., Southam, G., Lau, W. M., & Yang, J. (2013). Shewanella oneidensis MR-1 bacterial nanowires exhibit p-type, tunable electronic behavior. *Nano Letters*, *13*(6), 2407–2411.

Liaw, B. Y., Jungst, R. G., Nagasubramanian, G., Case, H. L., & Doughty, D. H. (2005). Modeling capacity fade in lithium-ion cells. *Journal of Power Sources*, *140*, 157–161.

Lohmann, N., Weßkamp, P., Haußmann, P., Melbert, J., & Musch, T. (2015). Electrochemical impedance spectroscopy for lithium-ion cells: Test equipment and procedures for aging and fast characterization in time and frequency domain. *Journal of Power Sources*, *273*, 613–623.

Madhi, I., Saadoun, M., & Bessais, B. (2012). Impedance spectroscopy study of porous ITO based gas sensor. *Procedia Engineering*, *47*, 192–195.

Malvankar, N. S., Vargas, M., Nevin, K. P., Franks, A. E., Leang, C., Kim, B. C., ... & Rotello, V. M. (2011). Tunable metallic-like conductivity in microbial nanowire networks. *Nature Nanotechnology*, *6*(9), 573–579.

Meera, R., Yadav, K. L., Amit, K., Piyush, K. P., Nidhi, A., & Jyoti, R. (2012). *Corrosion Engineering*, *3*, 286–292.

Mostert, A. B., Powell, B. J., Pratt, F. L., Hanson, G. R., Sarna, T., Gentle, I. R., & Meredith, P. (2012). Role of semiconductivity and ion transport in the electrical conduction of melanin. *Proceedings of the National Academy of Sciences, T. 109*(23), C. 8943–8947.

Muroga, T., & Pint, B. A. (2010). Progress in the development of insulator coating for liquid lithium blankets. *Fusion Engineering and Design*, *85*(7–9), 1301–1306.

Nakajima, H., Kitahara, T., & Konomi, T. (2010). Electrochemical impedance spectroscopy analysis of an anode-supported microtubular solid oxide fuel cell. *Journal of the Electrochemical Society*, *157*, B1686–B1692.

Nakajima, H., Konomi, T., Kitahara, T., & Tachibana, H. (2008). Electrochemical impedance parameters for the diagnosis of a polymer electrolyte fuel cell poisoned by carbon monoxide in reformed hydrogen fuel. *Journal of Fuel Cell Science and Technology*, *5*, 041013.

Nakayama, S. (2009). Mechanistic study by electrochemical impedance spectroscopy on reduction of copper oxides in neutral solutions. *SEI Technical Review*, 62–68.

Nakayama, S., Kimura, A., Shibata, M., Kuwabata, S., & Osakai, T. (2001). Voltammetric characterization of oxide films formed on copper in air. *Journal of the Electrochemical Society*, *148*(11), B467–B472.

Nakayama, S., Shibata, M., Notoya, T., & Osakai, T. (2004). Voltammetric Determination of Cupric and Cuprous Oxides Formed on Copper. *SEI Technical Review-English Edition*, 22–25.

Nakayama, S., Kaji, T., Shibata, M., Notoya, T., & Osakai, T. (2007). Which is easier to reduce, Cu2O or CuO? *Journal of The Electrochemical Society*, *154*(1), C1–C6.

Nakayama, S., Kaji, T., Notoya, T., & Osakai, T. (2008). Mechanistic study of the reduction of copper oxides in alkaline solutions by electrochemical impedance spectroscopy. *Electrochimica Acta, 53* (9), 3493–3499.

Nakayama, S., Shibata, M., Notoya, T., & Osakai, T. (2002). On standardizing to voltammetric determination of cupric and cuprous oxides formed on copper. *Bunseki Kagaku, 51*(12), 1145–1151.

Neto, A. C. Guinea, F. Peres, N. M. Novoselov, K. S. & Geim, A. K. (2009). The electronic properties of graphene. *Reviews of Modern Physics, 81*, 109.

Ni, S. Y., Chang, J., Chou, L., & Zhai, W. Y. (2007). Comparison of osteoblast-like cell responses to calcium silicate and tricalcium phosphate ceramics in vitro. *Journal of Biomedical Materials Research, 80*, 174–183.

Niu, G., Zhou, M., Yang, X., Park, J., Lu, N., Wang, J., … & Xia, Y. (2016). Synthesis of Pt–Ni octahedra in continuous-flow droplet reactors for the scalable production of highly active catalysts toward oxygen reduction. *Nano Letters, 16*(6), 3850–3857.

Nykvist, B., & Nilsson, M. (2015). Rapidly falling costs of battery packs for electric vehicles. *Nature Climate Change, 5*, 329–332.

Okamoto, A., Hashimoto, K., & Nakamura, R. (2012). Long-range electron conduction of Shewanella biofilms mediated by outer membrane C-typecytochromes. *Bioelectrochemistry, 85*, 61–65.

Opitz, A., Badami, P., Shen, L., Vignarooban, K., & Kannan, A. M. (2017). Can Li-ion batteries be the panacea for automotive applications? *Renewable & Sustainable Energy Reviews, 68*, 685–692.

Palmas, S., Ferrara, F., Vacca, A., Mascia, M., & Polcaro, A. M. (2007). Behavior of cobalt oxide electrodes during oxidative processes in alkaline medium. *Electrochimica Acta, 53*(2), 400–406.

Pan, J., Thierry, D., & Leygraf, C. (1996). Electrochemical impedance spectroscopy study of the passive oxide film on titanium for implant application. *Electrochimica Acta, 41*(7–8), 1143–1153.

Pirbadian, S., Barchinger, S. E., Leung, K. M., Byun, H. S., Jangir, Y., Bouhenni, R. A., … & Gorby, Y. A. (2014). Shewanella oneidensis MR-1 nanowires are outer membrane and periplasmic extensions of the extracellular electron transport components. *Proceedings of the National Academy of Sciences, 111*(35), 12883–12888.

Rawat, M., Yadav, K. L., Kumar, A., Patel, P. K., Adhlakha, N., & Rani, J. (2012). Structural, dielectric and conductivity properties of Ba 2+ doped (Bi0. 5Na0. 5) TiO3 ceramic. *Advanced Materials Letters, 3*(4), 286–292.

Reardon, P. N., & Mueller, K. T. (2013). Structure of the type IVa major pilin from the electrically conductive bacterial nanowires of Geobacter sulfurreducens. *Journal of Biological Chemistry, T. 288* (41), C. 29260–29266.

Richter, H., Nevin, K. P., Jia, H., Lowy, D. A., Lovley, D. R., & Tender, L. M. (2009). Cyclic voltammetry of biofilms of wild type and mutant Geobacter sulfurreducens on fuel cell anodes indicates possible roles of OmcB, OmcZ, type IV pili, and protons in extracellular electron transfer. *Energy & Environmental Science, 2*(5), 506–516.

Saadoun, M., Boujmil, M. F., El Mir, L., & Bessais, B. (2009). Nanostructured zinc oxide thin films for NO_2 gas sensing. *Sensor Letters, 7*, 1–6.

Sahoo, S., Pradhan, D. K., Choudhary, R. N. P., & Mathur, B. K. (2012). Dielectric relaxation and conductivity studies of $(K_{0.5}Bi_{0.5})(Fe_{0.5}Nb_{0.5})O_3$ ceramics. *Advanced Materials Letters, 3*, 97–101.

Saito, T., Abe, T., Fujinaga, K., Miyao, M., Kuroishi, M., Hiwatashi K., Ueno A. (2005). 9th CONFERENCE, Solid oxide fuel cells: (SOFC-IX), Quebec City, *In Proceedings- Electrochemical Society Pv, 1*(7), 133–140, Pennington, NJ.

Samadani, E., Farhad, S., Scott, W., Mastali, M., Gimenez, L. E., Fowler, M., & Fraser, R. A. (2015). Empirical modeling of lithium-ion batteries based on electrochemical impedance spectroscopy tests. *Electrochim Acta, 160*, 169–177.

Savkina, R., Smirnov, A., Kirilova, S., Shmid, V., Podolian, A., Nadtochiy, A., & Korotchenkov, O. (2019). Charge-carrier relaxation in sonochemically fabricated dendronized CaSiO–SiO_2–Si nanoheterostructures. *Applied Nanoscience, 9*(5), 1047–1056.

Schmid, T., Sebesta, A., Stadler, J., Opilik, L., Balabin, R. M., & Zenobi, R. (2010, February). Tip-enhanced Raman spectroscopy and related techniques in studies of biological materials. In *Synthesis and Photonics of Nanoscale Materials VII* (Vol. 7586, p. 758603). International Society for Optics and Photonics San Francisco, California, United States: SPIE LASE.

Sharma, S., Rai, R., Hall, D. A., & Shackleton, J. (2012). Nonlinear ferroelectric and dielectric properties of $Bi(Mg_{0.5}Ti_{0.5})$ O_3-$PbTiO_3$ perovskite solid solutions. *Advanced Materials Letters, 3*, 92.

Singh, N. K., Kumar, P., & Rai, R. (2011). Comparative study of structure, dielectric and electrical behavior of Ba $(Fe_{0.5}Nb_{0.5})O_3$ ceramics and their solid solutions with $BaTiO_3$. *Advanced Materials Letters, 2*, 200–205.

Stankovich, S., Dikin, D. A., Piner, R. D., Kohlhaas, K. A., Kleinhammes, A., Jia, Y., & Ruoff, R. S. (2007). Synthesis of graphene-based nanosheets via chemical reduction of exfoliated graphite oxide. *Carbon, 45*, 1558–1565.

Subramanian, P., Pirbadian, S., El-Naggar, M. Y., & Jensen, G. J. (2018). Ultrastructure of Shewanella oneidensis MR-1 nanowires revealed by electron cryotomography. *Proceedings of the National Academy of Sciences of the United States of America, 115*, E3246–E3255.

Surridge, N. A., Jernigan, J. C., Dalton, E. F., Buck, R. P., Watanabe, M., Zhang, H., … Murray, R. W. (1989). The electrochemistry group medal lecture. Electron self-exchange dynamics between redox sites in polymers. *Faraday Discussions of the Chemical Society, 88*(1), 1–17.

Tengvall, P., Elwing, H., Sjöqvist, L., Lundström, I., & Bjursten, L. M. (1989). Interaction between hydrogen peroxide and titanium: A possible role in the biocompatibility of titanium. *Biomaterials, 10*(2), 118–120.

Tomaev, V. V., Moshnikov, V. A., Miroshkin, V. P., Gar'Kin, L. N., & Zhivago, A. Y. (2004). Impedance spectroscopy of metal-oxide nanocomposites. *Glass Physics and Chemistry, 30*(5), 461–470.

Waag, W., Fleischer, C., & Sauer, D. U. (2013). On-line estimation of lithium-ion battery impedance parameters using a novel varied- parameters approach. *Journal of Power Sources, 237*, 260–269.

Wilson, J. R., & Barnett, S. A. (2008). Solid oxide fuel cell Ni–YSZ Anodes: Effect of composition on microstructure and performance. *Electrochemical and Solid-State Letters, 11*(10), B181–B185.

Woodman, J. L., Jacobs, J. J., Galante, J. O., & Urban, R. M. (1983). Metal ion release from titanium-based prosthetic segmental replacements of long bones in baboons: A long-term study. *Journal of Orthopaedic Research, 1*(4), 421–430.

Xia, Y., Zou, J., Yan, W., & Li, H. (2018). Adaptive tracking constrained controller design for solid oxide fuel cells based on a wiener-type neural network. *Applied Science, 8*(10), 1758.

Yang, W. S., Noh, J. H., Jeon, N. J., Kim, Y. C., Ryu, S., Seo, J., & Seok, S. I. (2015). High-performance photovoltaic perovskite layers fabricated through intramolecular exchange. *Science, 348*, 1234–1237.

Yoon, Y., Jo, J., Kim, S., Lee, I. G., Cho, B. J., Shin, M., & Hwang, W. S. (2017). Impedance spectroscopy analysis and equivalent circuit modeling of graphene oxide solutions. *Nanomaterials, 7*(12), 446.

Zhu, Y., Murali, S., Cai, W., Li, X., Suk, J. W., Potts, J. R., & Ruoff, R. S. (2010). Graphene and graphene oxide: Synthesis, properties, and applications. *Advanced Materials, 22*, 3906–3924.

4 Analytical Techniques for Characterization of Oxide-Based Materials

Iraida N. Demchenko

4.1 ANALYTICAL TECHNIQUES FOR CHARACTERIZATION OF OXIDE-BASED MATERIALS

4.1.1 INTRODUCTION

The ability to control the physical properties of novel materials, by controlling the crystallographic structure, arrangement of atoms inside the sample's volume and along the surface taking into account point defects, is of crucial importance nowadays from both fundamental and applied research points of view. In order to understand the nature of matter, we need to address ourselves to its building blocks – atoms. These, in turn, consist of nuclei and electrons with the electronic structure ultimately determining all properties of matter. It is, therefore, natural to anticipate that knowledge of electronic structure of different systems together with the ability to describe and predict electronic structure of new systems will give a boost to science and technology. Among the ways to reach such information, X-ray spectroscopy techniques stand considerably out due to their capabilities to provide detailed information on material electronic structure and thus helping us to construct the informational bridge between the structural and electronic properties of wide class of materials. As is well known, X-ray methods could be classified according to various interaction processes occurring between radiation and substance. Generally, the following groups of methods can be distinguished: X-ray spectroscopy methods, methods based on diffraction and inelastic X-ray scattering methods. This chapter focuses on discussing X-ray spectroscopy methods such as X-ray absorption near edge structure (XANES)/X-ray emission spectroscopy (XES)/resonant inelastic X-ray scattering (RIXS) and presents some of their important examples, where the unique features of the photon "*in-/out*" process are exploited. In a nutshell, X-ray absorption spectroscopy (XAS) is a tool for probing the unoccupied density of electronic states of a material using energy tunable high intensity X-ray radiation. The measurement observes the fraction of X-ray photons that are absorbed by the specimen. Photons are absorbed by atomic core energy level (valence) electrons which are then promoted to higher previously unoccupied energy levels (conduction). The decay of electrons from these unstable states results in the emission of a photon of energy equivalent to the energy difference between the core level and excited state. XAS is an extremely powerful tool for measuring density of states because it is both element and site specific. XAS measurements can distinguish between the different binding energies of specific elements, as well as different bonding sites for the same element. Near the absorption edge, the dependence of the absorption coefficient exhibits a fine structure. Two types of fine X-ray absorption structure are distinguished: the near-edge (XANES) and the extended oscillatory structure (EXAFS) expressed by the function of, so-called, oscillations EXAFS, χ. Using the XANES method, it is possible to study the electronic structure of matter. The symmetry and energies of vacant molecular orbitals in molecules or electron bands lying above the Fermi level in solids could be determined. Information on the oxidation state of the absorbing atom and on the symmetry of its coordination sphere could be obtained as well. The EXAFS analysis provides more extensive information about the local environment of the absorbing atom, namely the type and number of nearest neighbors, as well as interatomic distances in a sphere with a radius of 5–6 Å. In addition to interatomic distances, the EXAFS data, taking into account multiple scattering of photoelectrons, as well as simultaneous processing the data for the absorption edges of several different elements in the examined compound provide information on the bond angles. The details could be found in Bunker (2010).

X-ray emission spectroscopy (XES) is a complementary technique that probes the occupied density of electronic states of a material (van Bokhoven & Lamberti, 2016). XES is also element and site specific. Emission spectroscopy can occur in two forms, either resonant inelastic X-ray emission spectroscopy (RIXS) or non-resonant X-ray emission spectroscopy (NXES). NXES probes the occupied density of states similarly to XAS for unoccupied. In this case, when an electron is excited to an unoccupied state, it leaves behind a core hole. This core hole may be filled by a previously excited electron or it may be filled by an electron from higher occupied states. This decay to fill the core hole may emit a photon as in the case of total fluorescence yield, but for NXES a spectrometer is used to select only those photons of energies corresponding to decays from occupied states. NXES will not be discussed further in this chapter since it is not a subject of our interest (for details of this technique see, for instance, de Groot and Kotani (2008)). RIXS functions by using the inelastic scattering of photons that are of a specific energy that corresponds to a resonance peak in an XAS spectrum. It is also element and site specific similar to XAS and XES. As it was mentioned above, RIXS is a *photon-in-photon-out* process where it is the energy and momentum change of the scattered photon that are measured observing excitations in the material caused by the transfer of energy. In other words, RIXS measures the energy and momentum differences between the occupied and unoccupied states. The RIXS method allows one to study the subtle features of the electronic structure (band structure) of solids: many electronic transitions that are not detected in ordinary absorption spectra or in non-resonant inelastic scattering spectra due to their low intensity are well manifested in RIXS spectra.

Let us note that most of the modern research in the field of X-ray spectroscopy is carried out using synchrotron radiation, since such studies require varying the energy of the incident radiation in a wide range. The basic concepts and properties of synchrotron radiation could be found in Attwood (1999).

As a complementary to X-ray spectroscopy methods, X-ray photoelectron spectroscopy (XPS)/resonant photoemission spectroscopy (RPES)(where photon “in-” photoelectron ‘out-’ process are exploited) and Rutherford backscattering spectrometry (RBS) (where an ion scattering technique is considered) techniques will be briefly described using oxide-based materials as examples. In methods of classical X-ray electron spectroscopy, the energy distribution of electrons knocked out of a sample by X-ray radiation is measured. There are several types of electron spectroscopy: XPS, Auger, spectroscopy of secondary electrons, etc. All methods of electron spectroscopy are used to study the surface, since the electron escape depth in the energy range typical of these methods does not exceed 50–100 Å. XPS and Auger methods are fundamental in determining the quantitative chemical composition of a surface (Hofmann, 2013). The positions of the lines in the photoelectron spectra correspond to the binding energies of the electronic levels, i.e. provide information on the electronic state of atoms on the surface (chemical shifts of core levels), as well as on the energy/electronic structure of the valence band. The effects that manifest themselves in photoelectron spectra and carry additional information include spin-orbit splitting of lines into multiplets, two-electron processes (low-energy "*shake-up*" and "*shake-off*" satellites), and inelastic photoelectron losses (characteristic bulk and surface plasmons).

Additional information in XPS can be obtained by "setting" the energy of the exciting X-ray radiation close to the edge of the resonant absorption (resonant photoemission, RPES). This method, in particular, is used to determine the contribution of individual atomic orbitals to molecular orbitals, which is important for describing the chemical bond. By varying the energy of the exciting radiation, the angle of incidence of the X-ray beam on the sample, or the angle of electron registration, it is possible to construct concentration profiles of the detected elements with depth in samples (making so call depth-profiles). For further details about this technique, refer to de Groot and Kotani (2008).

The last technique that will be discussed in this chapter is RBS. It is a quantitative, non-destructive method to analyze the depth profiles of the atoms of a solid (for details see Chu (1978)). A collimated beam of mono-energetic particles (for instance, He^+-ions) is accelerated by the Pelletron and steered into the scattering chamber, where the samples are mounted on a 3-axis goniometer. A small fraction of the impinging ions will backscatter (> 90°). The backscattered ions build up an energy spectrum out of which three parameters can be extracted: (i) the energy transfer during the collision, which depends on mass of a bombarding ion and the mass of the scattering atom allowing the mass of this atom to be determined; (ii) information about the depth profile, since additional energy loss occurs due to electronic stopping, and (iii) the concentration of the atoms in the sample, according to the Rutherford scattering cross section formula. The energy of the backscattered ions is measured with two Silicon Surface Barrier detectors, positioned/located at a fixed angle of 12° with respect to the incoming ion beam (for optimal mass resolution) and at a variable angle (for improved depth resolution). When the incoming beam is aligned with a major crystallographic direction of a single crystalline target, there is a drastic decrease of the backscattered yield. This is due to the fact that the incoming ions are guided into the channel or plane, which decreases the probability of direct collisions, i.e. a phenomenon called ion beam channeling. From these channeling measurements, information is obtained on the crystalline quality of the lattice (i.e. defects), the elastic strain, and the azimuthal orientation in film/crystal.

The chapter is organized as follows. After a short introduction presented in this section, the applicability of X-ray spectroscopy (both XANES and XES (including RIXS)) for the CdO films is presented in Section 4.1.2. Application of joint XANES, XES, and RIXS methods to estimate direct/indirect band gap size of materials will be discussed. The importance of XPS as a tool for observation of magnetic anisotropy of MgO/Co system with/without gold interlayer is discussed in Section 4.1.3. Section 4.1.4 presents usefulness of RPES spectroscopy to extract photoemission response of the localized Rare Earth (RE) impurity levels from the host electronic band structure using Yb implanted ZnO films as an example. Overall, the examples demonstrated in this chapter show efficiency of X-ray spectroscopic methods to obtain important information about the electronic structure of wide class of oxide-based materials confirming the fact that these techniques are relevant to a broad variety of scientific tasks in material science.

4.1.2 Concurrent Application of Soft X-Ray Emission and Absorption Spectroscopies: Direct Measurement of the Band Positions; Estimation of Direct and Indirect Bandgaps in Semiconductors

With the arrival of high-brilliance third-generation synchrotron sources, a new approach to study the bulk band structure has been established, namely to utilize RIXS. In RIXS, an electronic Raman scattering process is used to select specific excitations of valence electrons into unoccupied conduction band states. In other words, a core electron is resonantly excited into an unoccupied state at a certain k value, and the resonant fluorescence decay of a valence electron with the same k value into the core hole is detected. The observed RIXS spectrum thus contains momentum-resolved information about the occupied and unoccupied electronic states, which can be analyzed based on the Kramers-Heisenberg formalism (Rovezzi & Glatzel, 2014). Since the reachable information depth is typically of the order of a few hundred nanometers, study of systems with poorly defined surface properties or protective cap layers becomes possible. Starting from the first resonant X-ray emission spectra gathered using synchrotron radiation, a vivid discussion of resonance effects in RIXS has been occurring. Today it is widely accepted that band structure effects can be readily observed, and that this conclusion is true also for materials with core excitons. The purpose of the example presented below was to demonstrate that RIXS indeed excellently describes the band structure of semiconductor (using CdO as an example) and the experimental spectra agree very well with *ab initio* DFT calculations based on band structures derived with various functionals.

4.1.2.1 Analysis of O K-Edge XANES for CdO Films Combined with a Soft-X-Ray Emission Investigation

In order to probe the bulk-sensitive occupied valence band electronic structure, near-threshold excitation X-ray emission spectra were recorded for CdO films for excitation energies at the Oxygen K-edge absorption threshold. The main idea of this investigation was to demonstrate that the XES experiments are unique since they can provide direct evidence to the band structure of a semiconductor, in particular whether a semiconductor has direct or indirect band gap, without further theoretical input. In a XES experiment the band-gap type can be determined by observing the emission spectra as a function of excitation energy. In case of a direct energy gap material, emission at the highest energy is expected for excitation energy in the vicinity of the absorption threshold (into the conductive band (CB) minimum, CBM). As the excitation energy increases, the emission should shift to a lower energy. For indirect band-gap materials, the opposite behavior is expected, i.e., a shift of the emission spectrum (namely, top of the valence band (VB) maximum, VBM) to higher energy with increasing excitation energy. In other words, as the excitation energy increases, the probing transitions get closer, in k space, to the top of the VB. Our XES/RIXS data clearly show this tendency. Figure 4.1 presents the O K-edge XES spectra recorded with two different excitation energies, which demonstrate that increasing the excitation energy from 530 (excitation energies just below the absorption threshold) to 534 eV (slightly above the first absorption resonance) shifts the top of the VB to higher energies at about 0.3 eV. This behavior results from a well-established k-selectivity effect, whereby restrictions on the intermediate state relaxation enhance the emission at the k point of the CB minimum relatively. Consequently, the selective appearance of a peak at 526 eV below the top of VB is a clear signature of an indirect band gap in CdO as opposed to a direct band-gap semiconductor which would have the opposite relative enhancement of spectral "weight" at the VBM.

As mentioned in Demchenko et al. (2010), there is a controversy in the identification of the size of CdO energy gap. The experimental direct band gap values, accepted in literature, are around 2.2–2.4 eV. The experimental values of indirect band gap represented in the literature vary from 0.55 and 0.84 eV (see Koffyberg, 1976; Köhler, 1972; Madelung, 1982) to the higher ones of about 1.2 and 1.9 eV (Altwein, Finkenrath, Koňák, Stuke, & Zimmerer, 1968; Dakhel & Henari, 2003; Kocka & Konak, 1971; Maschke & Rössler, 1968). The same level of ambiguity exists in the reference data concerning the theoretical prediction of the indirect gap values, where the different ones are obtained using different approaches: linear combination of atomic orbitals (1.18 eV (Γ-L) and 1.12 eV (Γ-Σ); augmented plane wave

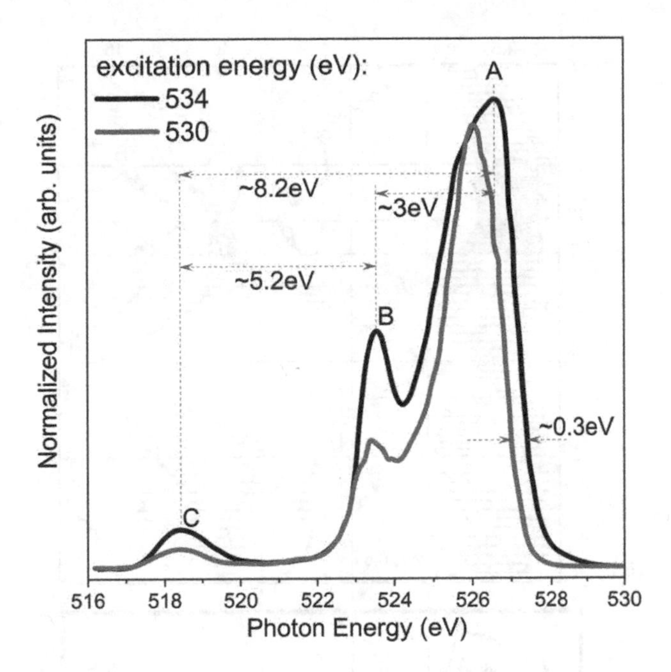

FIGURE 4.1 Oxygen 2p→1s XES spectra for CdO film with different excitation energies: 530 (red) and 534 eV (black) lines.

(0.8 eV (Γ-L) and 1.2 eV (Γ-Σ) and 1.11 eV (Γ-L) and 0.95 eV (Γ-Σ); local density approximation (1.7 eV); GGA+U+Δ and HSE03 +GW (0.68 eV). Right up until our publication (Demchenko et al., 2010), to the best of our knowledge, there were no reports available on the immediate evaluation of indirect gap values for CdO directly from combination of X-ray emission and absorption spectroscopy, except some attempts to indirectly estimate it (McGuinness, Stagarescu, & Ryan, 2003). In order to fill in this gap in knowledge, RIXS measurements were performed to probe transitions at different points in the Brillion zone in the CdO film. In resonant excitation, the core electron is promoted to a bound state in the CB, so the generated core hole recombines with the electrons from the higher electronic level (according to dipole transition rule), giving rise to the emission of a photon with the energy hv less than the energy of incident radiation, hv. For comparison, non-resonant excitation occurs when the incoming radiation promotes a core electron to the continuum.

When a core hole is created in this way, it could recombine through one of the several different decay paths. Since the core hole is refilled from the sample's high-energy free states, the decay and emission processes must be treated as the separate dipole transitions. It is in contrast to RIXS, where the events are concerned with each other and must be treated as a single scattering process.

The RIXS spectra (see Figure 4.4 in Demchenko et al. (2010)) were recorded in 0.5 eV step through the threshold region up to the first absorption peak at about 533 eV, as indicated by arrows on the XAS spectrum (see Figure 4.2(b)), and then for the resonances 536.51, 541.7, and 553.5 eV. An elastic emission peak in the threshold-excited XES is used for correspondence with the XAS photon energy scale. The key experimental observation is the relative loss of intensity near the VB maximum and apparent shifts to lower energy as excitation energy is tuned closer to the CB minimum threshold (at about 528 eV). Nonlinearity of the VBM profile (see region 525–527 eV) in comparison with peak positioned at around 523.2 eV is clearly seen on the intensity map (see Figure 4.3(a)).

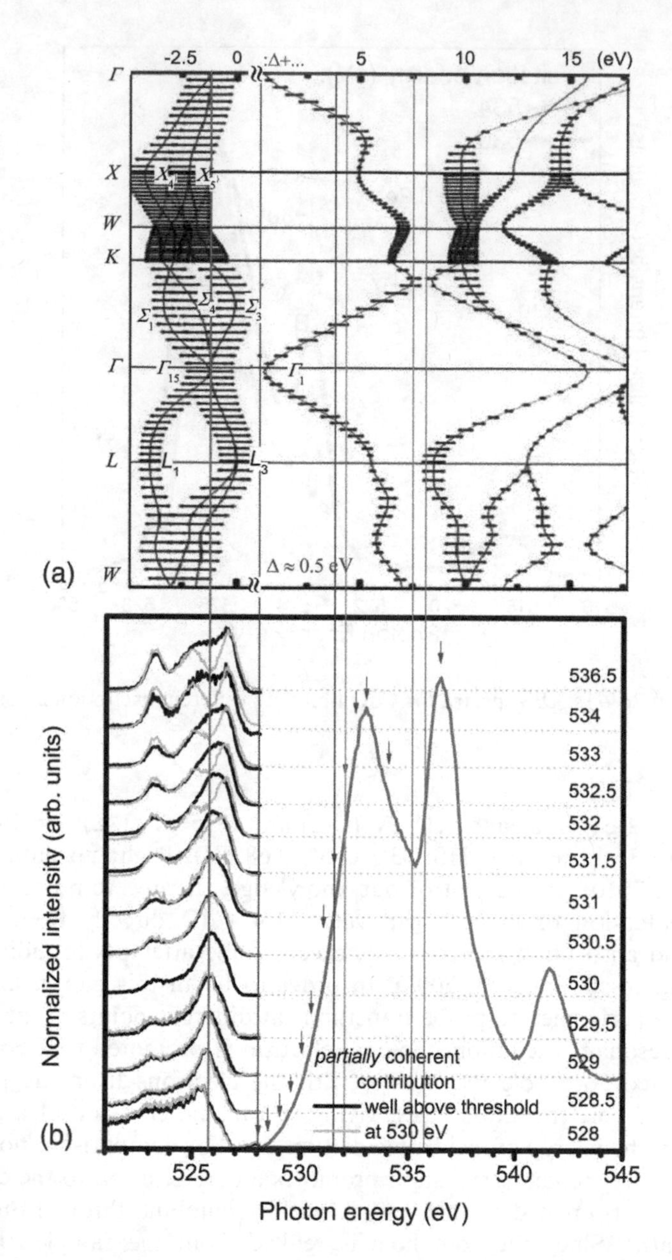

FIGURE 4.2 (Color online) Energy-aligned comparison of (a) theoretical HSE03-GW band structure of CdO Madelung (1982) to (b) the partially coherent contribution of resonant XES spectra normalized to their maxima and vertically offset. To obtain partially coherent contribution, two different approaches were applied: subtraction from RIXS data a XES spectrum well above the absorption threshold – black line, and XES spectrum at 530 eV – light gray line (see text for details). Additionally, the oxygen K-edge experimental XANES spectrum is shown on the right (dark gray line).

Reprinted with permission from Demchenko et al. (2010).

To understand the energy shifts and RIXS line shape changes, a procedure for subtracting the residual "incoherent" fraction component of the RIXS spectra has been proposed and used for enhancing the band-structure effects of the coherent fraction component

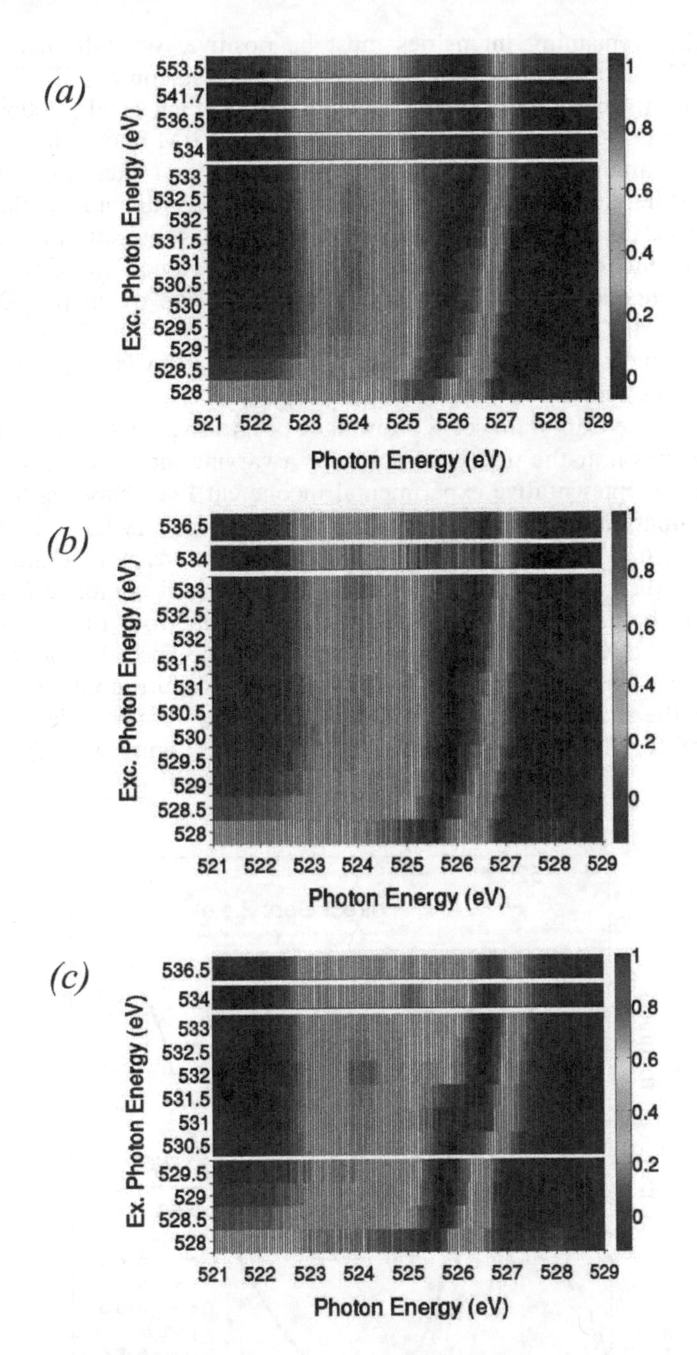

FIGURE 4.3 (Color online) (a) Intensity maps of normalized RIXS and partially coherent fractions of XES corresponding to (b) standard and (c) intermediate approaches (see text for details). Intermediate procedure works well for visualization of branching dispersion of occupied states.

Reprinted with permission from Demchenko et al. (2010).

(Lüning, Rubensson, Ellmers, Eisebitt, & Eberhardt, 1997). This involves usage of a high-energy excited spectrum well above the threshold as a representative of the incoherent X-ray emission with k mixing via intermediate state relaxations and scattering processes. This incoherent spectrum is then scaled and subtracted from the closer-to-threshold excited spectra with the restriction that remaining intensities must be positive. We call such a procedure as a standard one. For indirect band-gap materials such as silicon and BeTe, this procedure has been applied with clear and understandable results (Eich et al., 2006; Eisebitt et al., 1998). However, for CdO the leading energy shift of the VBM above threshold is problematic for these data analysis procedure (namely for a few first excitation energies spectra) since even with different scaling either a negative intensity dip above the 527 eV VBM results from subtraction of a high energy XES spectrum or subtraction has no effect on the spectra profile (an incoherent contribution is still there). Hence, there is a quandary that while there is 0% incoherent fraction between 527 and 527.5 eV in the 528 eV threshold spectrum, there is still clearly an incoherent shoulder from 526.2 to 527 eV extending above the 525.5 eV peak that one would like to subtract off. It is still visible after any type of subtraction taken for the estimation of pure coherent fraction.

We propose that the CdO data set is showing a progressively varying partial k mixing of initial and final states near the threshold and thus a varying incoherent line shape. The failure to find a single representative experimental incoherent line shape results from an over-simplified approximation that breaks down, probably, for the very large 4 eV CB dispersion of CdO. The single incoherent line shape subtraction works well for Si and BeTe where the CB lowest branch dispersion is <1 eV (Refs. 19 and 20) and complete k mixing is easier even near the threshold. From this scenario, we then claim from Figures 4.2 and 4.4 that the VB-CB gap in the XES-XAS threshold spectrum has no special meaning due to partial k mixing. Since the standard incoherent subtraction procedure could not result in completely removing the incoherent fraction and could not give fully a demonstration of the band structure of (Γ_{15}-L and Γ_{15}-Σ) branching effect (see Figure 4.2 (a)), it was decided,

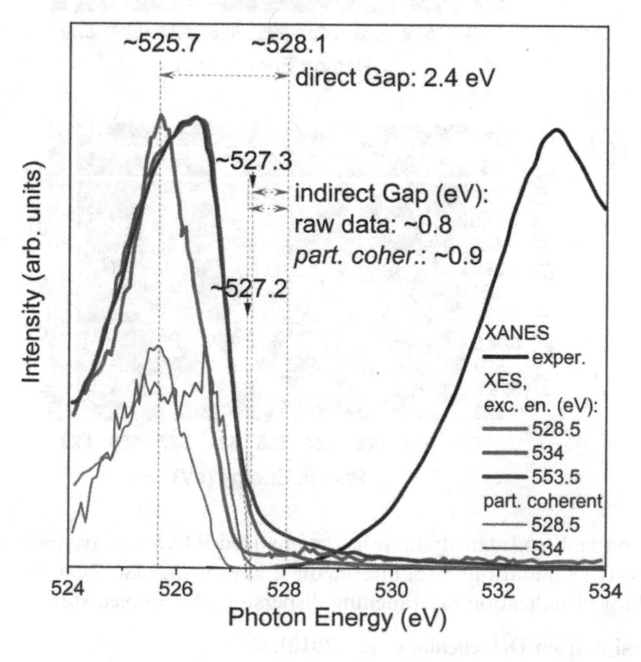

FIGURE 4.4 (Color online) Zoomed threshold region of oxygen 2p→1s XES and 1s→2p XANES spectra for CdO film with quantitative identification of direct and indirect band gaps. The excitation energies were 528.5 (thick red), 534 (thick green), and 553.5 (thick blue) eV. Red and green thin lines correspond to *partially* coherent fraction (standard procedure): 528.5 and 534 eV, respectively (see text for details, curves of the smaller intensity). XES energy scale was aligned to XAS by using elastic peak on emission spectra.

then, in parallel to the standard approach described above, how to arbitrarily enhance the excitation energy line shape changes by subtracting off an intermediate XES spectrum at 530 eV. In Figure 4.2(b) we present the resulting, partially coherent, emission spectra with comparison to the band structure (Figure 4.2(a)) taken from Piper et al. (2008). The partially coherent contributions corresponding to standard procedure are denoted by black lines, whereas light gray lines correspond to subtraction of an intermediate XES spectrum at 530 eV excitation energy (called here as an intermediate). In Figure 4.3 we present intensity maps corresponding to standard and intermediate partially coherent fraction of XES along with RIXS for a better visualization of band dispersion branches away from the (Γ_{15} point (see Figure 4.3(a)). To align the calculated band structure with respect to the experimental data, the VB Γ_{15} high-symmetry point was matched with the VBM of XES corresponding to excitation energy at 528.5 eV, so the CBM was shifted from the theoretical value of 1.9 eV to the optical experimental value at about 2.4 eV.

The estimated direct band gap value for CdO films, from our own optical absorption measurements, by an extrapolation of the square of the absorption coefficient comes to ~2.4 eV. Qualitatively, the aligned band structure agrees quite well with our results. This comparison highlights the low-energy and high-energy dispersions of the VB intensity with a good correspondence to the X and L high-symmetry points that lie at about 0.6 eV below and at about 1.0 eV above the Γ-point Γ_{15} energy. The Cd 5*s* orbital with even symmetry and the O 2*p* orbital with odd symmetry cannot mix at the Γ point, so the translation symmetry constraints are such that the Cd 5*p*-O 2*p* interaction is fully antibonding. Taking this into account and comparing our PDOS distribution with the presented band structure (Figure 4.2(a)) we are able to conclude that the orbital character of the lowest energy CB branch is mostly of Cd 5*s*-O 2*p* σ^*. As is well known, the VB upper bands have predominantly O 2*p* character.

4.1.2.2 Evaluation of Band-Gap Values

In Figure 4.4 the overlap of selected XES and XAS spectra (with the excitation energies equal to 528.5, 534, and 553.5 eV) in photon energy scale is presented. Two curves corresponding to partially coherent contributions (standard ones) are added to the plot (at 528.5 and 534 eV of excitation energy, curves of the smaller intensity) for more precise estimation of energy gap. Starting from the excitation energy of 534 eV, the edge of VB remains at the same energy and only the high energy tail increases its intensity. It should be noticed that the VB maximum movement occurs without any constant energy separation from the elastic peak position when the excitation energy increases. This suggests that the VBM for excitation energy of 528 eV cannot be associated with a constant (Raman-type) loss feature. To be firmly convinced, the 528.5 eV excitation energy spectrum was assumed to be a determinative one during our estimation of band-gap value. Taking into account that the CB minimum lies at 528.1 eV (point where absorption starts, obtained by taking the first nonzero values of the XAS spectrum first derivative), and the maximum of XES spectrum with the lowest excitation energy (528.5 eV) lies at 525.7 eV, we can evaluate the band-gap value to be 2.4 eV (Γ_1–Γ_{15} gap on the band structure). This is in good agreement with the direct band gap value presented in the literature and coincident with our own optical results.

Next, the difference between the CBM and the top of VB (at about 527.2 eV close to the L point) with the highest excitation energy is equal to about 0.9 eV which is in quite good agreement with the reported indirect band gap of 0.84 eV. Moreover, as is well known, the procedure of the indirect band-gap value estimation in case of thin CdO films by means of optical absorption spectra faces essential difficulties in view of weak signal and influence of Fabry-Pérot oscillations on the pre- and near-edge region. In order to directly measure the indirect gap of CdO by optical absorption, a thick CdO film (0.6 μm) was grown with the aim to make the indirect gap absorption more significant. Electrical property of this thick

CdO film is comparable to the thin CdO studied here. Linear extrapolation of the square root of the absorption coefficient gave a value of indirect band gap at about 0.95 eV which agrees well with our results discussed above.

Thus, combination of Figures 4.2 and 4.4 allows us to conclude that for investigated CdO thin films – (i) the direct energy gap at the Γ point is ≈2.4 eV, (ii) an indirect gap is ≈0.9 eV which implies dispersion along Γ-L and Γ-K, and (iii) lower energy dispersion is along Γ-X. Furthermore, the overall theoretical VB bandwidth is in good agreement with the lower energy XES peak at about 518 eV and overall 4–5 eV band width. Such clear and distinct indirect band gap and valence band dispersion signatures in CdO RIXS were not observed in previous X-ray emission studies, in part, because of the worse energy resolution and since the threshold excitation region down to 528 eV was not probed.

4.1.3 Assessing Factors Influencing Magnetic Anisotropy Type in Co/MgO System with Gold Interlayer

A phenomenon of perpendicular magnetic anisotropy (PMA) in F/MO_x interfaces, where F represents a ferromagnetic metal, M stands for a diamagnetic metal, and MO_x marks a nonmagnetic oxide (i.e. isolator) might be having an exciting future in novel devices for spintronics. Fundamental origin of PMA in such systems and a role of interfacial orbital hybridization require further investigations though (Daalderop, Kelly, & Shuurmans, 1994; Nakajima et al., 1998). The PMA appears when the interface anisotropy energy overcomes the magnetostatic and volume energy contribution to the free energy of the magnetic layer. This type of magnetic anisotropy, so-called interface or surface anisotropy, was predicted as a result of lowering symmetry at the surface or interface. However, the development of PMA in materials based on F/MO_x interfaces is still problematic due to incomplete understanding of its causes. Some researchers declare that PMA can be created only through a hybridization of F 3*d* and O 2*p* orbitals at the F/MO_x interface, while others show that placing an appropriate underlying nonmagnetic material is critical for developing PMA (Chen, Ma, & Wang, 2007; Gladczuk, Aleshkevych, Lasek, & Przyslupski, 2014; Manchon et al., 2008; Yang et al., 2011; Zhang et al., 2014). Studies of the electronic structures of F/MO_x linked together with magnetic measurements and theoretical studies should, hopefully, lead to a full understanding of PMA in such systems.

4.1.3.1 XPS as a Tool for Studies of the F/MO_x Interface

XPS is one of the primary tools used to analyze the interfaces utilizing either conventional X-ray tubes or complex synchrotron sources. These studies are frequently accompanied by sputtering to investigate depth dependence of XPS signals. That, however, may lead to unambiguous results due to the fact that an interpretation of XPS data for buried interfaces recorded in combination with ion sputtering procedure should be performed with special care as sputtering itself can seriously affect the interlayer structure (Chen et al., 2014). One should remember that ion sputtering, even when using noble gas ions, generates a large number of artifacts in subsurface region, as for instance, atomic mixing and knock-on implantation, preferential sputtering, bond breaking, phase formation, segregation, radiation-enhanced diffusion, roughness formation, etc. Such effects have been studied over the last decades and critical reviews of their influences on surface analytical techniques were published (Briggs & Seah, 1990; Demchenko et al., 2017; Hofmann, 1998; Lam, 1988; Oswald & Reiche, 2001). Taking into account the knowledge gathered within experimental observation of electronic structure modification due to sputtering procedure, it was decided to abandon it and study possible electronic structure modification at Co/MgO interface after addition of the thin layer of Au without sputtering procedure.

The samples containing Co/MgO interface with and without a thin gold interlayer in-between were grown onto a plane sapphire substrate at room temperature by molecular beam epitaxy. Complete details of their growth procedure can be found in Gladczuk et al. (2014). The thicknesses of each layer in the samples were identified as Mo (20 nm)/Au (20 nm)/Co (1.8 nm)/Au (0.3 nm)/MgO (2 nm) (Sample 1) and Mo (20 nm)/Au (20 nm)/Co (1.8 nm)/MgO (5 nm) (Sample 2). Details of XPS experimental set up, ferromagnetic resonance measurements, and software utilization could be found in Demchenko, Syryanyya, et al. (2018a). The HR XPS spectra for Sample 1 (with the gold interlayer) are shown in Figure 4.5.

The Au $4f_{7/2}$ photopeak maximum, located at 84 eV, was taken for calibration of energy scale (see Figure 4.5(a)). The Au $4f_{5/2}$ peak is overlapped with the Mg 2s states

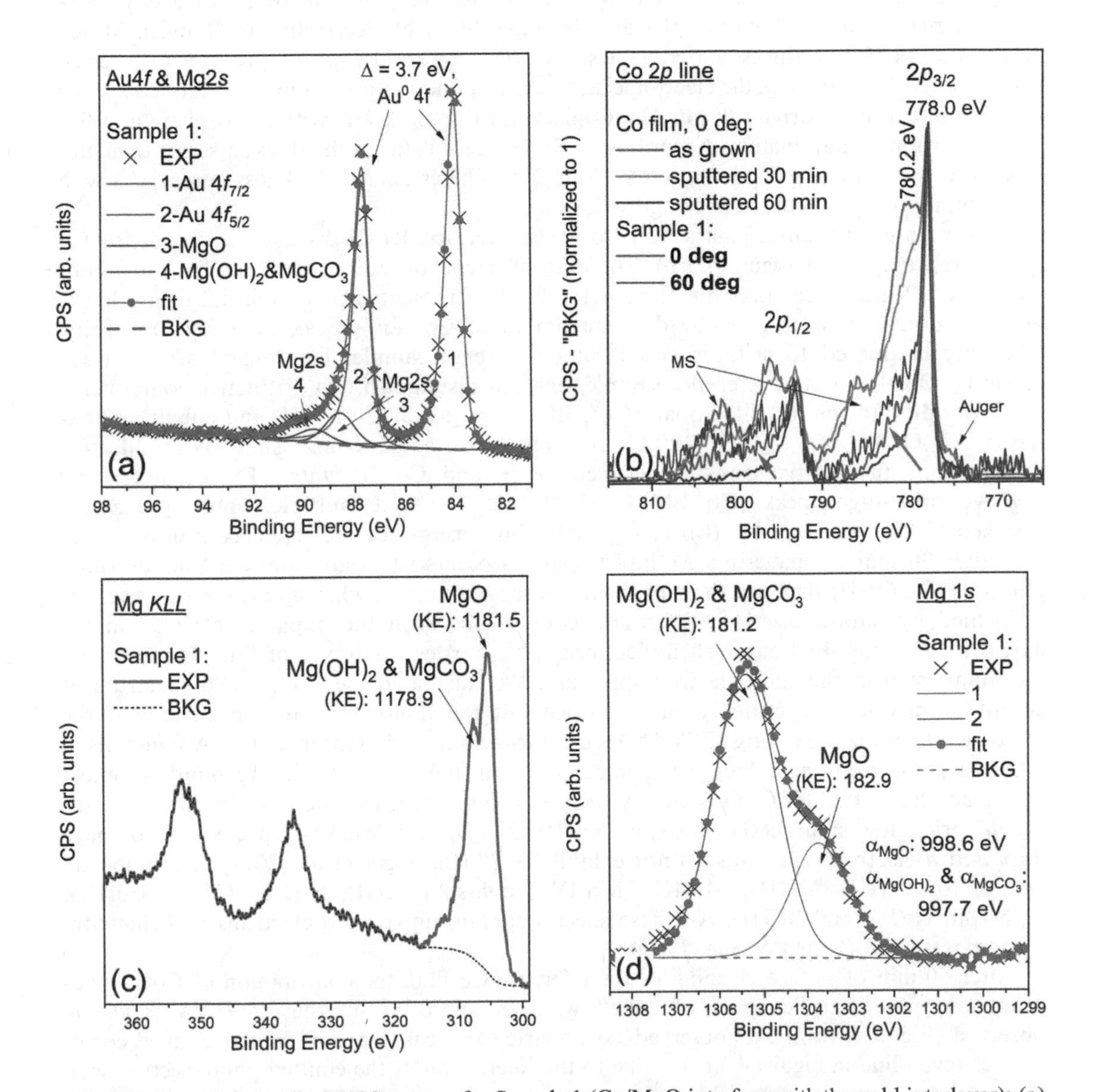

FIGURE 4.5 (Color online) XPS spectra for Sample 1 (Co/MgO interface with the gold interlayer): (a) split of Au 4*f* and Mg 2*s* states; (b) Co 2*p* states compared to metallic cobalt (after and before surface cleaning, see text for details); (c) Auger Mg KLL; and (d) Mg 1s line.

corresponding to various oxides of magnesium (marked as "3" and "4") with binding energies (BE) of 88.6 and 89.3 eV. The obtained values agree well with the data presented in literature (see, for instance, Vincent Crist, 1999), in which the Mg 2s peak positions of $Mg(OH)_2$, $MgCO_3$, and MgO are listed at the energies 89.2 eV, 89.3 eV, and 88.6 eV, respectively. Motivated by the uniqueness of peak shapes and positions within Auger spectra, which is useful for both elemental identification and chemical state analyses, a detailed analysis of Auger Mg KLL line (Figure 4.5(c)) in conjunction with Mg 1s XPS peak (Figure 4.5 (d)) was performed. The kinetic energy (KE) values of the Auger Mg KLL features and of the Mg 1s states are depicted in Figure 4.5(c) and (d), respectively. The estimated difference between photoelectron (Mg 1s, Figure 4.5(d)) and Auger (Mg KLL, Figure 4.5(c)) peak positions provides the so-called Auger parameter (α). It is particularly useful for chemical state analysis and can be used without interference of surface charging. The derived Auger parameters for magnesium in both samples are 998.6 and 997.7 eV. According to Bouvier, Mutel, and Grimblot (2004), the estimated values represent MgO and $Mg(OH)_2$ & $MgCO_3$ species, respectively. Consequently, the electronic states corresponding to the above mentioned species reflect interaction of originally pure MgO phase (in the top layer) with atmosphere and following carbon contamination. According to SESSA calculations, the thicknesses of that sublayer and the following MgO one are ~ 20 ± 2 Å. The estimated thickness agrees well with the nominal one predicted from the growth process.

As our interest focuses on the top Co/MgO interface, let us discuss the results for Co $2p$ line represented in Figure 4.5(b). The data of deconvolution of Co $2p_{3/2}$ lines for Samples 1 and 2 are summarized in Table 4.1. The data presented in Figure 4.2(b) are background subtracted and normalized to maximum of Co^0 $2p_{3/2}$ peak intensity for clarity. They are compared to metallic cobalt film (reference sample) before and after surface cleaning. On the pristine reference sample, one can distinguish two distinctive components corresponding to the metallic cobalt (Co^0, BE of $2p_{3/2}$: 778.0±0.15 eV) and cobalt monoxide (Co(II)O, BE of $2p_{3/2}$: 780.2±0.15 eV). The observed "chemical shift" is an effective indicator of the charge transfer between O $2p$ and Co $3d$ states. Furthermore, one observes an Auger peak (Co LMM: 777.0±0.15 eV) and multiplet splitting features (marked "MS" in Figure 4.5 (b)) at higher binding energy for the reference sample before and after 30 min of sputtering. Multiplet splitting occurs when an atom contains unpaired electrons (for Co(II) there are three of them). In these instances, when a core electron vacancy is formed by photoionization, there is a coupling between the unpaired electron in the core with the unpaired outer shell electrons. This creates a number of final states, which are manifested in the photoelectron spectrum. We calculated the Co $2p$ XPS spectrum of cobalt monoxide compound taking into account the multiplet coupling effect and the covalency hybridization using CTM4XAS code (Stavitski & de Groot, 2010). A comparison of the experimental and calculated spectra is shown in Figure 4.6. Strong multiplet effect at the $3d^7$ final state for CoO is clearly seen. It is important to note that, in the first transition series, low-spin Fe(II), low-spin Ni(II), Cr(VI), and Mn(VII) species do not have unpaired d electrons and thus do not exhibit "MS" (Biesinger et al., 2011). The opposite is true for Cr(III), Mn(II), Mn(III), Mn(IV), high-spin Fe(II), Fe(III), Co(II), Co(III), high-spin Ni(II), and Ni(III). All these species contain unpaired d electrons and therefore demonstrate MS (Gupta & Sen, 1975).

After 60 min of surface cleaning of the reference Co film, the contribution of CoO disappears and pure metallic phase of cobalt with $2p$ spin-orbit splitting 14.97 eV is clearly observed (Figure 4.5(b)). The observed asymmetric Co^0 peak shape of sputter-cleaned cobalt surface (cyan line in Figure 4.5(b)) is due to the interaction of the emitted photoelectron with the conduction electrons available in conductive/metallic samples. These shake-up-like events generate a tail on the higher binding energy side of the main peak instead of discrete shake-up satellites (Briggs, 2003). The comparison of the Co $2p$ states of Sample 1 to reference

TABLE 4.1
Cobalt $2p_{3/2}$ spectral fitting parameters for metallic cobalt Sample 1 and Sample 2. The energy calibration was done for Au $4f_{7/2}$ line at 84 eV

Name	Description	Position (eV) ± 0.15eV	FWHM (eV)	L. Sh.	Area (%)	St. Dev. (%)
Co^{O}, *Sputtered*	Auger LMM	777.0	3.12	GL(30)	12.27	0.06
	$2p_{3/2}$	778.0	0.75	LA (1.2,5,5)	70.75	0.05
	Plasmon 1	781.0	3.28	GL(30)	9.91	0.01
	Plasmon 2	783.0	3.28	GL(30)	7.08	0.01
Sample 1	Auger LMM	777.0	3.12	GL(30)	10.80	0.07
Au/Co/Au(0.3 nm)/MgO Co^{O} components	$2p_{3/2}$	778.0	0.75	LA (1.2,5,5)	64.86	0.15
	Plasmon 1	781.0	3.28	GL(30)	9.08	0.02
	Plasmon 2	783.0	3.28	GL(30)	6.49	0.02
Co(II)O components	$2p_{3/2}$	779.9	2.3	GL(30)	4.38	0.07
	MS 1	782.2	2.6	GL(30)	2.41	0.04
	MS 2	786.4	3.7	GL(30)	1.97	0.09
Sample 2	Auger LMM	776.94	3.12	GL(30)	7.7	0.16
Au/Co/MgO Co^{O} components	$2p_{3/2}$	777.94	0.75	LA (1.2,5,5)	58.05	0.21
	Plasmon 1	780.94	3.28	GL(30)	8.13	0.03
	Plasmon 2	782.94	3.28	GL(30)	5.80	0.02
Co(II)O components	$2p_{3/2}$	779.84	2.3	GL(30)	10.16	0.08
	MS 1	782.14	2.6	GL(30)	5.60	0.04
	MS 2	786.34	3.7	GL(30)	4.57	0.04

sample (with varied geometry of the XPS signal acquisition) manifests overlaying of minor cobalt oxide component with the major contribution of metallic cobalt. By deconvolution of the Co $2p_{3/2}$ peak, the fraction of Co(II) in Sample 1 is determined to be 9.8%. Here, the CoO fraction is calculated as the ratio of CoO phase to the metallic cobalt phase in the cobalt layer. The amount of each phase was determined estimating areas of the respective photoelectron sub-peaks during de-convolution of XPS spectra normalized to 100%. This indicates that despite a thin gold interlayer between cobalt and magnesium oxide layers, some amount of cobalt atoms is bonded to oxygen. This observation allowed us to suggest that gold layer grown on the cobalt top interface is in the form of non-coalescing islands; in other words, the top gold interlayer is not continuous. Thus, during deposition of the MgO layer, oxygen atoms from the MgO combine with the neighbouring Co atoms, leading to a formation of CoO at the Co/MgO interface in areas between the gold islands. Estimated by SESSA software thickness of CoO at metallic cobalt interface is ~7 ± 2 Å. It is worth noticing here that Au/Co/Au (0.3 nm)/MgO sample reveals PMA, see Figure 4.7(a).

Before general discussion of the influence of gold interlayer (between Co/MgO) on magnetic anisotropy of Sample 1, let us briefly mention the XPS results from the sample without gold interlayer between cobalt and magnesium oxide layers. It was shown that the fraction of cobalt oxide phase in this sample is about two times larger compared to the sample with gold interlayer. That means that oxygen atoms from MgO combine with the neighbouring Co

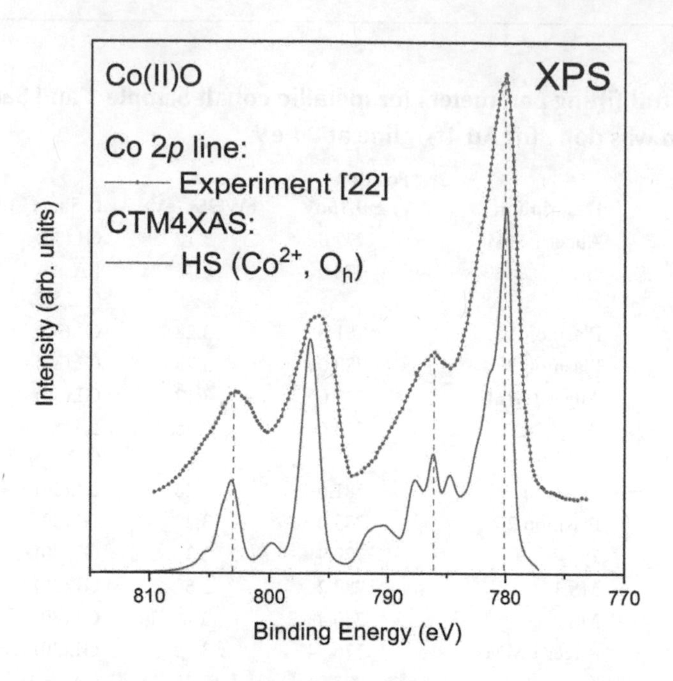

FIGURE 4.6 (color online) Charge transfer multiplet calculations of Co^{2+} in octahedral crystal field (red line) together with the experimental (circles) Co 2p XPS spectrum from CoO taken from (Biesinger et al., 2011). The used parameters values Δ are =2.5 eV (charge transfer), U_{dd} = 7.0 eV, and V(eg) = 2.0 eV.

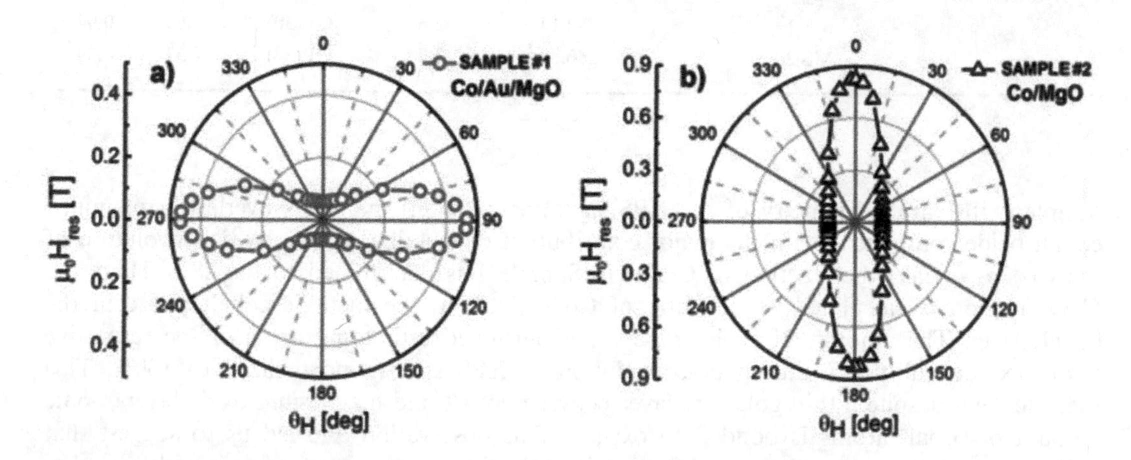

FIGURE 4.7 (color online) The polar angle dependence of the resonant field as a function of the angle between the direction of external magnetic field H_{ext}, and the normal to the sample surface for (a) Sample 1: Au (20 nm)/Co (1.8 nm)/Au (0.3 nm)/MgO (2 nm) and (b) Sample 2: Au (20 nm)/Co (1.8 nm)/ MgO (5 nm).

atoms leading to a formation of CoO at the Co/MgO interface. Estimated by SESSA software thickness of CoO interlayer at metallic cobalt interface for that sample is bigger compared to Sample 1 and is equal to ~10 ± 2 Å. By the way, Sample 2 manifests *in-plane* magnetic anisotropy (IMA, see Figure 4.7(b)).

4.1.3.2 Complementary Ferromagnetic Resonance (FMR) Investigations

The FMR resonance field (H_{res}) for Samples 1 and 2 is shown in Figure 4.7(a) and (b), respectively. In case of Sample 2 (with Co/MgO interface), a maximum ($\mu_0 H_{res}$= 0.73T) and minimum ($\mu_0 H_{res}$= 0.17T) of H_{res} was observed for perpendicular and parallel orientation of the external magnetic field, respectively (Figure 4.7(b)). Those extrema of $\mu_0 H_{res}(\theta_H)$ indicate an easy axis of magnetization in the plane of the magnetic layer. However, the structure with the gold incorporated at Co/MgO interface (Sample 1) has the maximum $\mu_0 H_{res} = 0.41\ T$ and minimum $\mu_0 H_{res} = 0.13\ T$ of H_{res} for parallel and perpendicular orientation of the H_{ext}, respectively, as an easy axis of magnetization is perpendicular to the sample surface. The results of the FMR investigation denote a significant enhancement of the surface anisotropy energy of cobalt layer (d_{Co} ~ 1.8 nm) as an effect of gold monolayer insertion between Co and MgO. The high surface anisotropy energy in Au/Co/Au/MgO structure, compared to the shape and magnetocrystalline magnetic anisotropy energy contributions orient the easy axis of magnetization perpendicular to the sample surface.

4.1.3.3 IMA vs PMA: General Remarks

The results previously published for Co/AlO_x system (Manchon et al., 2008) strongly suggest that the onset of PMA is related to the appearance of a significant density of interfacial Co–O bondings at the Co/AlO_x interface. However, the here-investigated Au/Co/MgO structure (Sample 2) reveals larger fraction of cobalt oxide compared to the sample with gold interlayer between Co and MgO (Sample 1) but, at the same time, IMA instead of PMA is observed. Consequently, there should be another factor explaining such an effect. First principle calculations for Fe/MgO and Co/MgO systems presented in Yang et al. (2011) make clear that in the case of ideal metal/isolator interfaces both systems reveal PMA with values of 2.93 and 0.38 erg/cm^2, respectively. That obviously differs for the investigated case since Sample 2 demonstrates IMA. The calculations for Fe/MgO showed that PMA weakens in the presence of interfacial disorder and lowers down to 2.27 and 0.98 erg/cm^2 for under- and over-oxidized cases, respectively (Yang et al., 2011). The over-oxidation of metal layer is detrimental to PMA (Khoo et al., 2013; Lacour et al., 2007; Schellekens et al., 2013) becausc the number of mixed states with both metal d_z^2 and oxygen p_z orbitals (which is critical to PMA at "metal/nonmagnetic oxide" interface) *is reduced* due to the local charge redistribution induced by additional oxygen atoms (see Figure 4.2 and the discussion in Zhang, Butler, and Bandyopadhyay (2003). This reduction is attributed to the split of the Co-d_z^2 and O-p_z hybridized states around Fermi level in the presence of additional oxygen. As the surface energy is decreased, the IMA in Sample 2 is observed (see Figure 4.7(a)).

Taking into account the above-mentioned, it is interesting to identify the PMA origin in Sample 1. The fitted surface anisotropy constant K_s (Gladczuk et al., 2014) for Au/Co/Au (0.3 nm)/MgO heterostructure is 1.6 erg/cm^2 (let us note that estimated value is higher then for Au/Co/MgO heterostructure (1.2 erg/cm^2)) and is approximately four times larger than theoretically predicted PMA value of 0.38 erg/cm^2 for ideal Co/MgO interface (Zhang et al., 2014). A decreased fraction of the cobalt monoxide (down to 9.8%) *and* an assumption of ideal Co/MgO interface do not explain fully the estimated value of K_s. A possible explanation (additional factor) of PMA existence in Sample 1 is the interfacial hybridization, i.e. a strong spin-orbit (SO) interaction, between the magnetic (cobalt) and nonmagnetic (gold) metals. For instance, several theoretical studies (Daalderop, Kelly, & den Broeder, 1992; Daalderop, Kelly, & Schuurmans, 1990; Daalderop et al., 1994; Gay & Richter, 1986; Wang, Wu, & Freemanet, 1993a, 1993b, 1994; Wu, Li, & Freeman, 1991) predicted that large SO coupling of Pd plays an important role for obtaining PMA in Co/Pd multilayers. In fact, there are plenty of reports regarding Co/Pd, Co/Pt, and Co/Au films possessing PMA (Carcia, 1988; Engel, England, Leeuwen, Wiedmann, & Falco, 1991;

Nakajima et al., 1998). All authors share the same opinion that a strong interfacial *d-d* hybridization produces an enhanced perpendicular Co orbital momentum, which causes PMA by SO coupling. Consequently, it is likely that *d-d* hybridization increases the surface energy (0.83 erg/cm^2 for Co/Au interface (Gladczuk et al., 2014) and plays an important role in developing PMA, as it appears for Sample 1. In other words, introduction of the gold interlayer at the Co/MgO interface induces the hybridization of Au 5*d* levels with 3*d* electrons of the ferromagnetic layer that generates/enhances PMA, in context of mixed states with both metal d_z^2 and oxygen p_z orbitals at Co/MgO interface.The results of XPS analysis presented here show clearly that some fraction of Co atoms at the Co/MgO interface is bonded to oxygen atoms. Moreover, an "oxidation zone" (thickness of CoO interlayer estimated by SESSA software) is approximately 1.5 times bigger for sample revealing IMA (Sample 1, without gold interlayer). With this in mind, the reaction of Co with oxygen atom through oxygen migration mechanism (Chen et al., 2014; Gilbert et al., 2016; Zhou et al., 2016) can be attributed via the redox reaction at the Co/MgO interface. It is clear that in Sample 1 the non-continuous gold interlayer between Co/MgO partially blocks the migration of oxygen atoms into the layer of cobalt. The discussed above "over-oxidation" of the Co/MgO interface turns out to be the only reason to explain decreasing of a surface energy leading to IMA for the sample without gold interlayer (Sample 2). The opposite is true for the sample with gold interlayer (Sample 1), namely, a metal–metal SO interaction plays a leading role in the manifestation of PMA. It is important to note also that according to Cagnon et al. (2001), other effects like interface roughness, magnetostriction, etc., all are not considered here, which may also come into play.

Overall, the obtained results identify a possibility of controlling the type of magnetic anisotropy in Co/MgO systems through addition of a gold interlayer, the fact that could be used in novel devices for spintronics.

4.1.4 RE-Implanted ZnO Before and After Thermal Annealing Probing by RPES and RBS

ZnO doped with RE elements system has experienced a close attention as it can be used in optoelectronics, information storage and photovoltaic devices (Chuang, Liou, & Woon, 2017; Jagadish & Pearton, 2006; Ratajczak et al., 2018). Usage of appropriate and effective processing techniques, such as atomic layer deposition (ALD) and ion implantation, can additionally increase this attention. ALD is a growth method suitable to cover large area substrates, thus the method can be applicable for low cost solar batteries production (Hurle, 1994). Ion implantation, on the other hand, is a convenient method for introducing RE ions into the crystal lattice with easily controllable ion concentration and its depth profile (Williams & Poate, 1984). The resultant system has, however, an extremely complex electronic band structure due to overlap of the RE $4f^n$ levels with the host VB, making detailed analysis a complicated endeavor. Typically, the main interest lays in measuring the energies of the 4*f* electrons relative to the band states, as well as the effects of the RE doping on the host lattice. Therefore, estimation of the binding energy of the VB maximum is required for every studied material. In this context, RPES is a useful tool for extracting photoemission response of the localized RE impurity levels from the host electronic band structure. We examined thin ZnO films grown by ALD and implanted with Yb to a fluence of 5e15 ions/cm^2 keeping in mind that such films can be utilized as down-converter material in Si-based solar cells. RPES was applied to study the electronic band structure of ZnO:Yb and extract the binding energy position of the Yb 4*f* electrons from the photoemission response of the system. The structure damage level in the examined system, lattice position, depth distribution of RE ions, and redistribution of the defects as a result of annealing, were probed by channeling RBS (RBS/c).

4.1.4.1 Quality of RE-Implanted ZnO Films Probed by RBS/c

Details of growth process of investigated samples and experimental set-up of applied techniques could be found in Demchenko, Ratajczak, Melikhov, Konstantynov, & Guziewicz (2019). The RBS/c results for the ZnO film implanted with Yb to a fluence of 5e15 at/cm^2 are shown in Figure 4.8. In this particular case, RBS/c spectra can be split into two independent regions: the lower energy part representing the scattering by host Zn atoms (Figure 4.8(a), left panel) and the higher energy part connected with the scattering by impurity atoms (Figure 4.8(a), right panel). Such a split emphasizes different information provided by the RBS data: the Zn signal reflects the radiation defect as effect of ion implantation, whereas the Yb signal shows the depth distribution of the ions and their lattice location.

The Zn related RBS/c spectra clearly reveal two damage peaks (see Figure 4.8(a)). A deeper one (located ~1260 eV) is the characteristic damage peak located close to the incident Yb ions depth range in ZnO (~33 nm) and its position is consistent with calculations performed by the Stopping and Range of Ions in Matter (SRIM) code (Ziegler, Biersack, & Ziegler, 2009) (see Figure 4.8 (b)). The second peak (between 1300 and 1325 eV) shows a strong damage close to the sample's surface due to the oxygen deficiency and enrichment in Zn interstitials (Zn_i) (Altwein et al., 1968; Ratajczak et al., 2018; Ratajczak, Mieszczynski, Prucnal, et al., 2017). After annealing, the deeper damage peak becomes narrower and the number of displaced Zn atoms seems to increase in the middle of this peak. This means that annealing does not lead to the lattice recovery but rather to a transformation and agglomeration of defects in the structure. The mobility of Zn atoms in the ZnO lattice during high temperature annealing is the primary reason for these phenomena (Turos et al., 2017).

Before annealing, the Yb signal of the aligned spectrum is slightly lower compared to the random one as can be seen in Figure 4.1(a), right panel. Based on this observation, it can be concluded that after implantation a small part of Yb ions is found in substitutional positions in

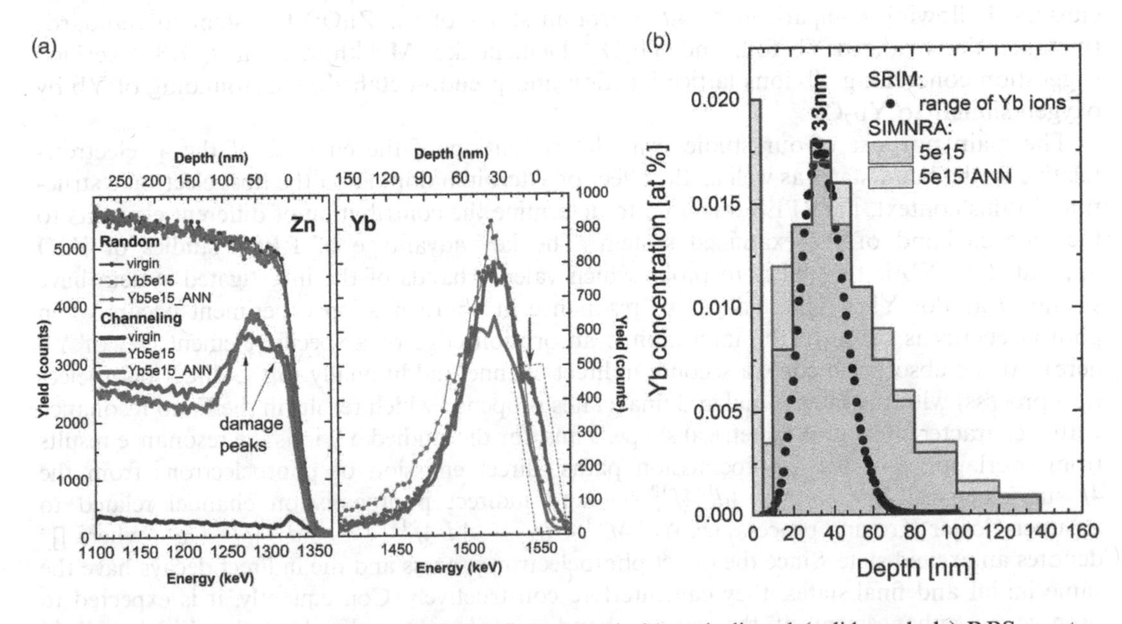

FIGURE 4.8 (color online) (a) The random (open symbols) and aligned (solid symbols) RBS spectra obtained for virgin ZnO films and after implantation with 150 keV Yb ions to a fluence of 5×10^{15} at/cm^2 (before and after annealing). (b) Yb ions depth distribution in ZnO matrix before and after annealing simulated by SIMNRA and SRIM codes.

ZnO matrix. This postulate was not confirmed by X-ray photoelectron spectroscopy studies for this fluence (not shown here) in the same way as for a sample with a higher fluence (shown in (Demchenko, Melikhov, et al., 2018b). In both cases, Yb 4*d* states corresponding to Yb^{2+} were not visible. This can presumably be explained by the fact that eventual Yb^{2+} concentration is below the sensitivity limit of this technique. During annealing all atoms become mobile. Consequently, a total rejection of originally substitutional Yb atoms into interstitial sites takes place. Moreover, partial precipitation of these ions on the surface region is followed as a result of a prolonged diffusion process (this issue will be further discussed below).

To illustrate migration of Yb ions in the host matrix, SIMNRA simulations were performed (see Figure 4.8(b)). These calculations were aimed to fit random RBS spectra shown in Figure 4.8(a), right panel. The Yb ions distribution calculated by the SRIM code is also added for comparison. As can be seen, the range of Yb atoms has been evaluated at depth of 33 nm, which is in a good agreement with both simulation results. Summarizing, analysis of the results presented in Figure 4.8 lets to the conclusion that during annealing of implanted samples two different processes affect redistribution of Yb ions. The first process is a diffusion that relies on replacement of the implanted Yb ions to interstitial positions. The second one is an out-diffusion: it relies on agglomeration of implanted ions on the surface as a result of a prolonged diffusion process. The latter process is seen in Figure 4.8(a), right panel, as a shoulder marked by the arrow.

4.1.4.2 Valence Band of Yb Doped ZnO Probed by RPES

Our previous X-ray photoelectron spectroscopy investigations of Yb 4*d* states in ZnO:Yb presented in Demchenko, Melikhov, et al. (2018b) clarified that the majority of ytterbium atoms (Yb^{3+}) in the system are bonded to oxygen and show an extended multiple structure indicating that one of the $4f^{14}$ electrons has been promoted to the valence level. This multiple structure can be attributed to the $4f^{13} \leftrightarrow 4d^9$ interaction. The Yb^{3+} ions in the ZnO host matrix are located in the interstitial positions and according to investigations presented in Ratajczak, Prucnal, Guziewicz, et al. (2017) form presumably pseudo-octahedron YbO_6 clusters. Following comparison of 4*d* ytterbium states of the ZnO:Yb system to standards (metallic Yb, oxidized Yb foil, and Yb_2O_3) Demchenko, Melikhov, et al. (2018b) verified suggestion concerning Yb ions lattice location and pseudo-octahedral surrounding of Yb by oxygen similar to Yb_2O_3.

The main purpose of our studies was determination of the energies of the 4*f* electrons relative to the band states as well as the effect of ytterbium doping on the host electronic structure. In this context, the RPES was used to determine the contribution of different elements to the valence band of the examined system. The key advantage of RPES studies of ZnO implanted by Yb is the ability to probe which valence bands of the investigated system have strong 4*f* and/or Yb weight. Fano-type resonance in photoemission experiment occurs when photon energy is tuned to the inter-atomic absorption edge of a specific element (N_5 of Yb here). At the absorption edge, a second, indirect channel (additionally to the direct photoelectron process) with the same initial and final states is opened which results in the Fano resonance with a characteristic anti-symmetrical shape. Thus, for the studied Yb ions the resonance results from overlapping of two photoemission paths: direct emission of photoelectrons from the 4*f* state, i.e. $4d^{10}4f^{13} + h\nu \rightarrow 4d^{10}4f^{12} + e^-$and indirect photoemission channel related to a Super-Coster-Kroning process, i.e. $4d^{10}4f^{13} + h\nu \rightarrow \left[4d^9 4f^{14}\right]* \rightarrow 4d^{10}4f^{12} + e^-$, where []* denotes an excited state. Since the direct photoelectron process and the indirect decays have the same initial and final states, they can interfere constructively. Consequently, it is expected to observe an enhancement of the valence band contributions related to the $Yb4d \rightarrow Yb4f$ absorption edge. Straightforward experimental observation of this process is shown in Figure 4.9(a) and (c). Here, the valence band photoemission spectra of ZnO implanted by ytterbium at various photon energies in the vicinity of the $Yb4d \rightarrow Yb4f$ absorption threshold, so-called energy distribution curves (EDC), are shown. The valence band spectra changes with photon energy and weak enhancement of the valence band features around binding energy of 7.5 eV below the Fermi level is clearly seen. The maximum intensity is observed at 4*d* absorption threshold energy (~182 eV photon energy). By taking selected difference spectra (the difference of energy distribution curves (ΔEDC)) for *on*- versus *off*-resonance (marked by green and black curves in Figure 4.9 (a, c)), the 4*f* weighted density of states (DOS) may be extracted which is shown in Figure 4.9 (b, d). This result supports a conclusion that the Yb 4*f* states are

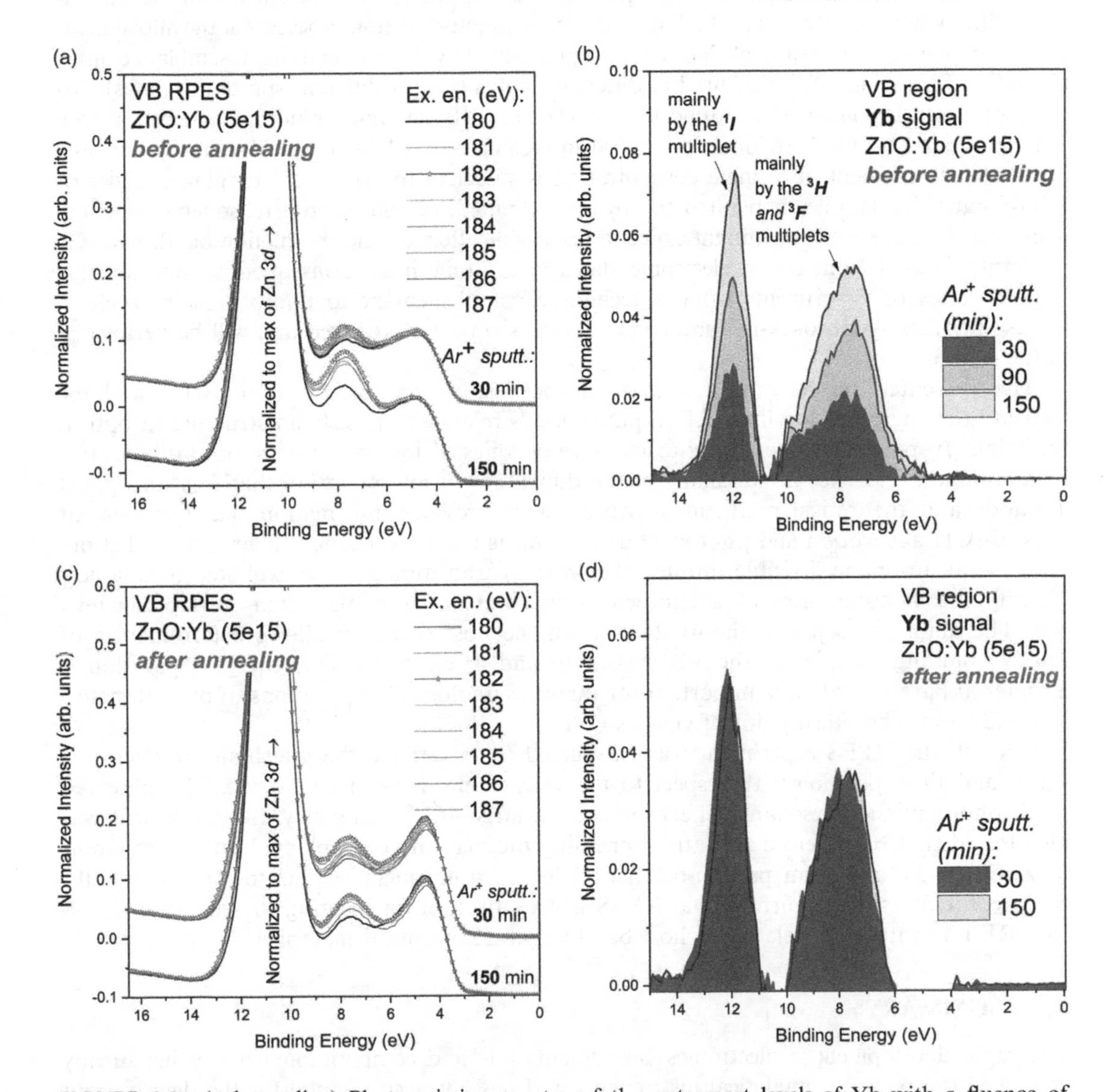

FIGURE 4.9 (color online) Photoemission spectra of the outermost levels of Yb with a fluence of 5×10^{15} ion/cm^2 in the ZnO (virgin) excited with different photon energies in the vicinity of the Yb 4*d*-absorbtion edge (a) *before* (sputtering time 30 min (top) and 150 min (bottom)) and (c) *after* annealing (sputtering time 30 min (top) and 150 min (bottom)). The energy distribution curves, EDC, of ZnO: Yb for 182 (ON, circle green) and 180 (OFF, black) incident photon energies are shown for each spectrum. The resonating components for Yb^{3+} state, the difference of energy distribution curves (ΔEDC), (b) *before* and (d) *after* annealing treatment taken across the Yb 4*d* → Yb 4*f* photoionization threshold.

positioned well below the bottom of the ZnO valence band, which is in a good agreement with a bar diagram showing the multiplet structures of Yb^{3+} calculated by Cox (1975). The 4*f* emission is dominated by two structures at around 7.5 and 11.7 eV binding energy. According to Cox (1975), the latter feature is mainly caused by the ^1I multiplet, whereas the former, lower positioned broader feature is mainly caused by ^3H, ^3F multiplets. Subsequent qualitative analysis of Yb DOS, namely intensity of the estimated signal, before and after annealing is in accordance with the RBS suggestions and the SRIM/SIMNRA simulations regarding the distribution of Yb dopant in the ZnO. The mean projected range of implanted ytterbium ions is about 33 nm. After annealing, some of the Yb ions migrate toward the surface of the investigated film, which is observed in the RBS spectra as an out-diffusion process. All the above mentioned explain an increase of Yb DOS signal intensity before and its resemblance after annealing for the samples measured at different depths (i.e. for different sputtering times). No concentration dependence was found for the 4*f* or the VB maximum binding energies for dose 5e15 ions/cm^2 and for 1e16 ions/cm^2 studied in Demchenko, Melikhov, et al. (2018b), suggesting that measurement of a single concentration is sufficient to extract the binding energies of Yb 4*f* states for samples implanted to any dose (limited, of course, to a reasonably low dosages). At the same time, one might expect a noticeable effect on the conduction band, which is primarily formed from cation electronic states in an ionic host. Consequently, to study this issue a series of experiments using a technique complementary to photoemission, such as inverse photoemission or X-ray absorption spectroscopy, are proposed and will be performed in the future.

The presented results exhibited that photoemission spectroscopy is a useful tool for locating the energy of localized RE impurity levels relative to host band structure in optical materials. In particular, the ability to measure energies of localized electronic states relative to a common reference avoids many of the difficulties in interpretation that exist in optical methods and theoretical predictions. Moreover, it provides information that complement excited-state absorption and photoconductivity measurements. Using synchrotron radiation, even "weak resonance" visible through the 4*d* to 4*f* transition may be exploited to uniquely identify the 4*f* component of a photoemission spectrum from the often-overlapping host VB. The ability to separate the 4*f* states from the host spectrum allows determination of relative binding energies of the VB maximum and 4*f* electrons. This energy separation is a material parameter that is important for many technological applications of optical materials and from theoretical point of view as well.

Overall, the RPES experiment was conducted to investigate the ytterbium 4*f* electronic states and their position with respect to the valence electrons of zinc oxide. The observed weak photoemission resonance effect indicates a large $4f^{14-\delta}$ occupancy. In addition, it was also found that both before and after annealing the majority of implanted ytterbium atoms in ZnO are 3+ and form pseudo-octahedral local arrangement similar to Yb_2O_3. Finally, the presented results confirmed that RPES is a useful tool for locating the energy of localized RE impurity levels relative to host band structure in optical materials.

4.2 SUMMARY

The rapid development in electronics, instrumentation, and computational power has greatly enhanced the power of analytical instruments and has given an impulse for the development of newer analytical techniques. These in turn lead to the development of solution techniques that allow a more advanced and more accurate characterization of impurity ions and, in connection with the analysis of host matrix, also allow the evaluation of environmental effects. The understanding of the local environment and electronic structure of the investigated elements at the atomic scale allows a better understanding of the mechanism for changes of material properties. Theoretically, the changes in the electronic structure caused by the presence of trace elements (dopants), point defects, quantum effects, etc. are difficult to discuss even qualitatively and hardly possible to discuss quantitatively within the framework of conventional solid state physics, as opposite to the case of defect-free crystals. Luckily, there are several experimental methods for analyzing the local environment of atoms and/or ions along with their electronic structure reorganization. Among these methods, X-ray spectroscopy (including XAS, XES, RIXS), and XPS including RPES, especially ones that use synchrotron radiation, occupy a special place being surface-sensitive techniques. This is because many of the processes occur on the surface and the presence of impurities or contaminations at the surface/bulk can affect the performance of the device. This has led to the need for surface/bulk characterization. Nuclear techniques like Rutherford backscattering spectroscopy, nuclear reaction analysis, and thin layer activation analysis can also provide valuable information in this endeavor. This chapter gave an overview of the basic principle and application of these techniques in materials characterization.

The use of newer materials as chemical sensors, detectors, catalysts and as critical components in electronics, computers, and energy conversion devices call for advanced analytical techniques for their chemical and structural characterization. The use of newer analytical techniques can provide a solution to these challenges.

ACKNOWLEDGEMENTS

The work was partially supported by the University of Warsaw, Department of Chemistry (Poland). The author acknowledges support from Elettra Sincrotrone (Italy) and Helmholtz-Zentrum Dresden-Rossendorf (Germany). The author is grateful to Dr. Renata Ratajczak of the National Centre for Nuclear Research (Poland) and to Dr. Yevgen Melikhov of Cardiff University (UK) for their invaluable help and insightful discussions during RBS/XPS experiments and data interpretation.

REFERENCES

Altwein, M., Finkenrath, H., Koňák, Č., Stuke, J., & Zimmerer, G. (1968). The electronic structure of CdO II. Spectral distribution of optical constants. *Physica Status Solidi B, 29*, 203–209.

Attwood, D. (1999). *Soft X-rays and extreme ultraviolet radiation: Principles and applications.* Cambridge: Cambridge University Press.

Biesinger, M. C., Payne, B. P., Grosvenor, A. P., Lau, L. W. M., Gerson, A. R., & Smart, R. S. C. (2011). Resolving surface chemical states in XPS analysis of first row transition metals, oxides and hydroxides: Cr, Mn, Fe, Co and Ni. *Applied Surface Science, 257*, 2717–2730.

Bouvier, Y., Mutel, B., & Grimblot, J. (2004). Use of an Auger parameter for characterizing the Mg chemical state in different materials. *Surface and Coatings Technology, 180–181*, 169–173.

Briggs, D. (2003). XPS: Basic principles, spectral features and qualitative analysis. In D. Briggs & J. T. Grant (Eds.), *Surface analysis by Auger and X-ray photoelectron spectroscopy* (p. 840). Chichester: IM Publications.

Briggs, D., & Seah, M. P. (Eds.). (1990). *Practical surface analysis in: Auger and X-ray photoelectron spectroscopy* (p. 657). Chichester: Wiley.

Bunker, G. (2010). *Introduction to XAFS: A practical guide to X-ray absorption fine structure spectroscopy.* Cambridge: Cambridge University Press.

Cagnon, L., Devolder, T., Cortes, R., Morrone, A., Schmidt, J. E., Chappert, C., & Allongue, P. (2001). Enhanced interface perpendicular magnetic anisotropy in electrodeposited Co/Au(111) layers. *Physical Review B, 63*, 104419.

Carcia, P. F. (1988). Perpendicular magnetic anisotropy in Pd/Co and Pt/Co thin-film layered structures. *Journal of Applied Physics, 63*, 5066–5073.

Chen, D., Ma, X. L., & Wang, Y. M. (2007). First-principles study of the interfacial structures of Au/MgO(001). *Physical Review B, 75*, 125409.

Chen, X., Feng, C., Wu, Z. L., Yang, F., Liu, Y., Jiang, S., … Yu, G.H. (2014). Interfacial oxygen migration and its effect on the magnetic anisotropy in Pt/Co/MgO/Pt films. *Applied Physics Letters, 104*, 052413.

Chu, W. K. (1978). Energy loss of charged particles. In J. Thomas & A. Cachard (Eds.), *Material characterization using ion beams* (Vol. 28, pp. 3–34). NATO Advanced Study Institutes Series (Series B: Physics). Boston, MA: Springer US.

Chuang, Y.-T., Liou, J.-W., & Woon, W.-Y. (2017). Formation of p-type ZnO thin film through coimplantation. *Nanotechnology, 28*, 035603.

Cox, P. A. (1975). Fractional parentage methods for ionisation of open shells of *d* and *f* electrons. In J. D. Dunitz, P. Hemmerich, R. H. Holm, J. A. Ibers, C. K. Jorgensen, J. B. Neilands, D. Reinen, R. J. P. Williams (Eds.), *Photoelectron spectrometry. Structure and bonding* (Vol. 24, pp. 59–81). Berlin and Heidelberg: Springer.

Daalderop, G. H. O., Kelly, P. J., & den Broeder, F. J. A. (1992). Prediction and confirmation of perpendicular magnetic anisotropy in Co/Ni multilayers. *Physical Review Letters, 68*, 682–685.

Daalderop, G. H. O., Kelly, P. J., & Schuurmans, M. F. H. (1990). First-principles calculation of the magnetic anisotropy energy of (Co)n/(*X*)m multilayers. *Physical Review B, 42*, 7270–7273.

Daalderop, G. H. O., Kelly, P. J., & Shuurmans, M. F. H. (1994). Magnetic anisotropy of a free-standing Co monolayer and of multilayers which contain Co monolayers. *Physical Review B, 50*, 9989–10003.

Dakhel, A. A., & Henari, F. Z. (2003). Optical characterization of thermally evaporated thin CdO films. *Crystal Research and Technology, 38*, 979–985.

de Groot, F., & Kotani, E. (2008). *Core level spectroscopy of solids.* Boca Raton, FL: CRC Press.

Demchenko, I. N., Chernyshova, M., Tyliszczak, T., Denlinger, J. D., Yu, K. M., Speaks, D. T., … Lawniczak-Jablonska, K. (2011). Electronic structure of CdO studied by soft X-ray spectroscopy. *Journal of Electron Spectroscopy Related Phenomena, 184*, 249–253.

Demchenko, I. N., Denlinger, J. D., Chernyshova, M., Yu, K. M., Speaks, D. T., Olalde-Velasco, P., … Lawniczak-Jablonska, K. (2010). Full multiple scattering analysis of XANES at the Cd L_3 and O K edges in CdO films combined with a soft-x-ray emission investigation. *Physical Review B, 82*, 075107.

Demchenko, I. N., Lisowski, W., Syryanyy, Y., Melikhov, Y., Zaytseva, I., Konstantynov, P., … Cieplak, M. Z. (2017). Use of XPS to clarify the Hall coefficient sign variation in thin niobium layers buried in silicon. *Applied Surface Science, 399*, 32–40.

Demchenko, I. N., Melikhov, Y., Konstantynov, P., Ratajczak, R., Turos, A., & Guziewicz, E. (2018b). Resonant photoemission spectroscopy study on the contribution of the Yb 4f states to the electronic structure of ZnO. *Acta Physica Polonica A, 133*, 907–909.

Demchenko, I. N., Ratajczak, R., Melikhov, Y., Konstantynov, P., & Guziewicz, E. (2019). Valence band of ZnO:Yb probed by resonant photoemission spectroscopy. *Materials Science in Semiconductor Processing, 91*, 306–309.

Demchenko, I. N., Syryanyya, Y., Melikhov, Y., Nittler, L., Gladczuk, L., Lasek, K., … Chernyshova, M. (2018a). X-ray photoelectron spectroscopy analysis as a tool to assess factors influencing magnetic anisotropy type in Co/MgO system with gold interlayer. *Scripta Materialia, 145*, 50–53.

Eich, D., Fuchs, O., Groh, U., Weinhardt, L., Fink, R., Umbach, E.,… Waag, A. (2006). Resonant inelastic soft x-ray scattering of Be chalcogenides. *Physical Review B, 73: 115212*.

Eisebitt, S., Lüning, J., Rubensson, J.-E., Settels, A., Dederichs, P. H., Eberhardt, W., … Tiedje, T. (1998). Resonant inelastic soft X-ray scattering at the Si L3 edge: Experiment and theory. *Journal of Electron Spectroscopy Related Phenomena, 93*, 245–250.

Engel, B. N., England, C. D., Leeuwen, R. A. V., Wiedmann, M. H., & Falco, C. M. (1991). Interface magnetic anisotropy in epitaxial superlattices. *Physical Review Letters, 67*, 1910–1913.

Gay, J. G., & Richter, R. (1986). Spin anisotropy of ferromagnetic films. *Physical Review Letters, 56*, 2728.

Gilbert, D. A., Olamit, J., Dumas, R. K., Kirby, B. J., Grutter, A.J., Maranville, B.B., … Liu, K. (2016). Controllable positive exchange bias via redox-driven oxygen migration. *Nature Communications, 7*, 11050.

Gladczuk, L., Aleshkevych, P., Lasek, K., & Przyslupski, P. (2014). Magnetic anisotropy of Au/Co/Au/MgO heterostructure: Role of the gold at the Co/MgO interface. *Journal of Applied Physics, 116*, 233909.

Gupta, R. P., & Sen, S. K. (1975). Calculation of multiplet structure of core p -vacancy levels. II. *Physical Review B, 12*, 15–19.

Hofmann, S. (1998). Sputter depth profile analysis of interfaces. *Reports on Progress in Physics, 61*, 827–886.

Hofmann, S. (2013). *Auger- and X-ray photoelectron spectroscopy in materials science* (Vol. 49). Springer Series in Surface Sciences. Berlin and Heidelberg: Springer-Verlag.

Hurle, D. T. J. (Ed.). (1994). *Handbook of crystal growth: Thin films and epitaxy: 3* (Vol. 3). Amsterdam, New York & North-Holland: Elsevier.

Jagadish, C., & Pearton, S. J. (Eds.). (2006). *Zinc Oxide bulk, thin films and nanostructures.* New York: Elsevier Science.

Khoo, K. H., Wu, G., Jhon, M. H., Tran, M., Ernult, F., Eason, K., ... Gan, C. K. (2013). First-principles study of perpendicular magnetic anisotropy in CoFe/MgO and CoFe/$Mg_3B_2O_6$ interfaces. *Physical Review B, 87*, 174403.

Kocka, J., & Konak, C. (1971). The structure of the indirect absorption edge of CdO. *Physica Status Solidi B, 43*, 731–738.

Koffyberg, F. P. (1976). Thermoreflectance spectra of CdO: Band gaps and band-population effects. *Physical Review B, 13*, 4470–4476.

Köhler, H. (1972). Optical properties and energy-band structure of CdO. *Solid State Communications, 11*, 1687–1690.

Lacour, D., Hehn, M., Alnot, M., Montaigne, F., Greullet, F., Lengaigne, G., ... Schuhl, A. (2007). Magnetic properties of postoxidized Pt/Co/Al layers with perpendicular anisotropy. *Applied Physics Letters, 90*, 192506.

Lam, N. Q. (1988). Ion bombardment effects on the near-surface composition during sputter profiling. *Surface and Interface Analysis, 12*, 65–77.

Lüning, J., Rubensson, J.-E., Ellmers, C., Eisebitt, S., & Eberhardt, W. (1997). Site- and symmetry-projected band structure measured by resonant inelastic soft x-ray scattering. *Physical Review B, 56*, 13147–13150.

Madelung, O. (Ed.). (1982). *Semiconductors: Physics of II-VI and I-VII compounds.* Berlin and Heidelberg: Springer-Verlag.

Manchon, A., Ducruet, C., Lombard, L., Auffret, S., Rodmacq, B., Dieny, B., ... Panaccione, G. (2008). Analysis of oxygen induced anisotropy crossover in Pt/Co/MO_x trilayers. *Journal of Applied Physics, 104*, 043914.

Maschke, K., & Rössler, U. (1968). The electronic structure of CdO I. The energy-band structure (APW method). *Physica Status Solidi B, 28*, 577–581.

McGuinness, C., Stagarescu, C. B., & Ryan, P. J. (2003). Influence of shallow core-level hybridization on the electronic structure of post-transition-metal oxides studied using soft X-ray emission and absorption. *Physical Review B, 68*, 165104.

Nakajima, N., Koide, T., Shidara, T., Miyauchi, H., Fukutani, H., Fujimori, A., ... Suzuki, Y. (1998). Perpendicular magnetic anisotropy caused by interfacial hybridization via enhanced orbital moment in Co/Pt multilayers: Magnetic circular x-ray dichroism study. *Physical Review Letters, 81*, 5229–5232.

Oswald, S., & Reiche, R. (2001). Binding state information from XPS depth profiling: Capabilities and limits. *Applied Surface Science, 179*, 307–315.

Piper, L. F. J., DeMasi, A., Smith, K. E., Schleife, A., Fuchs, F., Bechstedt, F., ... Munoz-Sanjosé, V. (2008). Electronic structure of single-crystal rocksalt CdO studied by soft x-ray spectroscopies and ab initio calculations. *Physical Review B, 77*, 125204.

Ratajczak, R., Guziewicz, E., Prucnal, S., Łuka, G., Böttger, R., Heller, R., ... Turos, A. (2018). Luminescence in the visible region from annealed thin ALD-ZnO films implanted with different rare earth ions. *Physica Status Solidi A, 215*, 1700889.

Ratajczak, R., Mieszczynski, C., Prucnal, S., Guziewicz, E., Stachowicz, M., Snigurenko, D., ... Turos, A. (2017a). Structural and optical studies of Pr implanted ZnO films subjected to a long-time or ultra-fast thermal annealing. *Thin Solid Films, 643*, 24–30.

Ratajczak, R., Prucnal, S., Guziewicz, E., Mieszczynski, C., Snigurenko, D., Stachowicz, M., ... Turos, A. (2017b). The photoluminescence response to structural changes of Yb implanted ZnO crystals subjected to non-equilibrium processing. *Journal of Applied Physics, 121*, 075101.

Rovezzi, M., & Glatzel, P. (2014). Hard x-ray emission spectroscopy: A powerful tool for the characterization of magnetic semiconductors. *Semiconductor Science and Technology, 29*, 023002.

Schellekens, A. J., Deen, L., Wang, D., Kohlhepp, J. T., Swagten, H. J. M., & Koopmans, B. (2013). Determining the Gilbert damping in perpendicularly magnetized Pt/Co/AlOx films. *Applied Physics Letters, 102*, 082405.

Stavitski, E., & de Groot, F. M. F. (2010). The CTM4XAS program for EELS and XAS spectral shape analysis of transition metal L edges. *Micron, 41*, 687–694.

Turos, A., Jóźwik, P., Wójcik, M., Gaca, J., Ratajczak, R., & Stonert, A. (2017). Mechanism of damage buildup in ion bombarded ZnO. *Acta Materialia, 134*, 249–256.

van Bokhoven, J. A., & Lamberti, C. (Eds). (2016). *X-ray absorption and x-ray emission spectroscopy: Theory and applications.* Chichester, West Sussex: John Wiley & Sons, Inc.

Vincent Crist, B. (1999). Handbooks of monochromatic XPS spectra. USA: XPS International, Inc.

Wang, D. S., Wu, R., & Freemanet, A. J. (1993a). State-tracking first-principles determination of magnetocrystalline anisotropy. *Physical Review Letters, 70*, 869–872.

Wang, D. S., Wu, R., & Freemanet, A. J. (1993b). Magnetocrystalline anisotropy of Co-Pd interfaces. *Physical Review B, 48*, 15886–15892.

Wang, D. S., Wu, R., & Freemanet, A. J. (1994). Magnetocrystalline anisotropy of interfaces: First-principles theory for Co-Cu interface and interpretation by an effective ligand interaction model. *Journal of Magnetism and Magnetic Materials, 129*, 237–258.

Williams, J. S., & Poate, J. M. (Eds.). (1984). *Ion implantation and beam processing*. Sydney: Academic Press.

Wu, R., Li, C., & Freeman, A. J. (1991). Structural, electronic and magnetic properties of Co/Pd(111) and Co/Pt(111). *Journal of Magnetism and Magnetic Materials, 99*, 71–80.

Yang, H. X., Chshiev, M., Dieny, B., Lee, J. H., Manchon, A., & Shin, K. (2011). First-principles investigation of the very large perpendicular magnetic anisotropy at Fe|MgO and Co|MgO interfaces. *Physical Review B, 84*, 054401.

Zhang, J. Y., Yang, G., Wang, S. G., Liu, Y. W., Zhao, Z. D., Wu, Z. L., … Yu, G. H. (2014). Effect of MgO/Co interface and Co/MgO interface on the spin dependent transport in perpendicular Co/Pt multilayers. *Journal of Applied Physics, 116*, 163905.

Zhang, X. G., Butler, W. H., & Bandyopadhyay, A. (2003). Effects of the iron-oxide layer in Fe-FeO-MgO-Fe tunneling junctions. *Physical Review B, 68*, 092402.

Zhou, X., Yan, Y., Jiang, M., Cui, B., Pan, F., & Song, C. (2016). Role of oxygen ion migration in the electrical control of magnetism in Pt/Co/Ni/HfO2 films. *The Journal of Physical Chemistry C, 120*, 1633–1639.

Ziegler, J. F., Biersack, J. P., & Ziegler, M. D. (2009). *SRIM- the stopping and range of ions in solids.* Chester, MD: SRIM Co. Retrieved from www.srim.org.

5 Low-Dimensional Structures of In_2O_3, SnO_2 and TiO_2 with Applications of Technological Interest

Javier Bartolomé, María Taeño, Antonio Vázquez-López, Félix del Prado, Miguel García-Tecedor, G. Cristian Vásquez, Julio Ramírez-Castellanos, Ana Cremades, and David Maestre

5.1 INTRODUCTION

The huge progress achieved in the development of new nanomaterials in the last decade has been driven to a large extent by the design and fabrication of novel semiconductor nanostructures that exhibit size-dependent properties (Alivisatos, 1996; Smith, & Nie, 2010). Reducing the size to nanoscale often offers advantages related to the use of a reduced amount of materials in high-performance devices, which enables the implementation of concepts such as portable, flexible, and low energy consumption or self-powering (Leung et al., 2018; Wang et al., 2014). Besides, the reduction of size also allows to improve properties and/or adding new ones achieving the design of devices based on smart nanomaterials, which are able to interact with external stimuli or their environment (Yoshida, & Lahann, 2008) with applicability in energy generation and storage (Greaney, & Brutchey, 2015; Wang, 2008), environmental monitoring and control, as well as in health-related fields (Marchesan, Kostarelos, Bianco, & Prato, 2015).

Low-dimensional structures, both in the micrometer and the nanometer scales, expose enormous surface to volume ratio, which is a key parameter to improve sensor performances (Huang, & Choi, 2007; Ramgir et al., 2010), and catalytic properties (Kaplan, Erjavec, Dražić, Grdadolnik, & Pintar, 2016; Kumar, Kumari, Karthik, Sathish, & Shankar, 2017), among others. Besides, the morphology could also be another parameter to develop new applications, as for example using micro or nanotubes, which peculiar geometry confers them improved performances (Bae et al., 2008; Yang et al., 2015) due to exposing a high amount of surface and the ability to load molecules inside acting as nano or micro-containers, among others. As an example, lab-on-a chip devices based on tubular nanostructures have been recently reported (Bobkov et al., 2019; Ghasemi et al., 2017). Finally, the dimensionality of the nano and microstructures is another issue in which a general nanomaterial classification is based. Nanoparticles, nanowires (NW), and bidimensional nanomaterials are paradigmatic exponents of 0D, 1D, or 2D nanostructures, and the combination of several dimensionalities in a complex material design allows also for achieving multifunctionality on smart materials.

During the late 90s, the research on nanomaterials was focused on growth issues, trying to optimize different techniques leading to high yield growth of nanostructures with an interest in the control of composition, morphology, and size. Later, once a certain maturity regarding the growth of nanostructures had been achieved, dopant elements were introduced to modify material properties such as conductivity, optical, sensing properties, and magnetism as well as to exploit the influence of the dopants on the growth mode. As an example, NW-based sensors offer, compared to the state-of-the-art corresponding to devices based on bulk materials, higher sensitivity allowing for lower levels of detection up to even single molecule sensing (Verardo, Agnarsson, Zhdanov, Höök, & Linke, 2019), lower working temperatures (e.g., for gas sensors (Zhang et al., 2014)), long-term stability for reliability in use, and a significantly reduced size and affordable cost, enabling the integration of nanosensors into many different devices and mobile and wearable systems (Akyildiz, & Jornet, 2014; Rai et al., 2014). In addition, they allow employing completely new sensing mechanisms as for example chemical sensors with optical readout (Clark, Hoyer, Philbert, & Kopelman, 1999; Rong, Tuttle, Reilly, & Clark, 2019).

Nowadays, although fundamental issues are still in focus, more applied envisions of the nanomaterials are driving the research topics toward the integration of nanostructures into devices and exploring new applications with challenging issues that are yet to be resolved. The potential development of environmentally clean, safe, and affordable energy harvesting devices requires exploring new materials and innovative strategies in order to develop and exploit systems with improved performance and broader applicability. Some aspects to be faced are related to engineering nanocomponents into nanoscale complex devices. For example, enabling communication capabilities among the nanocomponents as well as with an external system (Almazrouei, Shubair, & Saffre, 2018), locomotion as needed in devices for diagnosis and treatments in medicine (Freitas, 2005; Patra, 2013), or even self-assembly or self-replication that could facilitate the ability to perform macroscopic tasks. Hybrid composites based on conducting polymers functionalized with inorganic nanomaterials are receiving increasing attention, and novel research lines based on these hybrid composites are under development as they are expected to play a leading role in smart, flexible, and low-cost energy storage devices.

The industry is now addressing the increasing importance of a new trend, "More than Moore" (MtM), where added value to devices is provided by incorporating functional nanomaterials that do not necessarily scale according to Moore's Law. In the recent documents, "International Technology Roadmap for Semiconductors" (ITRS, International Technology Roadmap for Semiconductors, 2015) and "A European Industrial Strategic Roadmap for Micro- and Nano-Electronic Components and Systems" (DG Connect, 2014), NW and nano-electronic components and devices are highlighted as a huge opportunity for MtM applications and societal challenges such as power, electric, and hybrid cars, for energy management and savings, for efficient and smart lighting, and low consumption radiofrequency components for mobile. Nonetheless, challenges still remain in the understanding of their fundamental properties and their functionalization, in order to face the advent of modern devices. Moreover, developing complex structures (Bueno, Maestre, Díaz, Pacio, & Cremades, 2018; García-Tecedor, Bartolomé, Maestre, Trampert, & Cremades, 2019; Ramos Ramón, Cremades, Maestre, Silva González, & Pal, 2017; Wang et al., 2016) and advanced nanocomposites (García-Tecedor et al., 2018; Del Prado et al., 2019; Li et al. 2011) by combining different materials and dimensionalities is an actual strategy to allow bandgap and properties engineering of nanomaterials.

Among the materials of choice, the family of semiconducting oxides (Wang, Nayak, Caraveo-Frescas, & Alshareef, 2016; Zhai et al., 2009) stands out due to their widely expanded applicability in a vast range of technological fields. The most interesting oxide semiconductors present high chemical stability, nontoxicity, biocompatibility, wide band gaps, polymorphism, simultaneity of high electrical conductivity and transparency in the visible and a relevant role of native defects (Lany et al., 2012; Zakutayev et al., 2012), together with the feasibility to get high-quality nanomaterials with tunable microstructure and morphology by means of cost-effective technology in comparison with traditional high-tech microelectronics fabrication. SnO_2, TiO_2, and In_2O_3 are intensively investigated paradigmatic examples, which have attained growing attention as they have become key materials in several applications of technological interest including microelectronic, energy harvesting and storage, optoelectronic, catalysis or gas sensing, among others (Goriparti et al., 2014; Kumar et al., 2017; Sberveglieri, Faglia, Groppelli, Nelli, & Perego, 1993; Tran et al., 2017). In the particular case of In_2O_3, there is a substantial interest in developing devices with a low content of In, considered as a critical element (Critical Raw Materials, 2017), without losing its high-performance characteristics, which also could be achieved by using nanomaterials.

The ability to synthesize nanomaterials based on TiO_2, SnO_2, and In_2O_3 has been widely demonstrated (Bae et al., 2008; Maestre, Cremades, & Piqueras, 2005; Vásquez et al., 2013; Yin, Cao, Luo, Hu, & Wei, 2009). TiO_2 and SnO_2 share the rutile structure, which is the most stable phase, and therefore, some similarities mainly related to the growth mechanism have been observed presenting an inherent anisotropy related to the tetragonal crystalline structure, which is of outmost relevance for the growth of elongated structures such as tubes or wires (Cremades et al., 2014; García-Tecedor et al., 2016). As a different case, the In_2O_3 shows an isotropic cubic structure, presenting growth mechanisms of complex nanostructures assisted by dislocations (Maestre, Haussler, Cremades, Jäger, & Piqueras, 2011).

One of the most important aspects to correlate and improve the physical properties of functional oxides is the appropriate choice of the synthesis method. In this sense, the methods for obtaining solids with tunable size and morphology, as well as controlled compositional homogeneity, are among the most attractive. Specifically, the preparation of chemical systems of several components, such as doped binary semiconductor oxides, requires precursor mixtures with a well-defined and homogeneous composition at the atomic scale. Normally, synthetic synthesis routes are used, which make it possible to obtain precursors of homogeneous composition at the atomic level, since they substantially reduce the diffusion paths, allowing working at lower temperatures. Simultaneously, these methods allow in turn obtaining smaller and more homogeneous particle sizes. However, it must be borne in mind that the reproducibility and scalability of the synthesis, as well as the relation morphology–properties are key factors for rapid implementation in technological applications.

However, the materials formed by nanoparticles present some problems inherent to their high specific surface area and reduced volume, such as high reactivity, spontaneous coagulation, and agglomeration of nanoparticles, which impose a compromise between temperature, duration, and other experimental conditions of the synthesis to obtain reduced particle size. For this reason, different synthetic strategies are developed to obtain functional oxides from different systems.

Hydrothermal synthesis is a complex preparation method that requires precise control of many variables (starting reagents, solvents, filling volume, pH, reaction time, and temperature) to obtain reproducible materials with the desired characteristics. For this reason, this method has been selected for the production of several transparent conductive oxide (TCO) (SnO_2, TiO_2, etc.) nanoparticles (Elouali et al., 2012; Shen et al., 2011; Suematsu et al., 2014), allowing some of the results described in the following sections to be compared with those obtained by the polymeric precursor method also known as Liquid–Mix method. Controlling the morphology and composition by the hydrothermal conditions is a great advantage, but it also implies the need to establish precisely those conditions; what makes an extensive and laborious task for the optimization of a hydrothermal route as a scalable method of synthesis.

On the other hand, the decomposition of complex organic metallic precursors, achieved in the Liquid–Mix method (Del Prado et al., 2018), is especially interesting. Its main advantage lies in the presence of the different metals in the precursor in the same stoichiometric relationship that the oxide presents and, furthermore, at molecular level, which facilitates homogenization and softer conditions of decomposition, giving rise to smaller and homogeneous particles. In addition to being reproducible and versatile, this method of synthesis allows us to scale the obtaining of nanomaterials at the industrial level and facilitate the technological use of the nanomaterials produced. This method is based on the formation of metal–organic complexes of the metallic precursors, using a bi- or tri-dentate organic chelating agent, such as citric acid; a polyalcohol, such as ethylene glycol, for the formation of ester bonds by polycondensation with the chelates, giving rise to an intermediate three-dimensional resin of the precursors. After drying, the pyrolysis of the organic residue is continued; this combustion, at relatively low temperatures, removes the organic part leaving the desired composition of the mixed oxide chemically combined in a pure, uniform, and finely divided state. An important consequence of the mixture of cationic precursors occurring at the atomic level is the considerable decrease in calcination temperatures. This method has proved to be very useful by controlling the different parameters that affect the synthesis. As an example, in the production of TiO_2, the control of the solvent/precursor ratios, the pH, the presence of acids with catalytic function, and the dopants used allow the preferential synthesis of TiO_2 in the anatase or rutile phase. Furthermore, this method allows a control over the anatase/rutile transformation that occurs in these oxides.

Another synthesis strategy for obtaining nanoparticles of very small size is the use of coprecipitation methods (Sagadevan et al., 2019) under mild conditions in aqueous medium. This method of synthesis allows, at the same time and maintaining the simplicity of the process, the incorporation of other metals as dopants, which leads to the modification of the properties. In this way, the promising catalytic, thermoelectric, and ionic conductance properties of these systems can be exploited.

Sol–gel methods for obtaining nano-oxides are also being developed (Ali et al., 2018). The nanostructures of these materials present the difficulty of having to reconcile highly energetic synthesis conditions with maintaining a small particle size. For this reason, the examples where nanoparticles are obtained are very scarce.

The growing demand for semiconductors with structures of reduced dimensionality is due fundamentally to their potential applications in technological devices, as well as their study for the understanding of the quantum phenomena that can occur within them. In order to synthesize micro- and nanostructures of semiconductor oxides with elongated morphologies such as wires, rods, ribbons, or needles, growth techniques based on physical, chemical, or catalytically activated processes are used such as vapor phase deposition. In addition to the vapor–liquid–solid method, which requires the presence of a catalyst governing the anisotropic growth of the nanostructures, by using the vapor–solid (VS) method (Maestre et al., 2005), which involves mechanisms of evaporation-solidification, neither catalyst nor external substrate are required. Unlike other methods, VS method results in the growth of a great variety of elongated micro- and nanostructures such as wires, rods, and tubes, and including also bidimensional objects as nanosheets, which can be developed on the surface of the precursor pellet driven by different mechanisms and by avoiding the presence of a catalyst at the end of the structures. This situation allows a better control of the morphology of the nanostructures depending on the parameters that define the thermal treatment such as temperature, time, and gas flow.

To summarize, different soft chemistry and physical methods can be used in order to obtain monodisperse nanoparticles, as well as one-dimensional or bidimensional structures (sol–gel, hydrothermal synthesis, complex precursors, vapor–solid, and vapor–liquid–solid), which improve the physical properties for technological applications.

This chapter aims to review and highlight recent developments in the applicability of low-dimensional structures of SnO_2, TiO_2, and In_2O_3. Selected examples of the noteworthy use of these semiconducting oxides in gas sensing devices, optical and mechanical resonators, waveguides, batteries, solar cells, white luminescence devices, and photocatalysis will be reported. Special attention will be focused on the use of undoped and doped nanoparticles, elongated nanostructures in different forms (wires, rods, tubes, etc.), and composite materials fabricated by diverse physical and chemical routes. Moreover, the exploitation of phase control, doping, and shape engineering in different applications will be also discussed in this chapter, where some examples of recent patents will be presented as well. Section 5.2 will be dedicated to In_2O_3, section 5.3 to SnO_2, and section 5.4 will be dealing with TiO_2 low-dimensional structures and related materials. Section 5.5 will be devoted to conclusions, which is followed by references.

5.2 INDIUM OXIDE LOW-DIMENSIONAL STRUCTURES AND APPLICATIONS

Indium oxide is mostly known for its application as transparent electrode, when combined with tin to form indium tin oxide (ITO). However, over the last two decades an increasing interest has arisen for its properties as semiconductor material on its own (Bierwagen, & Speck, 2010), giving rise to a variety of new applications. There is a lot of available information on the use of In_2O_3 and ITO as transparent electrode, which can be found elsewhere (see for instance Hofmann, Cloutet, & Hadziioannou, 2018), and therefore this section will be focused on the work done on low-dimensional In_2O_3 structures with different applications such as gas sensing, field emission transistor, and as an active material in optics (nonlinear optics, harmonic generation, light confinement).

Indium oxide crystallizes in essentially three different phases, corresponding to the space groups $I2_13$, *Ia*-3 (cubic), and *R*-3 (rhombohedral) (Karazhanov et al., 2007). The most stable phase at ambient pressure and temperature is the *Ia*-3 phase (hereinafter called bcc-In_2O_3 phase), which has bixbyite structure and a unit cell of 80 atoms with 8 formula units per unit cell (see Figure 5.1a). It consists of two types of In atoms in octahedral and trigonal prismatic coordination, and one type of O atoms, located at Wyckoff positions 8b,

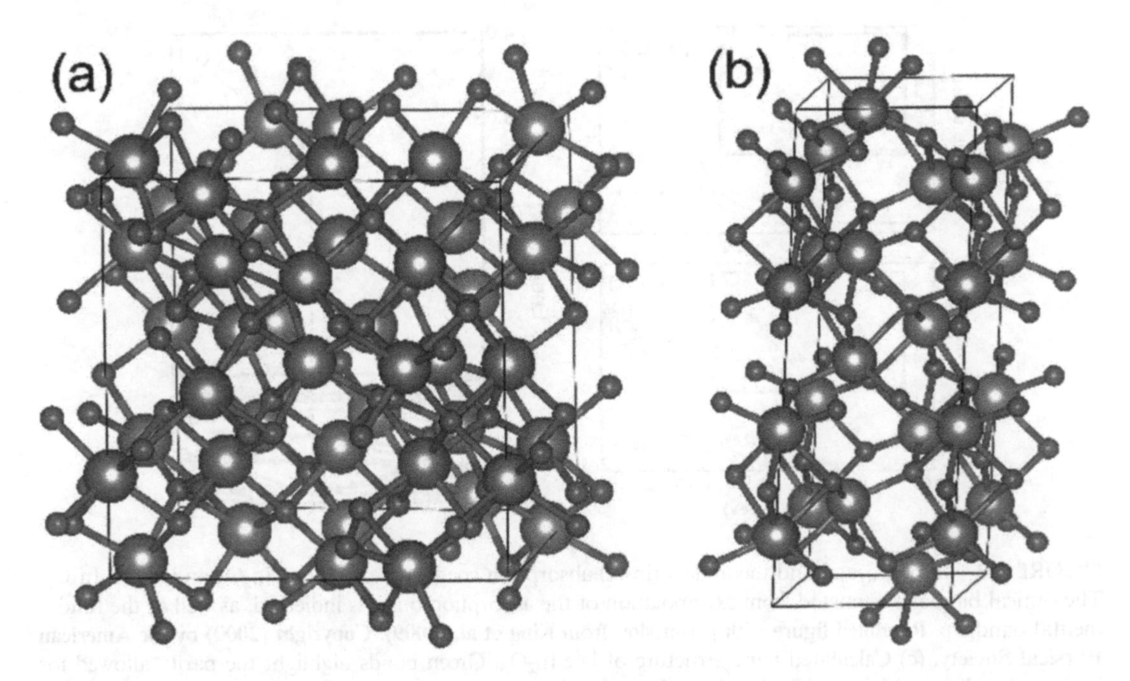

FIGURE 5.1 Schematic of the two most common phases: (a) bixbyite bcc-In_2O_3 and (b) corundum rh-In_2O_3.

Images were rendered by VESTA software (Momma, & Izumi, 2011).

24d, and 48e, respectively. The structure may be regarded as a 2 × 2 × 2 superstructure of fluorite, with ordered removal of O from one-fourth of the anion sites.

The rhombohedral modification *R*-3 (hereinafter rh-In_2O_3) crystallizes with corundum structure (Figure 5.1b), with two formula units per unit cell and only one type of In atoms in trigonal biprism coordination and one type of O atoms, occupying 12c and 18e Wyckoff positions, respectively. This phase is stable only at high pressures, but it can be stabilized in ambient conditions by epitaxial growth on selected substrates, such as rhombohedral Al_2O_3 (0001), and carefully tuning the growth conditions (Wang et al., 2008; King et al., 2009). It is worth noting that XRD patterns from rh-In_2O_3 resemble closely those produced by poorly oxidized bcc-In_2O_3, where a combination of metallic In and bcc-In_2O_3 peaks can be present, thus great care has to be taken before ascribing them to the rhombohedral modification. The $I2_13$ phase has been very rarely reported (Zachariasen, 1927) and experimental information is extremely limited; therefore, this phase will not be further discussed here.

Owing to the large predominance of the bcc-In_2O_3 phase, this section will focus essentially on its properties and applications, while references to rhombohedral phase will be explicitly stated. The bcc-In_2O_3 phase has a fundamental bandgap of 2.7–2.9 eV, (Bourlange et al., 2008; Walsh et al., 2008); however, for long time its value was believed to be close to 3.7 eV. The reason for this is the parity-forbidden nature, within the dipole approximation, of the direct transitions between the valence band maximum and the conduction band minimum, which causes a strong shift in the absorption onset of the Tauc plot that appears as a very wide "optical bandgap" of around 3.7 eV (Figure 5.2). Thus, while the electronic properties are essentially determined by the fundamental bandgap, the optical bandgap dominates the optical properties of indium oxide. This decoupling is largely the cause of the excellent

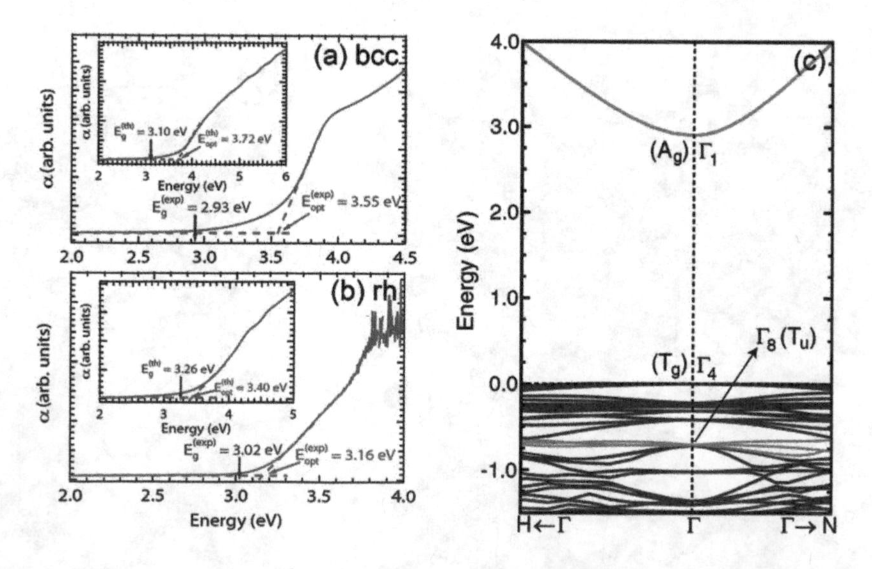

FIGURE 5.2 Experimental and theoretical (inset) absorption coefficient of (a) bcc-In_2O_3 and (b) rh-In_2O_3. The optical bandgap estimated from extrapolation of the absorption onset is indicated, as well as the fundamental bandgap. **Reprinted figure with permission from King et al. (2009). Copyright (2009) by the American Physical Society.** (c) Calculated band structure of bcc-In_2O_3. Green bands highlight the parity allowed for band-to-band transitions, with the first allowed valence bands lying 0.8 eV below the valence band maximum.

Reprinted figure with permission from Walsh et al. (2008). Copyright (2008) by the American Physical Society.

capabilities of indium oxide and its doped version ITO as transparent yet highly conductive electrode.

Degenerated doping may also cause some modifications on the optical bandgap due to the appearance of bandgap renormalization effects, which cause shrinkage of the fundamental gap, and the well-known Burstein–Moss shift (Walsh, Da Silva, & Wei, 2008). The latter is produced when the Fermi level lies above the conduction band minimum, leaving no empty states to be occupied by the excited electrons, which therefore require more energy to transit from the valence band maximum to empty states in the conduction band.

The electronic properties of indium oxide are highly influenced by its intrinsic defect structure. The presence of oxygen vacancies (V_O) are ubiquitous to the majority of oxide semiconductors and play a key role in several of their physical properties, from their electronic transport to their luminescence spectrum or their ionic diffusion. Doubly ionized oxygen vacancies in indium oxide are generally considered the origin of the high intrinsic conductivity of indium oxide, as first stated by De Wit (1973). Theoretical calculations performed by different authors (Ágoston, Albe, et al., 2009; Ágoston, Erhart, et al. 2009; Lany, & Zunger, 2007) have confirmed that these kind of defects have indeed the lowest formation energy, although, there is still some controversy on the energy levels they introduce in the gap. However, it is generally accepted that these are shallow levels, as measured experimentally (Bierwagen, & Speck, 2010). V_O have also been attributed as the origin of the blue-green emission of In_2O_3, centered at approximately 2.5 eV, which is compatible with the shallow donor-like level picture. Oxygen deficiency is even more pronounce at the surface, where V_O have been reported to have exceptionally low formation energies (Lany, & Zunger, 2007), giving rise to an electron accumulation layer which can dominate the conduction of In_2O_3 thin films. This is in contrast to its doped versions such as Sn:In_2O_3 (ITO) or Zn:In_2O_3 (IZO) where an electron depletion layer is found at the surface (Bartolomé, Maestre, Cremades, Amatti, & Piqueras, 2013; King et al., 2008).

A more comprehensive review of the physical properties of indium oxide by Bierwagen can be found in Bierwagen (2015).

5.2.1 Optics-Related Applications of Low-Dimensional In_2O_3 Structures

One of the earliest proposed applications for this material beyond its use as transparent electrode was in smart windows (Hamberg, & Granqvist, 1986) as wavelength selective filter. It is based on the large reflectivity in the infrared (IR) region presented by heavily Sn-doped indium oxide, combined with an excellent transparency in the visible range. An ITO thin-film coating can efficiently block the electromagnetic radiation in the range of 0.7–50 μm, which corresponds to the main spectral range emission of the black body at room temperature, thus hindering the radiative thermal energy transfer (Granqvist, 2007). The high IR reflectance of heavily doped indium oxide is caused by the large carrier concentration, which is proportional to the square of the plasma frequency, ω_p, within the Drude model. Thus, ω_p and in turn the reflectivity in the IR region can be tuned by varying the carrier concentration of the films as shown in Figure 5.3 (Granqvist, 2007; Hamberg, & Granqvist, 1986). A different approach not involving the deposition of thin films is the use of ITO microparticle dispersions on laminated glass, which has been effectively used as heat blockers on window panels (patent EP 1 698 599 B1) (Hagiwara, Nakagawa, Fukatani, Yoshioka, & Hatta, 2015).

The relatively high ω_p of heavily doped indium oxide is also interesting from the point of view of nonlinear optics (Liberal & Engheta, 2017). For a lossless material, any change in refractive index, n, depends on the variation of dielectric permittivity, ε, as $\Delta n^2 = \Delta\varepsilon^2/4\varepsilon$, and thus it is maximized when ε tends to zero, a condition that is reached at the plasma frequency (Alam, Leon, & Boyd, 2016). This is usually referred as epsilon-near-zero (ENZ)

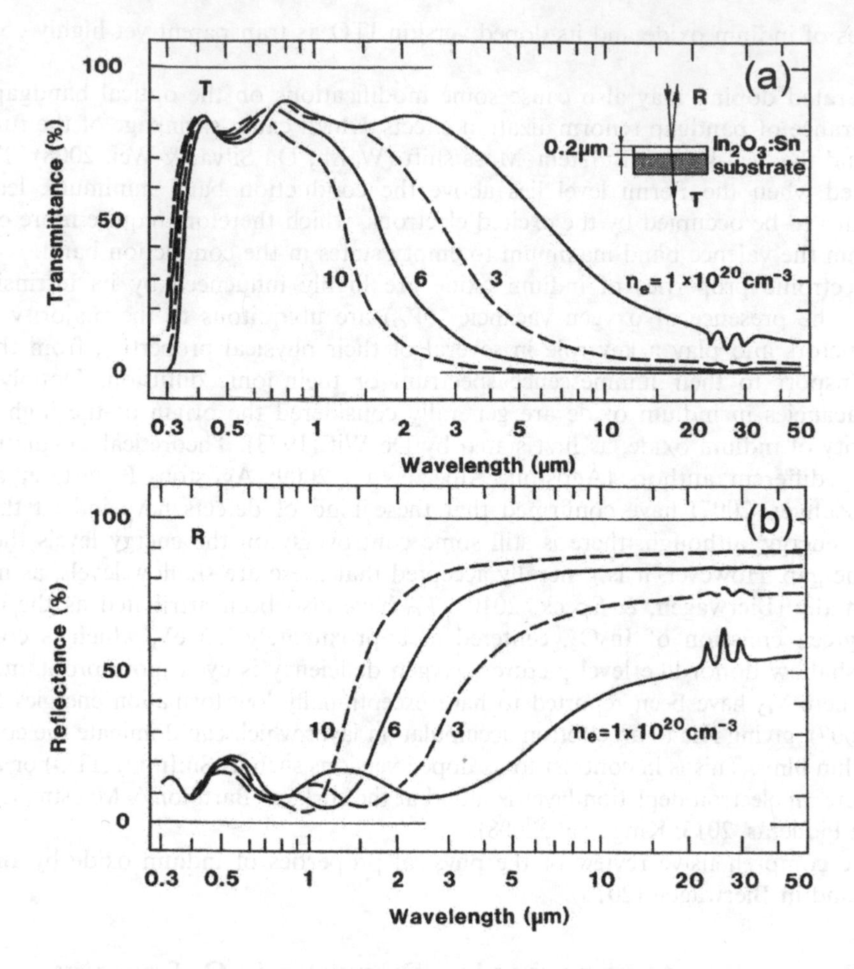

FIGURE 5.3 Theoretical spectral transmittance and reflectance of a 200 nm Sn:In_2O_3 thin film. The n_e values indicate the electron density used for the calculations.

Reprinted from Hamberg and Granqvist (1986) with permission from AIP Publishing.

condition. Hence, both indium oxide and ITO are expected to have strong nonlinear optical properties in the near-infrared light range (Alam et al., 2016). This phenomenon has been recently studied in ITO thin films, showing a strong (up to $\Delta n \sim 0.72$) and ultrafast (up to $\tau_{recovery}$ ~360 fs) third-order nonlinear response (Alam et al., 2016), and exploited in second (Capretti, Wang, Engheta, & Dal Negro, 2015) and third-harmonic (Capretti, Wang, Engheta, & Negro, 2015) light generation using telecommunication IR wavelengths as source light. The high transparency of ITO thin films in the visible range enables an efficient extraction of the generated light, while the ENZ condition reached at telecom wavelengths allows a complete integration with Si technology.

One of the problems of working in ENZ conditions is the larger intrinsic material losses at this wavelength range. Although ITO actually presents a negative nonlinear attenuation constant at ENZ (Alam et al., 2016), intrinsic losses can be further minimized using intrinsic In_2O_3 instead of ITO in the visible range, where In_2O_3 is highly transparent. In this case, the nonlinear behavior does not rely on the high sensitivity of the refractive index in the ENZ regime, but it is originated by a strong modification of interband optical transitions upon intense carrier pumping. Contrary to the previous examples, this process produces a negative variation of the refractive index (up to $\Delta n \sim -0.09$) and has been demonstrated on undoped indium oxide nanorod arrays (Guo, Chang, & Schaller, 2017). Figure 5.4 shows the In_2O_3 nanorod array obtained by vapor–liquid–solid process and the resulting modification of the array transmittance spectra at different transient times. These processes may find applications in the field of nano-photonics, plasmonics, or as saturable absorbers.

Light confinement and guiding is another application of indium oxide low-dimensional structures. As previously mentioned, the linear refractive index, n, of In_2O_3 is dependent on its plasma frequency, which can be widely varied depending on doping and V_O concentration, and therefore on its growth conditions. However, the refractive index in the visible range of undoped In_2O_3 lies around n = 1.9–2.0 (Bartolome, Cremades, & Piqueras, 2013; Dong et al., 2009), which is high enough to support light confinement through either Fabry-Pérot (FP) resonances or whispering gallery modes (WGM) in open dielectric resonator (ODR) structures. An ODR consists essentially in a structure made of a high n material with one or several dimensions close in size to the wavelength of the visible-IR light, which induces light confinement in these dimensions through multiple reflections in the inner walls of the ODR (see Figure 5.5). In the case of FP resonances, these are achieved through a large difference in n at the ODR interface, while for WGM resonances are obtained through total internal reflections.

Light confinement in the range of the visible light has been obtained for a variety of structures, from octahedral microcrystals (Dong et al., 2010) to elongated microrods with hexagonal or square cross sections (Bartolome et al., 2013; Dong et al., 2009). The photoluminescence of In_2O_3 is employed as a source of visible light, making use of its wide emission band centered at 600 nm, which spreads from the near-infrared to the blue range (Figure 5.5c). Thus, these structures can be used as both passive (optical cavities) and active (gain medium) materials in solid-state lasers. Quality factors (a measure of the energy stored by the cavity) as high as 350 have been obtained in highly crystalline hexagonal microrods (Bartolome et al., 2013). Because light confinement on elongated microstructures is

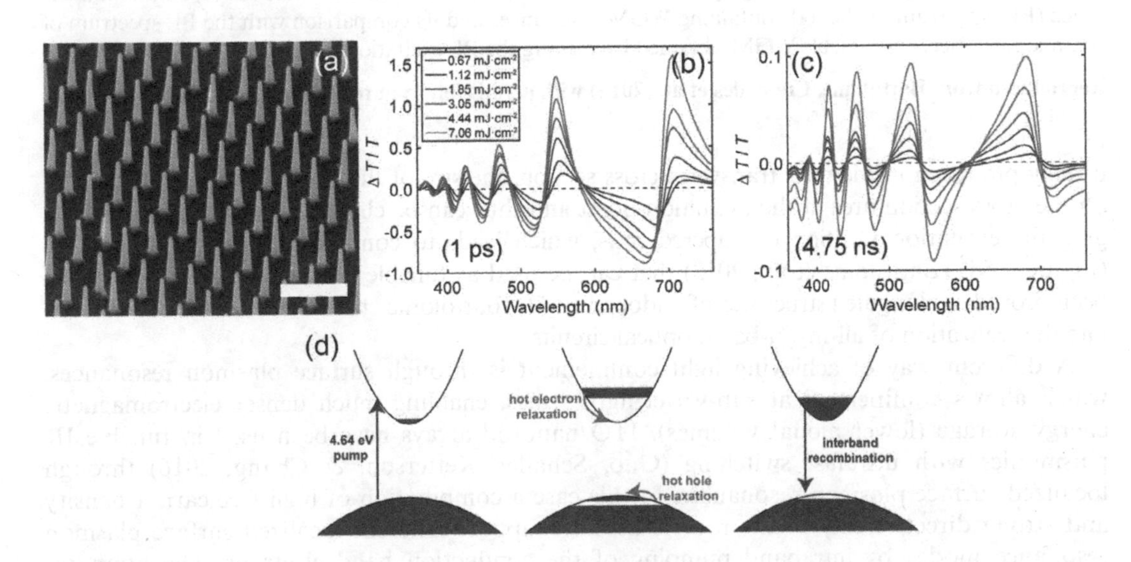

FIGURE 5.4 (a) Scanning electron micrograph of In_2O_3 nanorod array. (b) and (c) Variation of its transmittance spectrum with laser fluence at two different transient times. (d) Schematic of the electronic process leading to the negative nonlinear transient process.

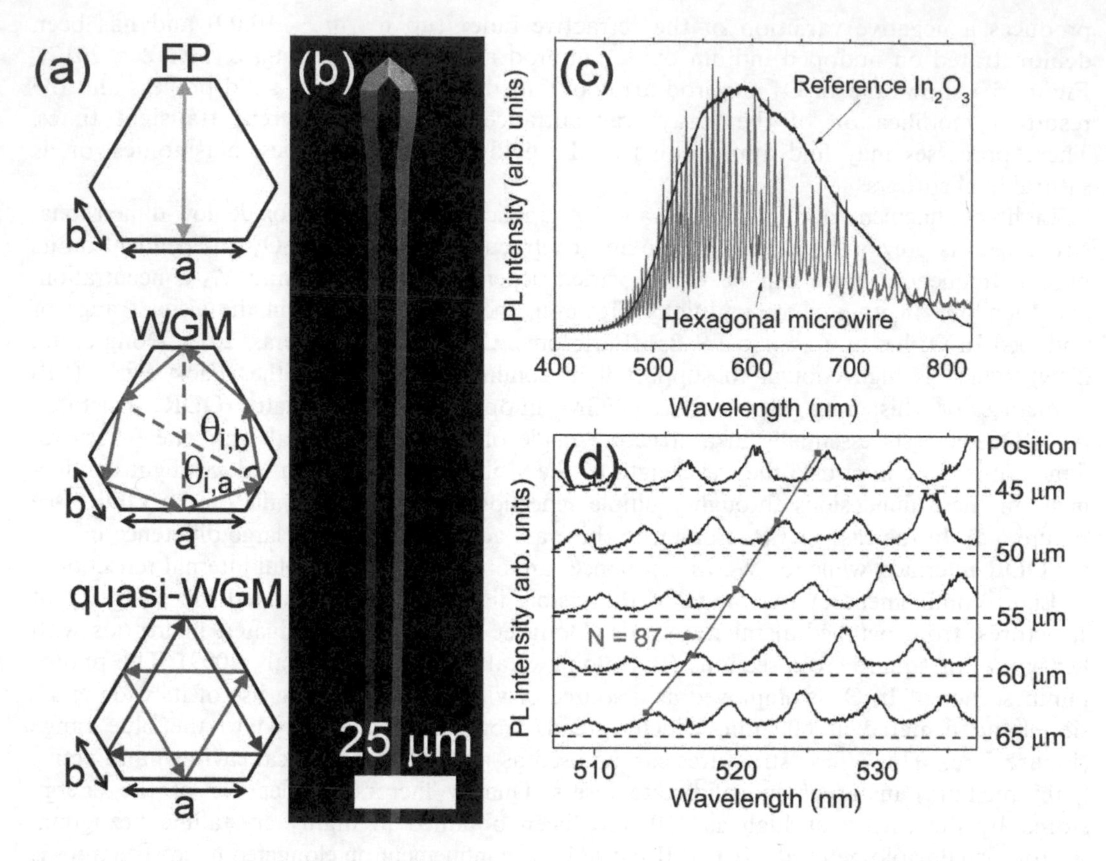

FIGURE 5.5 Visible light confinement in an In_2O_3 rod. (a) Different possible resonances in a hexagonal cavity. (b) Scanning electron micrograph of an In_2O_3 rod with hexagonal cross-section. (c) Photoluminescence (PL) spectrum of the rod containing WGM resonances and its comparison with the PL spectrum of ceramic reference. (d) Tunable WGM obtained by varying the PL excitation point along the wire.

Reproduced from Bartolome, Cremades et al. (2013) with permission from the Royal Society of Chemistry.

usually produced inside their transverse cross section, the size of the optical cavity depends only on the cross section area at the excitation point and thus can be changed continuously by changing the excitation position on tapered rods, which leads to continuously tunable resonators (Figure 5.5d) (Bartolome et al., 2013) that can be used as tunable filters. Waveguiding has also been proved in elongated structures of undoped In_2O_3 (Bartolome et al., 2013), which could facilitate the realization of all-In_2O_3-based optical circuits.

A different way of achieving light confinement is through surface plasmon resonances, which allows confinement at sub-wavelength scales, enabling much denser electromagnetic energy storage (lower modal volumes). ITO nanorod arrays have been used in tunable IR plasmonics with ultrafast switching (Guo, Schaller, Ketterson, & Chang, 2016) through localized surface plasmon resonances. In this case a combination of high free carrier density and strong directionality of the rods is used to support different localized surface plasmon resonance modes by intraband pumping of the conduction band electrons. The nonparabolic nature of ITO conduction band induces a red shift of the rods' plasmon frequency at high fluence pumping, which translates into a tunable mid infrared transmission modulation through tailoring of the sample geometry. Localized surface plasmonics on ITO nanorods find application in sensing, nanolasers, or nano-antennas.

5.2.2 Sensors with In_2O_3 as Active Material

One widespread application of low-dimensional semiconductor oxides is the field of sensor devices. There exist a large variety of different oxide-based sensors depending on their sensing mechanism and/or what is being sensed. Conductometric sensors are by far the most common type of oxide-based sensors and they work on the principle of sample conductance change upon exposure to chemical analytes. Other kind of sensors are field-effect transistor sensors, whose channel carrier concentration is sensitive to the presence of different chemical species, optical sensors, which are based on the change of refractive index or spectral absorption of the material, or force sensors, which measure either the change in the material stiffness or inertial mass under analyte adsorption. Several extensive reviews on the fundamentals and new advances of In_2O_3 thin (and thick) film conductometric sensors have already been published (see for instance Korotcenkov, Brinzari, & Cho, 2016, 2018). Therefore, this subsection will be focused on the recent progress of other low-dimensional In_2O_3-based sensors such as nanoparticles, nanowires, or rods.

Because the surface electron accumulation layer of indium oxide is produced by its large surface oxygen substoichiometry, any structure with increased surface to volume ratio, such as nanoparticles, nanowires, nanoribbons, or nanorods, would be extremely sensitive to oxidizing and/or reducing gases such as O_3, NO_x, NH_3, or ethanol (Rombach et al., 2016). These structures have the additional advantage of highly reduced time responses as compared to thin or thick films. The reason is that adsorption processes are produced directly on the surface of the structures, so no diffusion is required for the analytes to reach the inner layers of the material. This has another implication, as surface desorption is highly enhanced in indium oxide by ultraviolet (UV) light illumination (Wang et al., 2012). Hence, UV-enhanced desorption can be used to further decrease the time response of indium oxide nanostructure-based sensors, allowing room temperature operation without heating the samples at the usual 150–300 °C, with the associated energy cost reduction.

0D (nanoparticles) (Alvarado et al., 2018; Elouali et al., 2012; Gao et al., 2016; Gu, Nie, Han, & Wang, 2015; Neri et al., 2005; Xu et al., 2015) and 1D (nanowires, ribbons, rods, etc.) (Du et al., 2007; Li et al., 2003; Liang, Kim, Yoon, Kwak, & Lee, 2015; Rout et al., 2006; Singh et al., 2010; Vomiero et al., 2007; Xing et al., 2015; Zhang et al., 2004) indium oxide conductometric and FET gas sensors have been investigated since the early 2000s. Some general insights on nanoparticle and nanowire gas sensing can be found in Huang and Choi (2007); Hung, Le, & Van Hieu, (2017); Ramgir et al. (2010); Zhang, Liu, Neri, & Pinna (2016). Gas sensing with indium oxide nanowires and nanoparticles has been pushed down to detection limits as low as few particles per billion (ppb) for highly oxidizing gases such as NO_x (Zhang et al., 2004), O_3 (Wang et al., 2007), or for acetone (Feng et al., 2015), with improved time response as compared to other thin or thick film counterparts. The sensing mechanisms of nanowire-based devices are, in principle, rather simple, as they usually consist of a single wire or a bunch of them connected in parallel, which change in conductivity due to the electron extraction/injection of the absorbed analytes; while on nanoparticle films analyte chemi- and/or physisorption at the grain boundaries usually dominate the transport processes, leading to more complex behaviors and interplays. However, some detailed analysis revealed, even in the early works of In_2O_3 nanowire gas sensors, that the situation is actually more complex. Zhang et al. (2003) reported that undoped In_2O_3 nanowire response to NH_3 depends on the intrinsic carrier concentration (determined by its V_O concentration) and thus on the Fermi level of the studied wire, with *n*-type (increase in resistance under oxidizing conditions) sensing behavior for high carrier concentrations, and *p*-type (resistance decrease) for low concentrations. The reason for this, according to Zhang et al. (2003), was the relative position of the Fermi level with respect to the surface states induced by the adsorbed NH_3, which determine the role of NH_3 molecules as either electron traps or donors (see Figure 5.6). Thus, careful tuning (through doping) of the Fermi level relative to the surface states

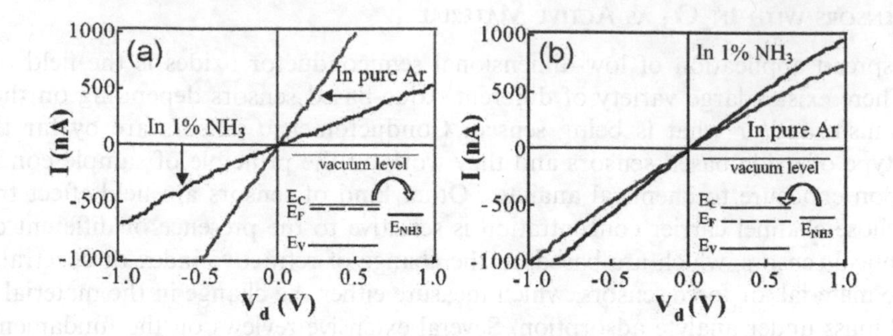

FIGURE 5.6 I–V_d curves for seemingly identical single In_2O_3 nanowire FET devices with high (a) and low (b) carrier concentrations, and their different behavior upon exposure to 1% of NH_3. Inset shows a schematic of the surface energy bands and the expected carrier flow.

Figure reprinted from Li et al. (2003) with permission from AIP Publishing.

introduced by the desired analytes should allow improvements on the selectivity and sensitivity of these devices (Singh et al., 2010). Similarly, an anomalous change from *p*-type to *n*-type sensing has been reported for In_2O_3 nanoparticles for increasing NO_x concentrations, which was explained by the competition between adsorbed NO_x^- and O^- species for the same adsorption sites (Xu et al., 2018). Each NO_x^- molecule displaces two O^- ions, which desorb releasing two electrons, with a net gain of one electron, leading to an increase in conductance as in a *p*-type material. At high NO_x concentrations, the majority of adsorbed O^- species are displaced and competition between both is no longer determining the material response, which thus behave as an *n*-type sensor. The presence of humidity and/or different operating temperature can also modify the sensing response of In_2O_3 to oxidizing or reducing agents (Korotcenkov et al., 2004).

Improved performances can be obtained also by functionalizing the structures with nanoparticle decorations of other inorganic compounds such as CuO (Liang et al., 2015), Bi_2O_3 (Park, Kim, Sun, & Lee, 2015), SnO_2 (Xu et al., 2015), WO_2 (Zachariasen, 1927; Feng et al., 2015), TiO_2 (Wu, Chou, & Wu, 2018), La_2O_3 (Zhan, Lu, Song, Jiang, & Xu, 2007), Au (Xing et al., 2015), Ag (Zhu, Chang, et al., 2016), or Pt (Neri et al., 2007), or with graphene/reduced graphene oxide (rGO) (Gu et al., 2015). The enhancement obtained with metallic nanoparticles is usually associated to their catalytic actions, as they accelerate the chemical reactions taking place on the surface of the matrix material (i.e., In_2O_3). However, in the case of metal oxide nanoparticles, the improvement is strongly related to the band bending at the heterojunction and its modification upon analyte exposure.

In_2O_3 nanoribbon arrays have also been used as biosensors for the detection of cardiac attack biomarkers (Liu et al., 2016). Liu et al. (2016) fabricated and patented (US2019120788A1) an In_2O_3 nanoribbon array in an FET configuration by a lithography-free shadow masking method (see Figure 5.7a). In_2O_3 nanoribbons demonstrated high sensitivity and negligible degradation to pH changes under wet conditions. These ribbons were subsequently functionalized by antibodies specific of three different biomolecules produced during acute myocardial infarctions which fixed the targeted antigens and, at the same time, induced an amplified pH change of the ribbons environment (Figure 5.7). These sensors showed excellent stability, reproducibility, and reusability.

A different kind of sensors consists in those based on mechanical or optical resonances. In_2O_3 optical resonators have already been briefly reviewed in the previous section, and the potential application of ODRs in sensing has been discussed elsewhere (Yang, & Guo,

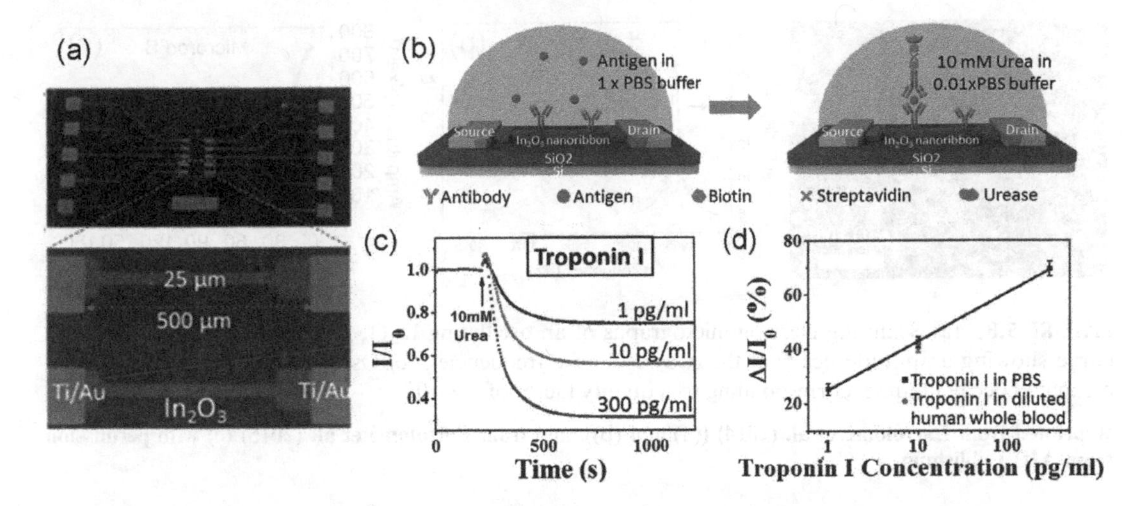

FIGURE 5.7 In_2O_3 nanoribbon biosensors. (a) Optical and scanning electron micrographs of the ribbons fabricated by shadow mask technique. (b) Scheme of antibody biofunctionalization and working principle. (c) and (d) Biosensor time response to 1, 10, and 300 pg mL^{-1} of cardiac troponin I antigens in phosphate-buffered saline, and average sensing results of three devices for the three concentrations in (c) and one concentration of troponin I in diluted human whole blood, marked as a red dot.

Figure adapted with permission from Liu et al. (2016). Copyright (2016) American Chemical Society. (US2019120788A1).

2006; Fan et al., 2011). However, to the best of the authors' knowledge, In_2O_3 optical resonators have not been employed for gas sensing yet, despite their promising results.

5.2.3 In_2O_3 Mechanical Resonators

Mechanical resonators, on the other hand, make use of their change in resonance frequency and/or quality factor upon molecule adsorption as transducing mechanism (Eom, Park, Yoon, & Kwon, 2011). These sensors can present extremely high sensitivities, allowing detection limits down to the single molecule. Most of the mechanical resonators consist of a single or double clamped beam of a selected material, typically Si or diamond, with cross sections in the range of the nanometers to few micrometers. Other configurations based on planar or more complex structures can also be found. The resonance frequency of the resonator depends fundamentally on the beam stiffness and its mass, and therefore the variation in resonance frequency due to single-molecule adsorption/desorption could be detected if the resonance peak is narrow enough, that is, if the quality factor is high enough. Indium oxide mechanical resonators have been obtained from single crystal microrods (Figure 5.8), presenting quality factors as high as 4×10^5 (Bartolomé, Cremades, & Piqueras, 2015; Bartolomé, Hidalgo, Maestre, Cremades, & Piqueras, 2014), comparable to other high-quality resonators, with the added advantage of high transparency in the visible and electrical conductivity. These resonators presented theoretical force detection limits of 10^{-16} $N/Hz^{1/2}$, opening the gate to ultrasensitive In_2O_3 mass sensors.

5.3 SnO_2-Based Nanostructures and Applications

Tin oxide is a wide bandgap semiconductor, with a direct gap of 3.6 eV at 300 K, usually presenting a rutile crystalline structure and high chemical stability. This material exhibits high electrical *n*-type conductivity due to the presence of native oxygen vacancies, as well as good

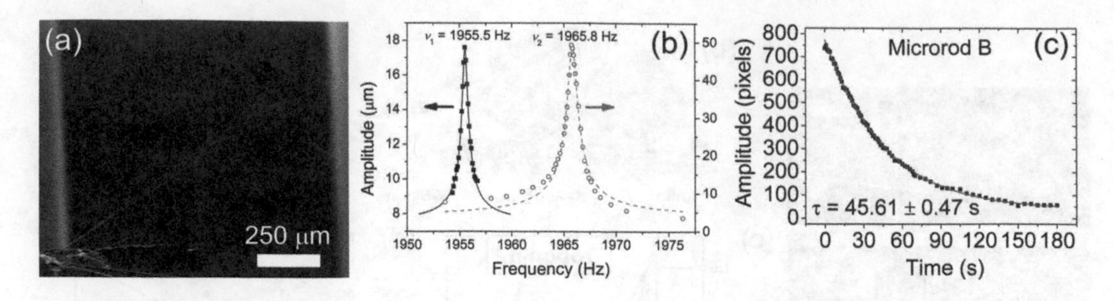

FIGURE 5.8 (a) Scanning electron micrographs of an oscillating In_2O_3 rod. (b) Oscillation amplitude curve showing amplitude peaks at the two resonance frequencies. (c) Oscillation decay curve with the estimated decay lifetime, corresponding to a quality factor of 4×10^5.

Reprinted from Bartolomé et al. (2014) ((a) and (b)), and from Bartolomé et al. (2015) (c) with permission from AIP Publishing.

transparency in the visible range. The combination of these properties includes tin oxide in the TCOs family, and together with its low cost and thermal stability, enables the functionalization of SnO_2 in different devices. In the last decades, SnO_2 has been extensively used in optoelectronic applications (Drevillon, Kumar, & Cabarrocas, 1989; Lee et al., 2004; Li, Zhou, Yu, Liu, & Zou, 2012), solar cells, LEDs, flat screens, Li-ion batteries (LIBs), catalysis, gas sensors, and smart windows, among others. Most of these applications require a control of the morphology and miniaturization of the materials, yielding in the micro- and nanoscale, where the surface to volume ratio increases considerably. Changes in size and morphology, as well as appropriate doping, can produce significant improvements in the optoelectronic properties of SnO_2 associated with the mobility of carriers, a large aspect ratio and effects of quantum confinement (Huang et al., 2009; Kar, & Patra, 2009; Kumar, Govind, & Nagarajan 2011; Zhao et al., 2008). In addition to SnO_2 nanoparticles (Aragón et al., 2016; Caetano, Meneau, Santilli, Pulcinelli, Magnani, & Briois, 2014; Del Prado et al., 2018; Suematsu et al., 2014), nanobelts (Wang, 2003), nanowires (Choi, Hwang, Park, & Lee, 2008; Wang, Lee, & Zeng, 2005), nanodiscs (Wang et al., 2005), or in the microscale morphologies as microrods (Choi, Akash, Sun, Wu, & Kim, 2013), microtubes (Del Prado et al., 2017; García-Tecedor et al., 2016; Garcia-Tecedor, et al., 2016; Maestre et al., 2005), or some more complex morphologies (Zhang et al., 2014) have been reported for SnO_2. Moreover, SnO_2 has also been used as a part of ternary compounds (e.g., ITO; FTO) and recently as a part of composite materials in combination with conductive polymers and carbonaceous materials.

Hence, different approaches have been followed during the past years to spread and improve the applicability of SnO_2 based on different micro- and nanostructures. In this section, some of the most relevant applications in the field of energy storage, solar cells, sensing, and optoelectronic devices are described.

5.3.1 Application of SnO_2 in Batteries

Currently, LIBs are considered as one of the most efficient resources for energy storage in portable devices (Goriparti et al., 2014; Marom, Amalraj, Leifer, Jacob, & Aurbach, 2011; Girishkumar, McCloskey, Luntz, Swanson, & Wilcke, 2010; Scrosati, & Garche, 2010; Kim et al. 2012; Goodenough, & Park, 2013; Etacheri, Marom, Elazari, Salitra, & Aurbach, 2011). However, in the last few years an increase in energy density has been required to achieve other applications that demand greater energy, such as electric vehicles (Etacheri et al., 2011;

Gallagher et al., 2014). There are still certain challenges for this technology to be the predominant one in high-performance devices, such as portable electronics or automatic electrical applications in the industry (Lavoie, Danet, & Lombard, 2017). An improvement in the performance of the batteries can be obtained by various factors, such as, for example, using materials with higher capacity than the commercial carbon-based electrodes, and optimizing the mechanical stability of anodes or cathodes (Goriparti et al., 2014; Nitta, & Yushin, 2014; Song, Kim, Kang, & Kim, 2017). Therefore, one of the main strategies that are being developed is the implementation of materials within the nanometric scale in the constituent electrodes of the LIBs, in order to avoid the disintegration of the electrode materials after a certain number of charging and discharging cycles, while improving the diffusion of Li through the battery. The use of nanomaterials has been proposed to solve the mechanical stresses due to the volumetric expansion and contraction that the electrodes suffer during the delithiation/lithiation processes, providing simultaneous advantages as the high relation surface to volume together with the presence of sites asset for the storage of Li, which facilitates reversibility in the insertion of Li^+ ions as well as the ionic diffusion of Li and electrons.

SnO_2 nanomaterials have been recently considered in LIB applications based on the high-moderate theoretical capacity (~782 mAh g^{-1}) and the low intercalation potential of Li-ions that SnO_2 exhibits. Nevertheless, its cyclability and final performance are substantially reduced, mainly to the volumetric expansion that the lattice suffers during the lithiation and de-lithiation processes. An alternative solution to this issue has been proposed employing low-dimensional SnO_2 as a part of a compound material that buffers these distortions, enhancing in this way the lifetime of the LIBs (Idota, Kubota, Matsufuji, Maekawa, & Miyasaka, 1997; Chen, & Lou, 2012; Nitta, & Yushin, 2014; Goriparti et al., 2014). Nanocomposites containing allotropic phases of carbon such as fullerenes and carbon nanotubes are reported to improve lithium storage properties compared to bare SnO_2 materials (Chen, & Xiong, 2013). Although graphene and GO composites had been getting into the spotlight due to their good properties for energy storage (Del Prado et al., 2019), it still remains an ongoing issue to implement them as part of commercial LIBs (Deng, Fang, & Chen, 2016). Several approaches are being evaluated in order to improve the capacity and cyclability of the SnO_2-graphene-based anodes in LIBs. As an example, Zhang et al. (2019) recently reported how the addition of ferrocene can improve the Li storage capability and enhance the reversible capacity of the SnO_2-based anodes, by suppressing Sn/SnO_2 agglomeration and hence retaining the nanoscale Sn/Li_2O interface. In their work a reversible capacity of up to 1084.5 mAh g^{-1} after 150 cycles was reported, as shown in Figure 5.9.

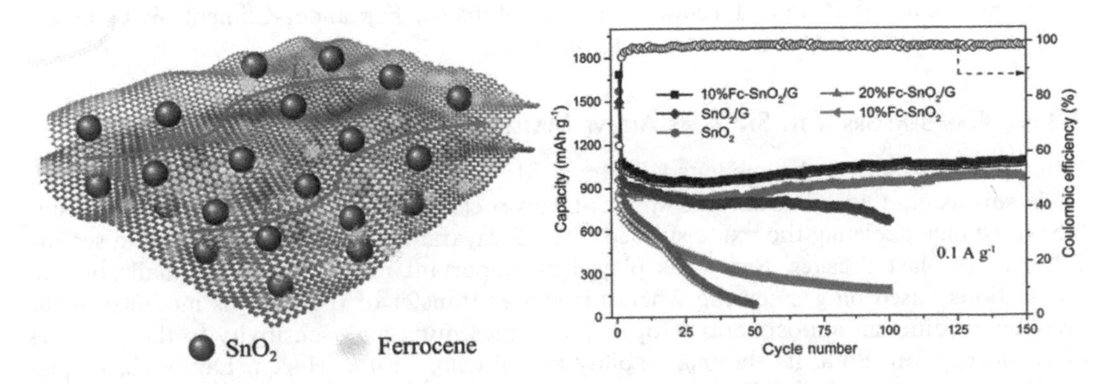

FIGURE 5.9 Scheme of the ferrocene SnO_2/graphene composite and cyclic performance as anode in a LIB. **Reprinted with permission from Zhang et al. (2019). Copyright (2019) American Chemical Society.**

The fact that GO, presenting a large amount of functional groups, owns a large surface to volume ratio makes it a promising material that can act as an absorbent, with improved performance in combination with nanoparticles due to the surface tension and electrostatically attraction that appears between both kind of materials (Huang, Boey, & Zhang, 2010). These features have motivated the investigation of GO-nanoparticles compounds and their implementation in different applications, specifically combining them with SnO_2 or TiO_2 nanoparticles focused on more effective energy storage (Chen, & Qin, 2014; Li, Lv, Lu, & Li, 2010; Lian et al., 2011).

5.3.2 SnO_2 Implementation in Solar Cells

SnO_2 has been widely used in the solar cell field, being employed as a transparent conductive layer usually doped with Fluor as fluorine doped tin oxide or alloyed with indium oxide as ITO. Even when the solar cells industry is still well dominated by silicon-based technologies, hybrid solar cells, combining organic and inorganic materials, seem promising. In this field, SnO_2 has been used as a passivation element of the silicon surface embedded in an organic host as the polymer PEDOT:PSS, which by itself possesses high conductivity and transparency in the visible range (Garcia-Tecedor, Maestre, Cremades, & Piqueras, 2017). Furthermore, SnO_2 also has an important role in the manufacture of transparent electrodes in photovoltaic devices of amorphous silicon (a-Si), CdTe and Gräztel cells (dye sensitized solar cells, DSSC) and as an electron selective layer in perovskite solar cells where it improves the carrier extraction (Bolzan, Fong, Kennedy, & Howard 1997; Kılıç, & Zunger, 2002; Gopel, 1985; Hagen, Lambrich, & Lagois, 1983; Zhu, Yang, et al., 2016; Fortunato, Ginley, Hosono, & Paine, 2007; Madou, & Morrison, 2012; Klein et al., 2010). Some of these promising SnO_2-based solar cells devices consist in hybrid materials, as the patent from the Schenzhen Xianjin Clean Power Tech. Research Co (CN109904331(A)) in which SnO_2 and rGO composites in form of thin films, with a 30–50 nm thickness, are produced and evaluated as electron transmission layer perovskites. Some other approaches are focused on an improvement of the passivation behavior of Si substrates by using hybrid organic—inorganic films formed by PEDOT:PSS and SnO_2 or TiO_2 nanoparticles, as described in the patent (patent published as ES2650213B2; EP3482426A1; WO2018010935A1; NO20161150A1; US2019319161A1). In that case charge carrier lifetimes of about 275 μs were achieved by adding a low concentration of SnO_2 nanoparticles to the conductive polymer, involving low-cost processing and the potential implementation in flexible substrates.

Nowadays, SnO_2 has been also employed as a hole mirror layer in heterojunction architectures, mainly with $BiVO_4$, for solar water splitting, improving notably the performance of the photoanodes (García-Tecedor, Cardenas-Morcoso, Fernández-Climent, & Giménez, 2019; Liang, Tsubota, Mooij, & van de Krol, 2011).

5.3.3 Gas Sensors with SnO_2 as Active Material

Detection of gases can be mediated by tin oxide, where the surface reaction between the oxide surface and adsorbed oxygen species involves charge transfer, which modifies the surface resistance depleting the oxide surface (Das, & Jayaraman, 2014), as described in section 5.2.2. In the last decades, SnO_2 has played an important role in the functionalization of applications based on gas sensing where the charge transfer of the SnO_2 is modified in the presence of different atmospheres (Gopel, 1985), presenting high sensitivity to the processes of chemisorption. Both, its thermal stability at low temperature (Hagen, Lambrich, Lagois, 1983) and easy ceramic fabrication have made SnO_2 a very cost- and performance-effective material in the sensing of gases such as H_2, H_2O, O_2, CO, and CO_2 (Choi et al., 2008; Sberveglieri et al., 1993; Zhang et al., 2014). Actually, SnO_2 is one of the most employed materials for gas sensing based on the resistance or conductance change due to gas adsorption (Das, & Jayaraman, 2014). The use of SnO_2 nanostructures allows achieving better surface-to-volume ratio and enhanced performance; hence the use of gas sensor based on SnO_2 nanomaterials has been greatly increased. The control of the size, morphology, doping, and defect distribution can tailor the electrical and chemical properties of the SnO_2 leading to variable selectivity and sensitivity to different gases. Gas sensors based on SnO_2 have been developed in the form of thin sheets, which exhibit good sensitivity but low response time (Choi et al., 2008). However, by reducing the size of the active materials by using morphologies such as nanoparticles, nanowires, nanotubes, and nanobelts, it has been possible to significantly increase their response in resistive sensors (Leite, Weber, Longo, & Varela, 2000; Ying, Wan, Song, & Feng, 2004).Many different approaches have been developed to increase SnO_2 sensing abilities, based on improving the following two main aspects: sensing sensibility (detection limit) and recovery time. One is the addition of noble metals (Au, Pt), which are known to enhance the rate of response (Lin, Li, Chen, & Fu, 2017; Moseley, 2017), as catalytic properties of metals lower the temperature range of gas response due to the formation of a Schottky barrier on the oxide surface. Also, *n*-type/*n*-type junctions have been used as gas sensors, as the combination of SnO_2 with TiO_2 or ZnO. On the other hand, the addition of dopants could inhibit the growth of nanoparticles while controlling conveniently the carrier concentrations, thus enhancing sensing abilities, as for Ni-doped SnO_2 (Lin et al., 2017). On the other hand, composites of tin oxide had been also used in gas sensing (Joshi et al., 2018) mainly because of its possibility to work under room temperature, lowering the operation temperature with improvements in sensibility and recent focus has been aiming in this direction (Joshi et al., 2018). As an example, SnO_2 quantum wire/rGO nanocomposites have been reported as selective gas sensors for H_2S, with remarkable sensitivity (Song et al., 2016), although still exhibiting long recovery times and high cost.

Great efforts have been devoted to patent continuous improvements in the sensitivity, selectivity, and miniaturization of SnO_2-based sensor devices with improved versatility. As an example, researches from the University of Korea recently patented an NO_2 sensing device usable at room temperature that consists of a multi-array chip based on 3D laminated structures which comprised SnO_2 nanoparticles and carbon nanotubes (Patent KR20180072980A).

5.3.4 Optical Applications of Low-Dimensional SnO_2

In the present section, a less exploited role of SnO_2 in waveguides, optical resonators, and white luminescence devices will also be presented. Microstructures of different TCOs as ZnO, In_2O_3, or Ga_2O_3 have been reported as optical resonators, with promising applicability in sensing, and lab-on-a-chip devices (Bartolome, et al., 2013; Huang et al., 2010; Lopez, Nogales, Mendez, & Piqueras, 2012). However, it is only recently that SnO_2 microstructures have been considered as optical resonators, which can be further exploited in future devices. The optical resonances observed in microcavities with regular sections are frequently attributed to FP or WGM (Born, & Wolf, 1980). FP resonances involve reflections between opposite flat ends of the cavity, while WG modes involved total internal reflections around the inner faces of the cavity. Each mode exhibits different properties such as the small mode volume for WGM presenting higher light confinement or the easy tuning of the resonance frequency in FP resonators; hence, mode selection could be a determining factor for specific applications. Microstructures with flat and smooth surfaces are crucial not only for the generation of optical resonances, but also, as it will be explained further, for achieving waveguiding behavior. Obtaining optical resonances and waveguiding behavior in the same material could bring up the possibility of achieving optical microsystems with a monolithic architecture, using the same material as light input element and optical microcavity. Therefore, a high control of the synthesis method used for the growth of the microstructures is required, as well as the ability of engineering the resonant modes through size, morphology, and doping. In addition, the quality of the cavities as optical resonators can be studied through the estimation of the finesse (F) and quality (Q) factors, which provide information about the resonator losses and stored photon energy, respectively. The behavior as optical resonators and waveguides of different undoped and Cr-doped SnO_2 microstructures as tubes, rods, and plates was recently reported (Garcia-Tecedor et al., 2017).

Figure 5.10 shows an SnO_2 microtube in which different optical resonances coexist. As explained before, the optical resonances depend on the morphology of the optical cavity. In this specific case, simultaneous FP and WGM resonances were detected due to the morphology of the tube, which is schematized in Figure 5.11. In that work, different morphologies were analyzed in order to tune the optical resonances.

In addition, the possible application of undoped and Cr-doped SnO_2 elongated microstructures with different morphologies were also reported as waveguides by Garcia-Tecedor et al. (2017). These microstructures were illuminated with a laser by one end and a bright light signal could be observed at the opposite end. Also the light can be guided toward both ends of the tube when illuminating at its center. This phenomenon confirms the waveguiding behavior with few losses along these structures. In the case of the microtubes, the light was guided along the lateral faces, therefore a red frame-feature was observed at the opposite end, as shown in Figure 5.12a. However, in the microrods, an intense red spot was found at the apex, marked with an arrow in Figure 5.12b, at hundreds of microns from the incidence of the laser. This light-guide behavior was observed even for the longest rods, with lengths in the range of millimeters. For the microplate, not only the final ending of the structure, but also the surrounding area was illuminated (Figure 5.12c), as a difference with the microrod. No scattered light can be appreciated along the structures, which is an indirect indication of low optical losses. Similar waveguiding behavior has been reported for microstructures of other semiconducting oxides as In_2O_3 and Ga_2O_3 (Bartolome et al., 2013; Lopez et al., 2012). It is important to point out that the guided light along the different microstructures is the laser itself, not the laser-induced luminescence of the material.

Among other applications, an increasing interest in the luminescent properties of GO in combination with TCOs such as ZnO (Williams, & Kamat, 2009) or transition metals (Cu) (Udayabhaskar et al., 2017) has recently emerged. Nevertheless, the luminescent properties of GO-SnO_2 nanoparticles have been scarcely studied and even lessen the possibility of synergetic effects between the SnO_2 and the GO matrix (Del Prado et al., 2019). These properties can be studied by means of luminescence techniques to get a deeper insight into the different processes involved as the charge transfer between the GO matrix and the nanoparticles. Even more, the luminescence of GO with metal oxides has been a motivation of study in the last decade (Chien et al., 2012; Cuong et al., 2010; Mei et al., 2010) in different fields with biomedical purposes or to develop optoelectronic devices (Eda et al. 2008; Mei et al., 2010). Nevertheless, the studies reported until the moment, related with GO and inorganic materials (Shen et al., 2011; Zhou, Zhu, Yang, & Li, 2011; Zou, Zhu, Sun, & Wang, 2011) are still scarce (Williams, & Kamat, 2009). In the majority of the cases, the luminescence has been explained applying different model or theories attending to the active emission as originating in the GO matrix. Usually, as a function of the fabrication processes and final components of the samples, isolated emissions in the near infrared range (Chien et al., 2012; Jiang et al., 2015), or alternatively in the yellow-green region (Jiang et al., 2015; Stengl et al., 2013) or in the blue or near-UV region (Eda et al., 2010; Jiang et al., 2015; Stengl et al., 2013) have been reported. The origin of these bands has been commonly associated with the aperture of a bandgap due to the functionalization of GO by different functional groups (Luo, Vora, Mele, Johnson, & Kikkawa, 2009; Mathkar et al., 2012), to the disorder of the GO containing the sp^2 and sp^3

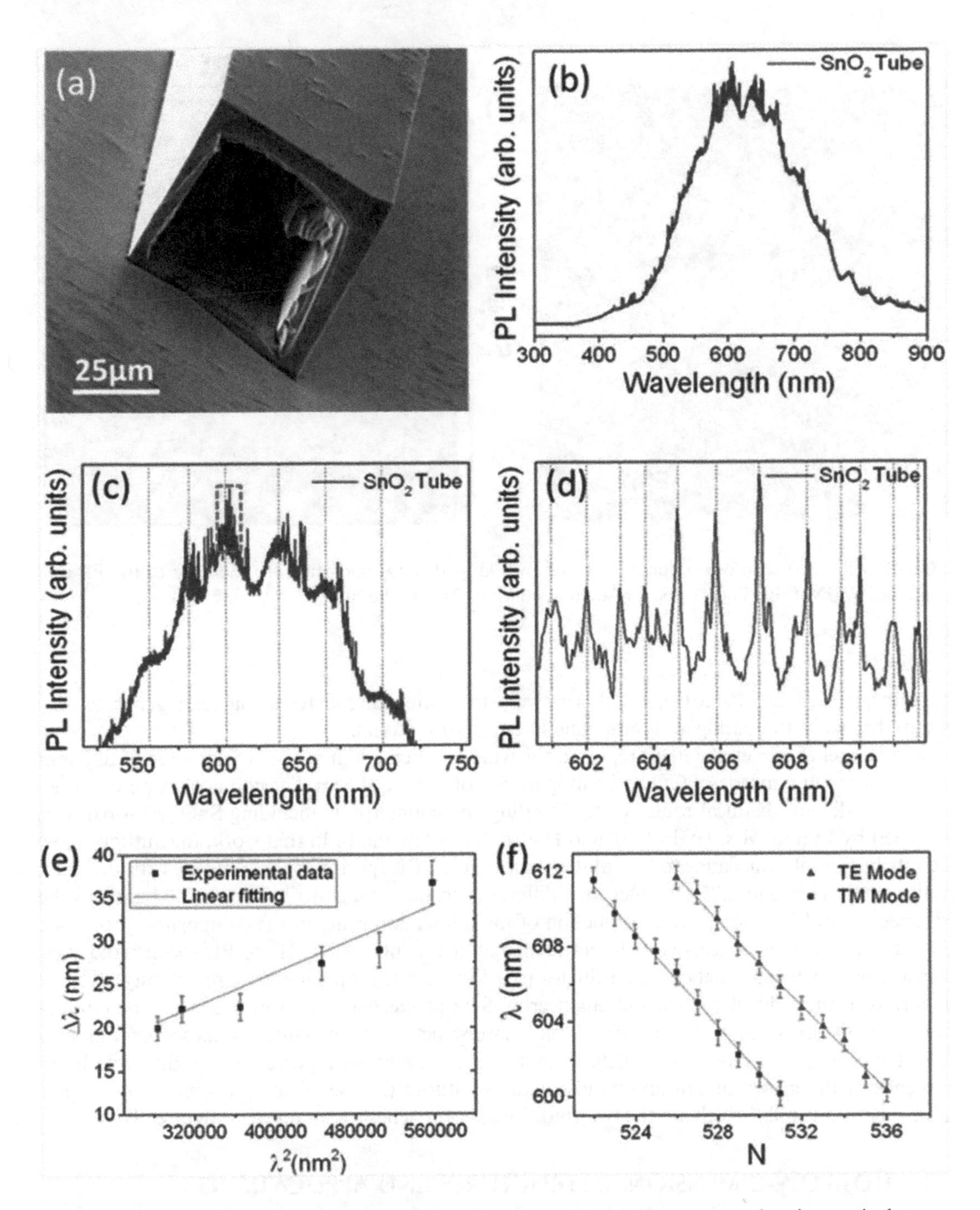

FIGURE 5.10 (a) SEM image of an undoped SnO_2 microtube. (b) PL spectrum showing optical resonances. (c) and (d) Amplified regions of the PL spectrum in (b) showing different resonant modes, where (d) corresponds to the marked region with a red rectangle in (c). (e) Linear fit $\Delta\lambda$ versus λ^2 corresponding to an FP mode. (f) Resonance wavelength λ as a function of the mode number N for the TE (triangular symbols) and TM (square symbols) polarizations corresponding to a WG mode.

Figures reprinted from Garcia-Tecedor et al. (2017). Reproduced with permission from IOP Publishing.

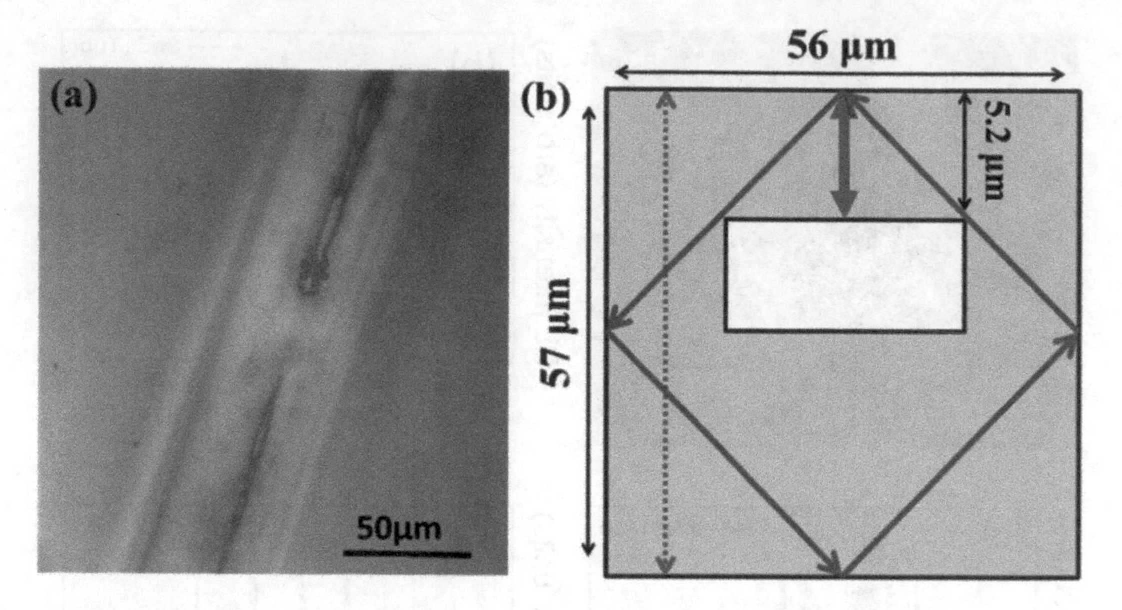

FIGURE 5.11 (a) Confocal optical image of an undoped SnO_2 microtube. (b) Scheme of the FP (red lines) and WGM (blue lines) optical resonances taking place in the tube.

domains (Chien et al., 2012), or alternatively to some other defects that can generate new energy bands in the bandgap in graphene like related material.

Lately, Del Prado et al. (2019) reported the wide band emission in the visible range of different composites with a matrix of GO and nanoparticles of SnO_2 as fillers. The composites were synthetized by different chemical routes or by blending the counterparts including SnO_2 nanoparticles obtained by Liquid-Mix, Hydrolysis, and Hydrothermal methods. In that work, the authors show a contribution of four bands to the total emission of the PL spectra of the samples, with the main peak centered around 2.25 eV, whereas a difference in the total contribution of the bands to the PL spectra could be observed as a function of the fabrication route of the composites. Moreover, the authors reported a near-white luminescence after the integration of the PL spectra and their transformation to chromaticity coordinates (see Figure 5.13), opening the opportunity to these materials in the field of science and engineering for optoelectronic devices. Recently, the fabrication of SnO_2-Li_2SnO_3 branched micro- and nanostructures with white luminescence was also reported (García-Tecedor et al., 2019). In that work, the authors reported the bright white luminescence of the grown micro- and nanostructures without the use of any phosphors or complex multi-quantum wells, which are the two traditional approaches to obtain white light emission.

5.4 TiO_2 LOW-DIMENSIONAL STRUCTURES AND APPLICATIONS

Titanium dioxide (TiO_2) is a semiconductor oxide with a wide and direct bandgap, having excellent electrical, electrochemical, and optical properties (Kumar et al., 2017; Longoni et al., 2017). The combination of high transparency in the visible range and electrical conductivity expands its applications further than the traditional catalysis implementation of TiO_2. In recent years, the potential of this material has been studied in various fields such as solar cells (Shogh, Mohammadpour, Zad, & Taghavinia, 2015), gas sensors (Tshabalala, Motaung, Mhlongo, & Ntwaeaborwa, 2016), and photocatalysis

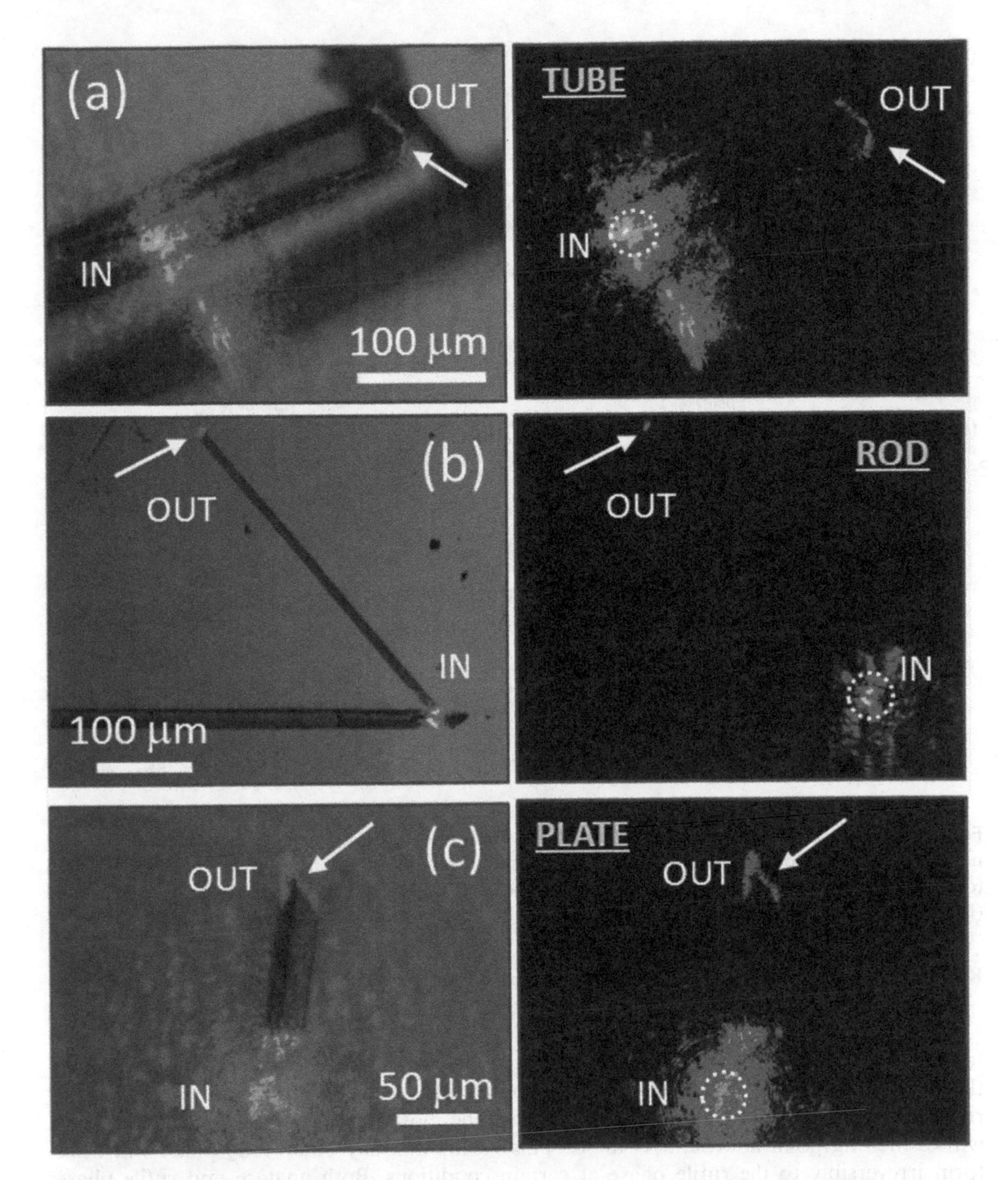

FIGURE 5.12 Bright field (left) and dark field (right) images from (a) tubes, (b) rods, and (c) plates. **Figures reprinted from Garcia-Tecedor et al. (2017).**

Reproduced with permission from IOP Publishing.

(Kaplan et al., 2016), among others, where the dimensions, morphology, and structural defects play an essential role.

TiO_2 presents three different crystalline structures: a rutile (tetragonal) stable phase and two metastable phases, known as anatase (tetragonal) and brookite (orthorhombic). Owing

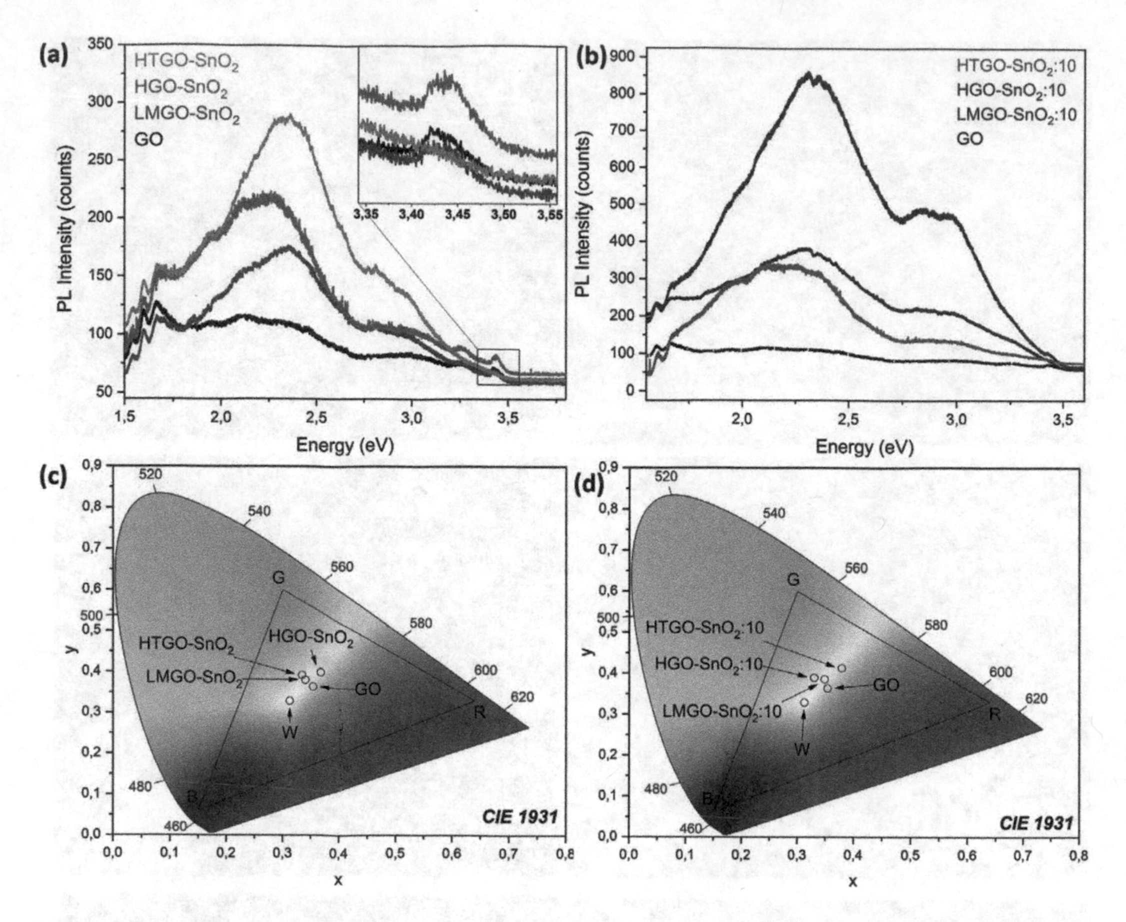

FIGURE 5.13 PL spectra acquired at T = 300 K corresponding to (a) composites with undoped SnO_2 nanoparticles and (b) composites with Li. PL spectrum from GO is also included as a reference. (c) and (d) CIE 1031 chromaticity coordinates of the undoped and Li-doped PL spectra from samples in (a) and (b), respectively. The black triangle denotes the RGB gamut with pure red, green, and blue in the vertices. Pure white (W) is also shown for comparison.

Reprinted from Del Prado et al. (2019), with permission from Elsevier.

to their physicochemical properties, stability at room temperature, and more developed synthesis methods so far, the most common and extensively used phases are the tetragonal rutile and anatase phases, being the rutile phase the most stable phase at high temperature, with a melting point at around 1840 °C. This means that any other TiO_2 phase will transform irreversibly to the rutile phase at certain conditions. Both anatase and rutile phases posses wide bandgaps (3–3.2 eV), are transparent to visible radiation, and exhibit good electrical conductive and high refractive index (n > 2), which make them appealing materials for electronic and optoelectronic applications such as solar cells, solar fuels, LIBs, and photocatalysis. TiO_2 also demonstrated excellent photocatalytic activity, which can be improved by using nanostructures with appropriate doping. Nowadays, TiO_2 is a go-to material for water purification or contaminant gas reduction due to its low cost and its high chemical stability; therefore, great efforts have been invested during the last decades to tune their physical and chemical properties using both doping and downscaling approaches. The selection of proper TiO_2 synthesis methods in order to control the chemical composition, crystal phase, crystal size, and morphology, as well as its structural defects is essential as these aspects are required in the development and improvement of different TiO_2-based applications (Liu et al., 2017; Usui, Yoshioka, Wasada, Shimizu, & Sakaguchi, 2015).

In this section, some of the most promising applications of TiO_2 low-dimensional structures in photovoltaics, optoelectronics, photocatalysis, gas sensing, and phase-control-based devices are described.

5.4.1 Photovoltaic Applications

One of the main applications of TiO_2 in the photovoltaic field is as active material of the Grätzel solar cells. The structure of one of these cells, also known as DSSCs, is shown in Figure 5.14. These TiO_2-based photovoltaic devices, which involve low cost and easy fabrication, are based on a photosensitized anode formed by a dye solution and TiO_2 nanostructures, a cathode and an electrolyte.

Control of morphology and crystal phase among other parameters is essential in order to achieve a good efficiency in these devices. DSSCs are usually fabricated using nanocrystalline TiO_2, mesoporous TiO_2, TiO_2 nanotubes, or hybrid anatase–rutile TiO_2 nanocrystalline material. However, recent studies have demonstrated a better efficiency for DSSCs using doped TiO_2. For example, the application of photoanodes prepared with N-doped TiO_2 nanotubes, improves the DSSC performance (Tran et al., 2017). Figure 5.15 shows TiO_2 nanotubes synthesized by a hydrothermal method and the corresponding *J–V* curves. The photoinduced current due to a xenon lamp used in the experiment increases around 30–40 % using N-doped TiO_2 as compared to the behavior of undoped TiO_2 electrodes due to the enhancement of the electron injection efficiency and the decrease of the dark current.

Doping TiO_2 has been a deeply explored field leading to enhanced photovoltaic behavior. Shogh et al. (2015) demonstrated a 26 % improvement in cell performance, obtained from Nd-TiO_2 photoelectrode compared to a pure one, due to the injection and transport enhancement in the doped photoelectrode. Most of the patents based on DSSC are focused on the achievement of better efficiency and improved efficiency by using different nanocrystalline TiO_2 and TiO_2-based composites (Fakharuddin, Jose, Brown, Fabregat-Santiago, & Bisquert, 2014).

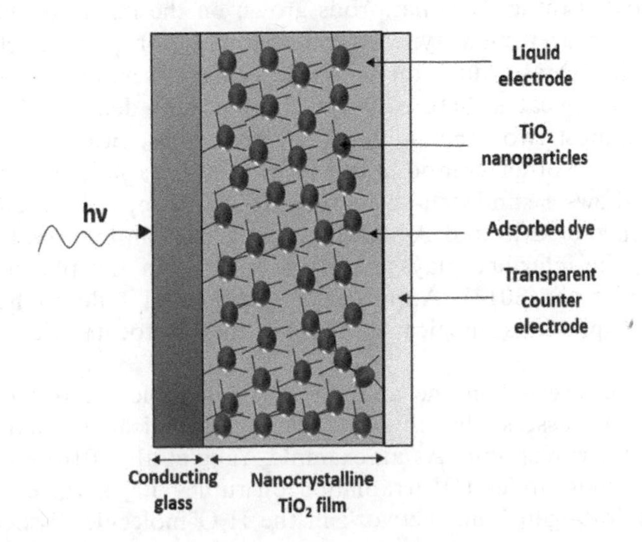

FIGURE 5.14 Scheme of a DSSC based on TiO_2 nanostructures.

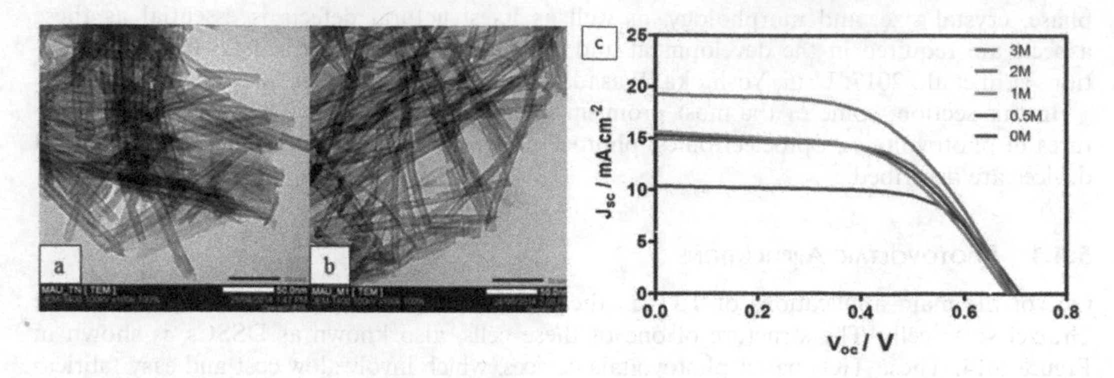

FIGURE 5.15 TEM images showing (a) undoped and (b) N-doped (0.5 M) TiO_2 nanotubes, and (c) the corresponding *J–V* curves.

Reprinted from Tran et al. (2017), with permission from Elsevier.

5.4.2 Photocatalytic Applications

TiO_2 has been studied as one of the most efficient environmental photocatalyst, being a good candidate for photodegradation of several pollutants or water splitting. Kaplan et al. (2016) synthesized a uniform anatase/rutile/brookite nanocomposite with excellent photocatalytic properties under UV light. It was determined that after 60 min of reaction, 94% conversion of Bisphenol A was achieved. On the other hand, doping TiO_2 usually can improve the response of this material. Demirci et al. (2016) studied the effect of Ag on the photodegradation of methylene blue under UV light. The results showed that Ag-doped TiO_2 have a degradation efficiency of 55% while undoped TiO_2 present a degradation efficiency around 36%.

The conversion of solar energy to hydrogen by means of water splitting is one of the most promising ways to achieve clean and renewable energy. Since the discovery of photocatalytic splitting of water on a TiO_2 electrode in 1972, there are numerous works focused on the study of the properties of this material under light illumination. Wang et al. (2016) fabricated rutile TiO_2 nanorods grown on the inner surface of arrayed anatase TiO_2 nanobowls as a new type of photoanodes for photoelectrochemical water splitting. The rutile/anatase junction improves charge separation; under solar light irradiation this hierarchical structure has a photocurrent density of 1.24 mA cm^{-2} at 1.23 eV, which is almost two times higher than pure TiO_2. Besides, a tuned heterojunction between TiO_2 and other compounds has been also exploited in the photocatalysis field. Figure 5.16 shows a sandwich-structure CdS/Au nanoparticles/TiO_2 nanorod array in which the presence of CdS and Au leads to enhanced photocurrent and light absorption. In this case, the interface plays a paramount role in the photocatalytic behavior, as reported by Li et al. (2014). Appropriate doping and tailored heterojunctions are among the most employed strategies to improve the photocatalytics response in TiO_2-based devices.

Recent efforts are invested in the analysis of single-molecule photocatalysis on TiO_2 surfaces in order to assess the involved physico-chemical mechanisms and design improved photocatalytic response. As an example, Yang et al. (2016) investigated photoinduced water dissociation under UV irradiation on rutile-TiO_2 surfaces and demonstrated the role played by hydrogen bond network in the H_2O molecule dissociation by H atom transfer.

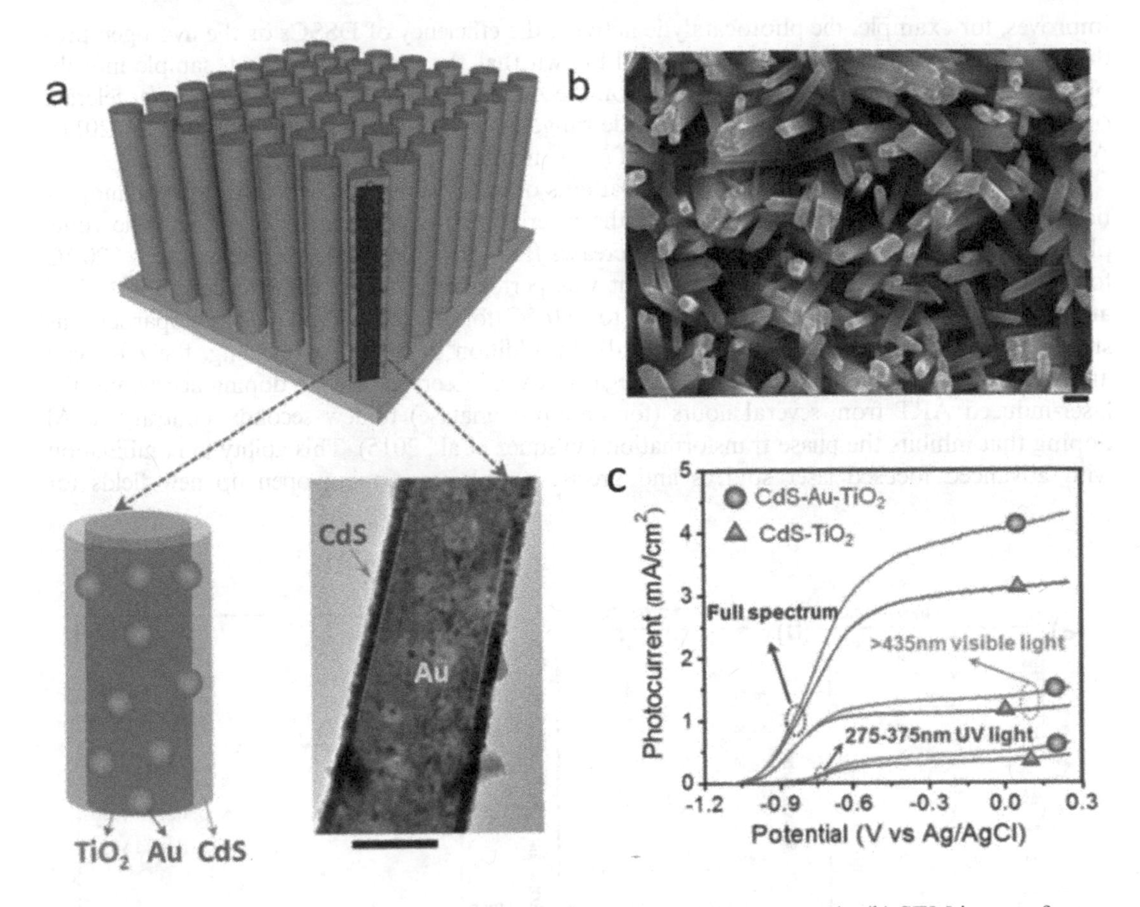

FIGURE 5.16 (a) Sandwich-microstructure formed by CdS/Au/TiO_2 nanorods. (b) SEM image of a top view from the nanorod array. (c) Photocurrent-applied potential ($J-V$) curves obtained by different irradiation conditions (full-spectrum, visible light (>430 nm), and ultraviolet (275−375 nm)).

5.4.3 Toward TiO_2 Phase-Control-Based Devices

In addition to the defect structure, doping, and morphology, most of the TiO_2 applications also depend on the crystallographic phase. Despite the fact that the band gap energy (E_g) for rutile and anatase phases is similar (E_g ~3.05 and ~3.2 eV at 300 K, respectively), their electrical, optical, and chemical properties are different due to different atomic arrangements and symmetries of the Ti and O atoms within the crystalline lattice. For example, rutile is the most thermally stable TiO_2 phase and highly resistant to chemical agents so it is suitable for operation in harsh environments such as in water splitting or photoelectrochemical applications, while anatase is well known because of the higher photocatalytic activity. However, bulk anatase transforms into the rutile phase at temperatures above ~700 °C, hindering its use in devices that operate at higher temperatures. So, great efforts have been invested to understand and manipulate the anatase to rutile phase transition (ART) temperature as well as developing methods to control precisely the anatase/rutile phase composition. Indeed, mixed anatase/rutile phase demonstrates advantages compared to their individual forms leading to a synergetic effect between the two phases that improves, for example, the photocatalytic activity, the efficiency of DSSCs or the hydrogen production in TiO_2-based solar fuels. It is well known that the synthesis method, sample morphology, and dimensions have critical impact on the ART, but doping can also promote/accelerate or inhibit/delay efficiently the ART in a wide range of temperatures (Hanaor, & Sorrell, 2011). Aluminum and iron doping are examples of such effects as shown in Figure 5.17.

Figure 5.17a and b show thermo-XRD patterns of undoped and 10 cat.% Al-doped nanoparticles as the temperature increases, where the asterisk marks the temperature when the rutile phase starts to appear. That temperature increases from 800 °C for undoped material to 920 °C for Al-doped samples. A similar experiment was performed for different concentrations of Al and Fe, decreasing the temperature down to 600 °C for 20 cat.% Fe-doped nanoparticles as shown in Figure 5.17c (Vásquez et al., 2014). In addition to thermal annealing, the ART can also be driven by laser irradiation. In this regard, it was observed that Fe doping accelerates the laser-induced ART from several hours (for undoped anatase) to few seconds, contrary to Al doping that inhibits the phase transformation (Vásquez et al., 2015). This ability in combination with advanced focused laser sources and precise sample positioning open up new fields for

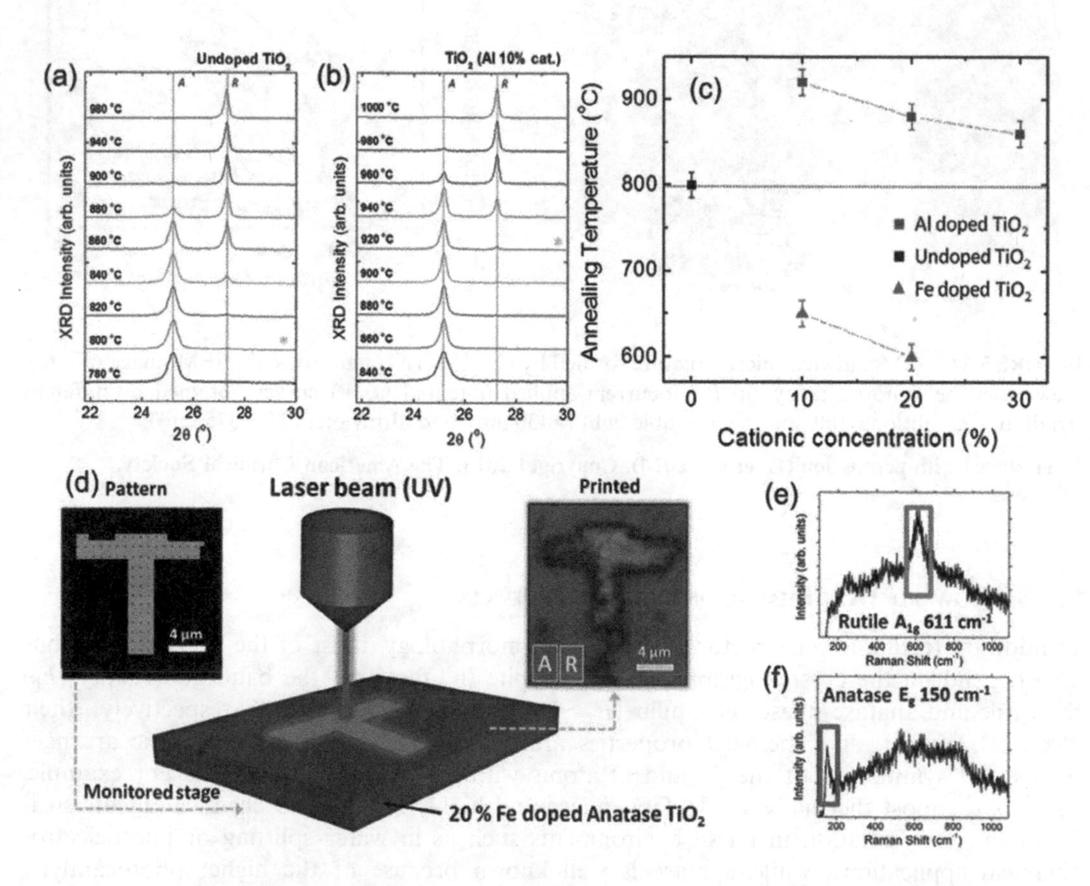

FIGURE 5.17 Thermo-XRD patterns on (a) undoped and (b) Al-doped TiO_2 nanoparticles. (c) Temperature at which the rutile phase starts to appear in XRD as a function of the concentration and Al and Fe dopants. Scheme of the laser patterning of rutile phase on 20 cat.% Fe-doped anatase. Raman spectra acquired on (e) laser irradiated and (f) unirradiated areas.

device fabrication, processing, and characterization, as shown in the scheme of Figure 5.17d for a laser-printed pattern (micropatterning) on Fe-doped anatase nanoparticles. The characteristic Raman spectra of rutile and anatase phases acquired on irradiated and nonirradiated areas are shown in Figures 5.17e and f, respectively, with the characteristic phonon lines marked by solid squares. This concept has potential applications in device processing since the phase composition could be spatially controlled at room temperature using appropriate laser power and dopant concentration (Benavides, Trudeau, Gerlein, & Cloutier, 2018; Vásquez et al., 2014; Wilkes, Deng, Choi, & Gupta, 2018). These results have also lead to two patents based on anatase to rutile phase transition (Patent ES2525393B2 (WO2016038230A1) (Vásquez et al., 2014) and Patent ES2525737B2 (WO2016046426A1)).

5.4.4 Applications of TiO_2-Based Composites

Research based on hybrid composites combining an organic material with nanostructured TiO_2 is gaining increasing attention due to their enhanced performance in different devices such as sensors, batteries, and solar cells, among others. In the field of LIBs, promising results have been reported by implementing TiO_2-based composites as anode material. Some of the main drawbacks of the pure TiO_2 anodes in the LIB, such as poor electronic conductivity and small lithium insertion and diffusion can be overcome when using composites (Wu, Chen, Hng, & Lou, 2012), in which diverse allotropic forms of carbon material (Graphite, Graphene, NTC, fullerenes) have been combined with TiO_2 nanostructures (Dahl et al., 2014). As an example, anatase TiO_2 nanospheres over graphene have shown six times higher cyclability than bare anatase (Li et al., 2011). Also rutile and anatase nanostructured TiO_2 graphene hybrid materials show enhanced Li-ion insertion/extraction, especially at high charge/discharge rates (Wang et al., 2009).

As previously mentioned, TiO_2 is the material of choice used for photoelectrochemical water splitting and still serves as the benchmark for photocatalyst/photoelectrode materials. The photocatalytic activity of TiO_2 can be also enhanced with the production of TiO_2 composites based on carbon allotropic phases (Wen et al., 2015). It has also been used to reduce GO to rGO due to the UV absorption of TiO_2 (Williams et al., 2008), where ethanol-suspended GO undergoes reduction as it accepts electrons from UV-irradiated TiO_2 particles. Composites formed by a combination of the inorganic TiO_2 and organic materials offer different configurations to achieve higher performance. Donor–acceptor systems can be created by blending TiO_2 with poly-3-hexylthiophene (P3HT) (Bai, Mora-Seró, De Angelis, Bisquert, & Wang, 2014).

TiO_2 nanoparticles have been also used as a passivation element of the silicon surface embedded in an organic host as PEDOT:PSS, which by itself possesses high conductivity and transparency to visible light (García-Tecedor et al., 2018) (patents published as ES2650213B2; EP3482426A1; WO2018010935A1; NO20161150A1; US2019319161A1). In that case, the hybrid composite formed using rutile TiO_2 nanoparticles in a concentration of 1 wt.% exhibits charge carrier lifetime of 160 μs, which converts this hybrid system in a promising candidate for the development of low cost and flexible solar cell devices. Figure 5.18 shows a scheme of the solar cell device and the corresponding Si-surface passivation results and current voltage measurements under dark or illumination conditions.

5.5 CONCLUSIONS

In order to face the advent of multifunctional modern devices, as well as to improve actual devices and explore novel functionalities, the development of low-dimensional, environmentally clean, easy to fabricate, scalable, and low-cost materials following innovative strategies is required. Wide-bandgap semiconducting oxides such as SnO_2, TiO_2, and In_2O_3 are considered to accomplish some of these requirements; therefore, great efforts are invested in their study, as

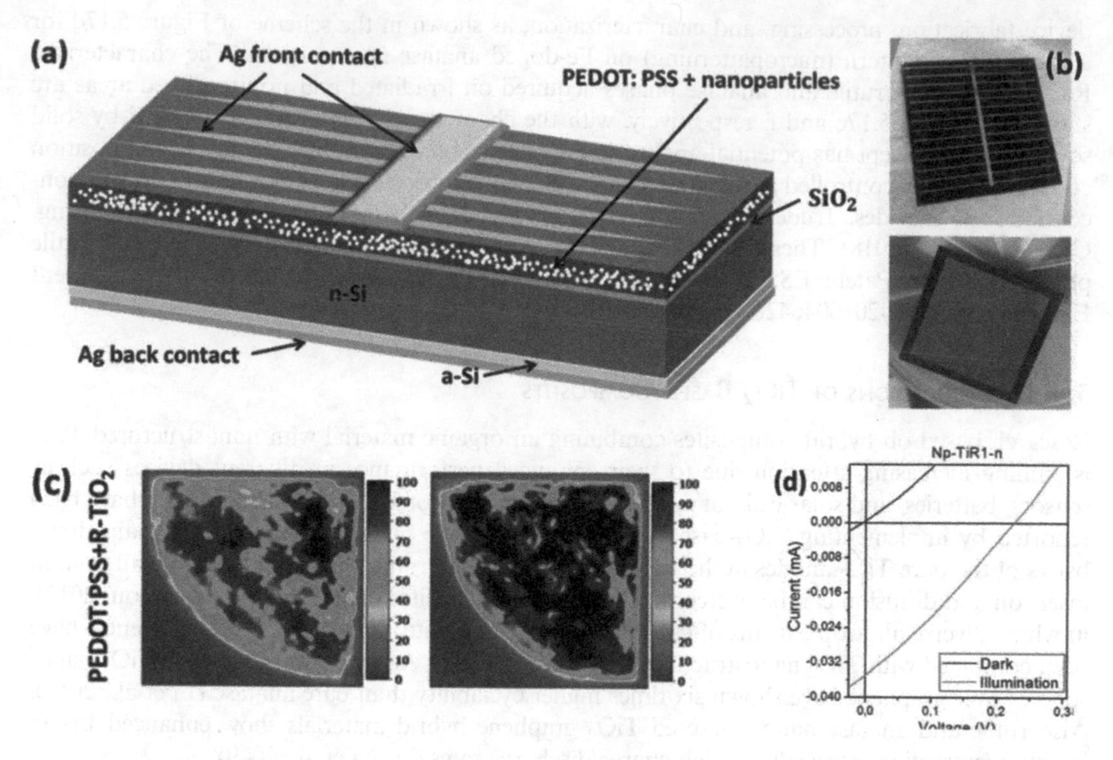

FIGURE 5.18 (a) Scheme of a photovoltaic device with hybrid composite formed by PEDOT:PSS and TiO_2 (or SnO_2) nanoparticles. (b) Images of the front and back side of the solar cell model. (c) PL images and QSS-PC measurements from composite sample containing TiO_2 nanoparticles in a 5 wt.% obtained by illumination from the PEDOT:PSS or from the a-Si:H side. (d) Current–voltage dependence under dark and illumination corresponding to the solar cell with the PEDOT:PSS-TiO_2 composite.

Figure reprinted from García-Tecedor et al. (2018). Reproduced with permission from IOP Publishing.

these materials will play a leading role in the development of sensors, photovoltaic devices, energy storage, photocatalysis, and optoelectronics devices, among others. In this chapter, different approaches based on size and morphological control, tailored composition by appropriate doping, tuned heterojunctions, and novel nanoarchitectures are reported and described in order to pave the way to broaden the applicability of these semiconducting oxides. Besides, the combination of these low-dimensional In_2O_3, SnO_2, or TiO_2 structures with organic materials such as graphene, carbon nanotubes, or conductive polymers is also described. Based on the synergy between their counterparts, these hybrid composites allow to reach breakthroughs in different fields of technological research; therefore, new generation of hybrid materials are being evaluated due to their easy processing, low costs, ability to be used in flexible substrates and scalability.

REFERENCES

Ágoston, P., Albe, K., Nieminen, R. M., & Puska, M. J. (2009). Intrinsic n-type behavior in transparent conducting oxides: A comparative hybrid-functional study of In_2O_3, SnO_2, and ZnO. *Physical Review Letters, 103*, 245501.

Ágoston, P., Erhart, P., Klein, A., & Albe, K. (2009). Geometry, electronic structure and thermodynamic stability of intrinsic point defects in indium oxide. *Journal of Physics: Condensed Matter, 21*, 455801.

Akyildiz, I. F., & Jornet, J. M. (2014). Electromagnetic wireless nanosensor networks. *Nano Communication Networks, 1*(1), 3–19.

Alam, M. Z., Leon, I. D., & Boyd, R. W. (2016). Large optical nonlinearity of indium tin oxide in its epsilon-near-zero region. *Science, 352*, 795–797.

Ali, I., Suhail, M., Alothman, Z. A., & Alwarthan, A. (2018). Recent advances in synthesis, properties and applications of TiO_2 nanostructures. *RSC Advances, 8*(53), 30125–30147.

Alivisatos, A. P. (1996). Perspectives on the physical chemistry of semiconductor nanocrystals. *Journal of Physical Chemistry, 100* (31), 13226–13239.

Almazrouei, E., Shubair, R. M., & Saffre, F. (2018). Internet of nanoThings: Concepts and applications, arXiv:1809.08914v1 [cs.ET] 21 Sep 2018.

Alvarado, M., Navarrete, È., Romero, A., Ramírez, J.L., & Llobet, E. (2018). Flexible gas sensors employing octahedral indium oxide films. *Sensors, 18*, 999.

Aragón, F. H., Huamaní Coaquira, J. A., Gonzalez, I., Nagamine, L. C. C. M., Macedo, W. A. A., & Morais, P. C. (2016). Fe doping effect on the structural, magnetic and surface properties of SnO_2 nanoparticles prepared by a polymer precursor method. *Journal of Physics D: Applied Physics, 49*, 155002.

Aravindan, V., Lee, Y. S., Yazami, R., & Madhavi, S. (2015). TiO_2 polymorphs in "rocking-chair" Li-ion batteries. *Materials Today, 18*(6), 345–351.

Archer, P. I., Radovanovic, P. V., Heald, S. M., & Gamelin, D. R. (2005). Low-temperature activation and deactivation of high-Curie-temperature ferromagnetism in a new diluted magnetic semiconductor: Ni^{2+}-doped SnO_2. *Journal of the American Chemical Society, 127*, 14479–14487.

Bae, C., Yo, H., Kim, S., Lee, K., Kim, J., Sung, M. M., & Shin, H. (2008). Template-directed synthesis of oxide nanotubes: Fabrication, characterization, and applications. *Chemistry of Materials, 20*, 756–767.

Bai, Y., Mora-Seró, I., De Angelis, F., Bisquert, J., & Wang, P. (2014). Titanium dioxide nanomaterials for photovoltaic applications. *Chemical Reviews, 114*(19), 10095–10130.

Bartolome, J., Cremades, A., & Piqueras, J. (2013). Thermal growth, luminescence and whispering gallery resonance modes of indium oxide microrods and microcrystals. *Journal of Materials Chemistry C, 1*, 6790–6799.

Bartolomé, J., Cremades, A., & Piqueras, J. (2015). High quality factor indium oxide mechanical microresonators. *Appl. Phys. Lett., 107*, 191910.

Bartolomé, J., Hidalgo, P., Maestre, D., Cremades, A., & Piqueras, J. (2014). In-situ scanning electron microscopy and atomic force microscopy Young's modulus determination of indium oxide microrods for micromechanical resonator applications. *Applied Physics Letters, 104*, 161909.

Bartolomé, J., Maestre, D., Cremades, A., Amatti, M., & Piqueras, J. (2013). Composition-dependent electronic properties of indium–Zinc–Oxide elongated microstructures. *Acta Materialia, 61*, 1932–1943.

Benavides, J. A., Trudeau, C. P., Gerlein, L. F., & Cloutier, S.G. (2018). Laser selective photoactivation of amorphous TiO_2 films to anatase and/or rutile crystalline phases. *ACS Applied Energy Material, 1*(8), 3607–3613.

Bierwagen, O. (2015). Indium oxide – A transparent, wide-band gap semiconductor for (opto)electronic applications. *Semiconductor Science and Technology, 30*, 024001.

Bierwagen, O., & Speck, J. S. (2010). High electron mobility In_2O_3(001) and (111) thin films with nondegenerate electron concentration. *Applied Physics Letters, 97*, 072103.

Bobkov, A., Varezhnikov, A., Plugin, I., Fedorov, F. S., Trouillet, V., Geckle, U., … Sysoev, V. (2019). The multisensor array based on grown-on-chip Zinc Oxide Nanorod Network for selective discrimination of alcohol vapors at sub-ppm range. *Sensors, 19*, 4265.

Bolzan, A. A., Fong, C., Kennedy, B. J., & Howard, C. J. (1997). Structural studies of rutile-type metal dioxides. *Acta Crystallographica Section B, 53*, 373–380.

Born, M., & Wolf, E. (1980). *Principles of optics* (6th ed.). Oxford: Pergamon.

Bourlange, A., Payne, D. J., Egdell, R. G., Foord, J. S., Edwards, P. P., Jones, M. O., … Hutchison, J. L. (2008). Growth of In_2O_3(100) on Y-stabilized ZrO_2(100) by O-plasma assisted molecular beam epitaxy. *Applied Physics Letters, 92*, 092117.

Bueno, C., Maestre, D., Díaz, T., Pacio, M., & Cremades, A. (2018). Fabrication of ZnO-TiO_2 axial micro-heterostructures by a vapor-solid method. *Materials Letters, 220*, 156–160.

Caetano, B. L., Meneau, F., Santilli, C. V., Pulcinelli, S. H., Magnani, M., & Briois, V. (2014). Mechanisms of SnO_2 nanoparticles formation and growth in acid ethanol solution derived from SAXS and combined Raman–XAS time-resolved studies. *Chemistry of Materials, 26*, 6777–6785.

Capretti, A., Wang, Y., Engheta, N., & Dal Negro, L. (2015). Comparative study of second-harmonic generation from epsilon-near-zero indium tin oxide and titanium nitride nanolayers excited in the near-infrared spectral range. *ACS Photonics, 2*, 1584–1591.

Capretti, A., Wang, Y., Engheta, N., & Negro, L. D. (2015). Enhanced third-harmonic generation in Si-compatible epsilon-near-zero indium tin oxide nanolayers. *Optics Letters*, *40*, 1500–1503.

Chen, J. S., & Lou, X. W. (2012). SnO_2 and TiO_2 nanosheets for lithium-ion batteries. *Materials Today*, *15*, 246–254.

Chen, J. S., & Xiong, W. L. (2013). SnO_2-based nanomaterials: Synthesis and application in lithium-ion batteries. *Small*, *9*, 1877–1893.

Chen, S. S., & Qin, X. (2014). Tin oxide-titanium oxide/graphene composited as anode materials for lithium-ion batteries. *Journal of Solid State Electrochemistry*, *18*, 2893–2902.

Chien, C. T., Li, S. S., Lai, W. J., Yeh, Y. C., Chen, H. A., Chen, I. S., … Chen, C. W. (2012). Tunable photoluminescence from graphene oxide. *Angewandte Chemie International Edition*, *51*(27), 6662–6666.

Choi, S.-W., Akash, K., Sun, G.-J., Wu, P., & Kim, S. S. (2013). NO_2-sensing performance of SnO_2 microrods by functionalization of Ag nanoparticles. *Journal of Materials Chemistry C*, *1*, 2834–2841.

Choi, Y.-J., Hwang, I. S., Park, J. G., & Lee, J. H. (2008). Novel fabrication of an SnO_2 nanowire gas sensor with high sensitivity. *Nanotechnology*, *19*, 095508.

Clark, H. A., Hoyer, M., Philbert, M. A., & Kopelman, R. (1999). Optical nanosensors for chemical analysis inside single living cells. 1. Fabrication, characterization, and methods for intracellular delivery of PEBBLE sensors. *Analytical Chemistry*, *71*(21), 4831–4836.

Cremades, A., Herrera, M., Bartolomé, J., Vásquez, G. C., Maestre, D., & Piqueras, J. (2014). On the thermal growth and properties of doped TiO_2 and In_2O_3 elongated nanostructures and nanoplates. *Physica B- Condensed Matter*, *453*, 92–99.

Critical Raw Materials. 2017. Retrieved from http://criticalrawmaterials.org/european-commission-publishes-new-critical-raw-materials-list-27-crms-confirmed/.

Cuong, T. V., Pham, V. H., Tran, Q. T., Hahn, S. H., Chung, J. S., Shin, E. W., & Kim, E. J. (2010). Photoluminescence and Raman studies of graphene thin films prepared by reduction of graphene oxide. *Materials Letters*, *64*, 399–401.

Dahl, M., Liu, Y., & Yin, Y. (2014). Composite titanium dioxide nanomaterials. *Chemical Reviews*, *114* (19), 9853–9889.

Das, S., and Jayaraman, V. (2014). SnO_2: A comprehensive review on structures and gas sensors. *Progress in Materials Science*, *66*, 112–255.

De Wit, J. H. W. (1973). Electrical properties of In_2O_3. *Journal of Solid State Chemistry*, *8*, 142–149.

Del Prado, F., Cremades, A., Maestre, D., Ramírez-Castellanos, J., González-Calbet, J. M., & Piqueras, J. (2018). Controlled synthesis of lithium doped tin dioxide nanoparticles by a polymeric precursor method and analysis of the resulting defect structure. *Journal of Materials Chemistry A*, *6*, 6299–6308.

Del Prado, F., Cremades, A., Ramírez-Castellanos, J., Maestre, D., González-Calbet, J. M., & Piqueras, J. (2017). Effect of lithium doping and precursors on the microstructural, surface electronic and luminescence properties of single crystalline microtubular tin oxide structures. *CrysEngComm*, *19*, 4321–4329.

Del Prado, F., Taeño, M., Maestre, D., Ramírez-Castellanos, J., González-Calbet, J. M., & Cremades, A. (2019). Effect of the synthesis method on the properties of lithium doped graphene oxide composites with tin oxide nanoparticles: Towards white luminescence. *Journal of Physics and Chemistry of Solids*, *129*, 133–139.

del-Castillo, J., Rodríguez, V. D., Yanes, A. C., & Méndez-Ramos, J. (2008). Energy transfer from the host to Er^{3+} dopants in semiconductor SnO_2 nanocrystals segregated in sol–Gel silica glasses. *Journal of Nanoparticle Research*, *10*, 499–506.

Demirci, S., Dikici, T., Yurddaskal, M., Gultekin, S., Toparli, M., & Celik, E. (2016). Synthesis and characterization of Ag doped TiO_2 heterojunction films and their photocatalytic performances. *Applied Surface Science*, *390*, 591–601.

Deng, Y., Fang, C., & Chen, G. (2016). The developments of SnO_2/graphene nanocomposites as anode materials for high performance lithium ion batteries: A review. *Journal of Power Sources*, *304*, 81–101.

DG Connect. 2014. A European industrial strategic roadmap for micro- and nano-electronic components and systems. A report to Vice President Kroes by the Electronic Leaders Group. Retrieved from https://ec.europa.eu/digital-single-market/en/news/european-industrial-strategic-roadmap-micro-and-nano-electronic-components-and-systems.

Dong, H., Chen, Z., Sun, L., Lu, J., Xie, W., Tan, H. H., … Shen, X. (2009). Whispering gallery modes in indium oxide hexagonal microcavities. *Applied Physics Letters*, *94*, 173115.

Dong, H., Sun, L., Sun, S., Xie, W., Zhou, L., Shen, X., & Chen, Z. (2010). Indium oxide octahedra optical microcavities. *Applied Physics. Letters, 97*, 223114.

Drevillon, B., Kumar, S., & Cabarrocas, P. R. I. (1989). In-situ investigation of the optoelectronic properties of transparent conducting oxide amorphous silicon interfaces. *Applied Physics Letters, 54*, 2088–2090.

Du, N., Zhang, H., Chen, B. D., Ma, X. Y., Liu, Z. H., Wu, J. B., & Yang, D.R. (2007). Porous indium oxide nanotubes: Layer-by-layer assembly on carbon-nanotube templates and application for room-temperature NH_3 gas sensors. *Advanced Materials, 19*, 1641–1645.

Du, N., Zhang, H., Chen, J., Sun, J., Chen, B., & Yang, D. (2008). Metal oxide and sulfide hollow spheres: Layer-by-layer synthesis and their application in Lithium-Ion battery. *Journal of Physical Chemistry B, 112*, 14836–14842.

Eda, G., Fanchini, G., & Chhowalla, M. (2008). Large-area ultrathin films of reduced graphene oxide as a transparent and flexible electronic material. *Nature Nanotechnology, 3*, 270.

Eda, G., Lin, Y. Y., Mattevi, C., Yamaguchi, H., Chen, H. A., Chen, I. S., … Chhowalla, M. (2010). Blue photoluminescence from chemically derived graphene oxide. *Advanced Materials, 22*, 505–509.

Elouali, S., Bloor, L. G., Binions, R., Parkin, I. V., Carmalt, C. J., & Darr, J. A. (2012). Gas sensing with nano-indium oxides (In_2O_3) prepared via continuous hydrothermal flow synthesis. *Langmuir, 28*, 1879–1885.

Eom, K., Park, H. S., Yoon, D. S., & Kwon, T. (2011). Nanomechanical resonators and their applications in biological/chemical detection: Nanomechanics principles. *Physics Reports, 503*, 115–163.

Etacheri, V., Marom, R., Elazari, R., Salitra, G., & Aurbach, D. (2011). Challenges in the development of advanced Li-ion batteries: A review. *Energy and Environmental Science, 4*, 3243–3262.

Fakharuddin, A., Jose, R., Brown, T. M., Fabregat-Santiago, F., & Bisquert, J. (2014). A perspective on the production of dye-sensitized solar modules. *Energy and Environmental Science, 7*, 3952–3981.

Fan, X., White, I. M., Shopova, S. I., Zhu, H., Suter, J. D., & Sun, J. (2011). Sensitive optical biosensors for unlabeled targets: A review. *Analytica Chimica Acta, 620*, 8–26.

Feng, C., Li, X., Ma, J., Sun, Y., Wang, C., Sun, P., … Lu, G. (2015). Facile synthesis and gas sensing properties of In_2O_3–WO_3 heterojunction nanofibers. *Sensors and Actuators B: Chemical, 209*, 622–629.

Fortunato, E., Ginley, D., Hosono, H., & Paine, D. C. (2007). Transparent conducting oxides for photovoltaics. *MRS Bulletin, 32*, 242–247.

Freitas, R. A. (2005). Nanotechnology, nanomedicine and nanosurgery. *International Journal of Surgery, 3*(4), 243–246.

Gallagher, K. G., Goebel, S., Greszler, T., Mathias, M., Oelerich, W., Eroglu, D., & Srinivasan, V. (2014). Quantifying the promise of lithium – Air batteries for electric vehicles. *Energy and Environmental Science, 7*, 1555–1563.

Gao, J., Wu, H., Zhou, J., Yao, L., Zhang, G., Xu, S., … Shi, K. (2016). Mesoporous In_2O_3 nanocrystals: Synthesis, characterization and NO_x gas sensor at room temperature. *New Journal of Chemistry, 40*, 1306–1311.

Gao, Y., Wang, S., Kang, L., Chen, Z., Du, J., Liu, X., Luo, H., & Kanehira, M. (2012). VO_2–Sb: SnO_2 composite thermochromic smart glass foil. *Energy and Environmental Science, 5*, 8234–8237.

García-Tecedor, M., Bartolomé, J., Maestre, D., Trampert, A., & Cremades, A. (2019). Li_2SnO_3 branched nano- and microstructures with intense and broadband white-light emission. *NanoResearch, 12*(2), 441–448.

García-Tecedor, M., Cardenas-Morcoso, D., Fernández-Climent, R., & Giménez, S. (2019). The role of underlayers and overlayers in thin film $BiVO_4$ photoanodes for solar water splitting. *Advanced Materials Interfaces, 0*, 1900299.

García-Tecedor, M., Del Prado, F., Bueno, C., Vásquez, G. C., Bartolomé, J., Maestre, D., … Piqueras, J. (2016). Tubular micro- and nanostructures of TCO materials grown by a vapor-solid method. *AIMS Materials Science, 3*(2), 434–447.

García-Tecedor, M., Karazhanov, S. Z., Vasquez, G. C., Haug, H., Maestre, D., Cremades, A., … Marstein, E.S. (2017). Silicon surface passivation by PEDOT: PSS functionalized by SnO_2 and TiO_2 nanoparticles. *Nanotechnology, 29*(3), 035401.

Garcia-Tecedor, M., Maestre, D., Cremades, A., & Piqueras, J. (2016). Growth and characterization of Cr doped SnO_2 microtubes with resonant cavity modes. *Journal of Materials Chemistry C, 4*, 5709–5716.

Garcia-Tecedor, M., Maestre, D., Cremades, A., & Piqueras, J. (2017). Tailoring optical resonant cavity modes in SnO_2 microstructures through doping and shape engineering. *Journal of Physics D: Applied Physics*, *50*, 415104.

Geng, X., Niu, L., Xing, Z., Song, R., Liu, G., Sun, M., … Liu, L. (2010). Aqueous-processable noncovalent chemically converted graphene–quantum dot composites for flexible and transparent optoelectronic films. *Advanced Materials*, *22*, 638–642.

Ghasemi, A., Amiri, H., Zare, H., Masroor, M., Hasanzadeh, A., Beyzavi, A., … Hamblin, M. R. (2017). Carbon nanotubes in microfluidic lab-on-a-chip technology: Current trends and future perspectives. *Microfluidics and Nanofluidics*, *21*(9), 156.

Girishkumar, G., McCloskey, B., Luntz, A. C., Swanson, S., & Wilcke, W. (2010). Lithium-air battery: Promise and challenges. *The Journal of Physical Chemistry Letters*, *1*, 2193–2203.

Goodenough, J. B., & Park, K. S. (2013). The Li-Ion rechargeable battery: A perspective. *Journal of the American Chemical Society*, *135*, 1167–1176.

Gopel, W. (1985). Chemisorption and charge-transfer at Ionic semiconductor surfaces – Implications in designing gas sensors. *Progress in Surface Science*, *20*, 9–103.

Goriparti, S., Miele, E., De Angelis, F., Di Fabrizio, E., Zaccaria, R. P., & Capiglia, C. (2014). Review on recent progress of nanostructured anode materials for Li-ion batteries. *Journal of Power Sources*, *257*, 421–443.

Granqvist, C. G. (2007). Transparent conductors as solar energy materials: A panoramic review. *Solar Energy Materials and Solar Cells*, *91*, 1529–1598.

Greaney, M. J., & Brutchey, R. L. (2015). Ligand engineering in hybrid polymer: Nanocrystal solar cells. *Materials Today*, *18*(1), 31.

Gu, F., Nie, R., Han, D., & Wang, Z. (2015). In_2O_3–graphene nanocomposite based gas sensor for selective detection of NO_2 at room temperature. *Sensors and Actuators B*, *219*, 94–99.

Guo, P., Chang, R. P. H., & Schaller, R. D. (2017). Transient negative optical nonlinearity of indium oxide nanorod arrays in the full-visible range. *ACS Photonics*, *4*, 1494–1500.

Guo, P., Schaller, R. D., Ketterson, J. B., & Chang, R. P. H. (2016). Ultrafast switching of tunable infrared plasmons in indium tin oxide nanorod arrays with large absolute amplitude. *Nature Photonics.*, *10*, 267–273.

Hagen, W., Lambrich, R. E., Lagois, J. (1983). Semiconducting gas sensors. In P. Grosse (eds) *Festkörperprobleme 23. Advances in Solid State Physics* (Vol 23, pp. 254–256). Berlin, Heidelberg: Springer.

Hagiwara, M., Nakagawa, T., Fukatani, J., Yoshioka, T. & Hatta, B. Tin-doped indium oxide microparticle dispersion, process for producing the same, interlayer for laminated glass having heat-ray blocking property produced with the dispersion, and laminated glass, pp. 1–33, United States Patent Application 20060225614.

Hamberg, I., & Granqvist, C. G. (1986). Evaporated Sn-doped In_2O_3 films: Basic optical properties and applications to energy-efficient windows. *Journal of Applied Physics*, *60*, R123.

Hanaor, D. A. H., & Sorrell, C. C. (2011). Review of the anatase to rutile phase transformation. *Journal of Materials Science*, *46*(4), 855–874.

Hofmann, A. I., Cloutet, E., & Hadziioannou, G. (2018). Materials for transparent electrodes: From metal oxides to organic alternatives. *Advanced Electronic Materials*, *4*, 1700412.

Huang, G., Bolanos Quinones, V. A., Ding, F., Kiravittaya, S., Mei, Y., & Schmidt, O. G. (2010). Rolled-up optical microcavities with subwavelength wall thicknesses for enhanced liquid sensing applications. *ACS Nano*, *4*, 3123–3130.

Huang, H., Lee, Y. C., Tan, O. K., Zhou, W., Peng, N., & Zhang, Q. (2009). High sensitivity SnO_2 single-nanorod sensors for the detection of H_2 gas at low temperature. *Nanotechnology*, *20*, 115501.

Huang, H., Ng, M., Wu, Y., & Kong, L. (2015). Solvothermal synthesis of Sb: SnO_2 nanoparticles and IR shielding coating for smart window. *Materials and Design*, *88*, 384–389.

Huang, H., Tian, S., Xu, J., Xie, Z., Zeng, D., Chen, D., & Shen, G. (2012). Needle-like Zn-doped SnO_2 nanorods with enhanced photocatalytic and gas sensing properties. *Nanotechnology*, *23*, 105502.

Huang, X., Boey, F., & Zhang, H. (2010). A brief review on graphene-nanoparticle composites. *COSMOS*, *6(2)*, 159–166.

Huang, X.-J., & Choi, Y.-K. (2007). Chemical sensors based on nanostructured materials. *Sensors and Actuators B*, *122*, 659–671.

Hung, C. M., Le, D. T. T., & Van Hieu, N. (2017). On-chip growth of semiconductor metal oxide nanowires for gas sensors: A review. *Journal of Science: Advanced Materials and Devices*, *2*, 263–285.

Idota, Y., Kubota, T., Matsufuji, A., Maekawa, Y., & Miyasaka, T. (1997). Tin-based amorphous oxide: A high-capacity lithium-ion-storage material. *Science, 276*, 1395–1397.

ITRS, International Technology Roadmap for Semiconductors. 2015 Edition. MORE MOORE. Retrieved from www.semiconductors.org/wp-content/uploads/2018/06/5_2015-ITRS-2.0_More-Moore.pdf.

Jiang, B., Zuo, J. M., Jiang, N., O'Keeffe, M. & Spence, J. (2003). Charge density and chemical bonding in rutile, TiO_2. *Acta Crystallographica Section A: Foundations of Crystallography, 59*(4), 341–350.

Jiang, K., Sun, S., Zhang, L., Lu, Y., Wu, A., Cai, C., & Lin, H. (2015). Red, green, and blue luminescence by carbon dots: Full-color emission tuning and multicolor cellular imaging. *Angewandte Chemie International Edition, 54*, 5360–5363.

Joseph, D. P., Renugambal, P., Saravanan, M., Raja, S. P., & Venkateswaran, C. (2009). Effect of Li doping on the structural, optical and electrical properties of spray deposited SnO_2 thin films. *Thin Solid Films, 517*, 6129–6136.

Joshi, N., Hayasaka, T., Liu, Y., Liu, H., Oliveira, O. N., & Lin, L. (2018). A review on chemiresistive room temperature gas sensors based on metal oxide nanostructures, graphene and 2D transition metal dichalcogenides. *Sensors and Actuators B: Chemical, 185*, 213.

Kaplan, R., Erjavec, B., Dražić, G., Grdadolnik, J., & Pintar, A. (2016). Simple synthesis of anatase/rutile/brookite TiO_2 nanocomposite with superior mineralization potential for photocatalytic degradation of water pollutants. *Applied Catalysis B: Environmental, 181*, 465–474.

Kar, A., & Patra, A. (2009). Optical and electrical properties of Eu^{3+}-doped SnO_2 nanocrystals. *Journal of Physical Chemistry C, 113*, 4375–4380.

Karazhanov, S. Z., Ravindran, P., Vajeeston, P., Ulyashin, A., Finstad, T. G., & Fjellvåg, H. (2007). Phase stability, electronic structure, and optical properties of indium oxide polytypes. *Physical Review B., 76*, 075129.

Kılıç, Ç., & Zunger, A. (2002). Origins of coexistence of conductivity and transparency in SnO_2. *Physical Review Letters, 88*, 095501.

Kim, H., Kim, S. W., Park, Y. U., Gwon, H., Seo, D-H., Kim, Y., & Kang, K. (2010). SnO_2/Graphene composite with high lithium storage capability for lithium rechargeable batteries. *Nano Research, 3*, 813–821.

Kim, T. H., Park, J. S., Chang, S. K., Choi, S., Ryu, J.H., & Song, H-K. (2012). The current move of Lithium Ion batteries towards the next phase. *Advanced Energy Materials, 2*, 860–872.

King, P. D. C., Veal, T. D., Fuchs, F., Wang, Ch. Y., Payne, D. J., Bourlange, A., … MacConville, C. F. (2009). Band gap, electronic structure, and surface electron accumulation of cubic and rhombohedral In_2O_3. *Physical Review B, 79*, 205211.

King, P. D. C., Veal, T. D., Payne, D. J., Bourlange, A., Egdell, R. G., & MacConville, C. F. (2008). Surface electron accumulation and the charge neutrality level in In_2O_3. *Phyical. Review Letters, 101*, 116808.

Klein, A., Korber, C., Wachau, A., Säuberlich, F., Gassenbauer, Y., Harvey, S. P., … Mason, T.O. (2010). Transparent conducting oxides for photovoltaics: Manipulation of fermi level, work function and energy band alignment. *Materials (Basel), 3*, 4892–4914.

Korotcenkov, G., Brinzari, V., & Cho, B. K. (2016). In_2O_3- and SnO_2-based thin film ozone sensors: Fundamentals. *Journal of Sensors., 2016*, 1–31.

Korotcenkov, G., Brinzari, V., & Cho, B. K. (2018). In_2O_3- and SnO_2-based ozone sensors: Design and characterization. *Critical Review in Solid State and Materials Sciences, 43*, 83–132.

Korotcenkov, G., Brinzari, V., Golovanov, V., Cerneavschi, A., Matolin, V., & Tadd, A. (2004). Acceptor-like behavior of reducing gases on the surface of n-type In_2O_3. *Applied Surface Science., 227*, 122–131.

Kumar, D. P., Kumari, V. D., Karthik, M., Sathish, M., & Shankar, M.V. (2017). Shape dependence structural, optical and photocatalytic properties of TiO_2 nanocrystals for enhanced hydrogen production via glycerol reforming. *Solar Energy Materials and Solar Cells, 163*, 113–119.

Kumar, V., Govind, A., & Nagarajan, R. (2011). Optical and photocatalytic properties of heavily F(-)-doped SnO_2 nanocrystals by a novel single-source precursor approach. *Inorganic Chemistry, 50*, 5637–5645.

Lany, S., Zakutayev, A., Mason, T. O., Wager, J. F., Poeppelmeier, K. R., Perkins, J. D., … Zunger, A. (2012). Surface origin of high conductivities in undoped In_2O_3 thin films. *Physical Review Letters, 108*, 016802.

Lany, S., & Zunger, A. (2007). Dopability, intrinsic conductivity, and nonstoichiometry of transparent conducting oxides. *Physical Review Letters, 98*, 045501.

Lavoie, Y., Danet, F., & Lombard, B. (2017). Lithium-ion batteries for industrial applications. *2017 Petroleum and Chemical Industry Technical Conference (PCIC)* (pp. 283–290). IEEE publisher.

Lee, J. S., Sim, S. K., Min, B., Cho, K., Kim, S-W., & Kim, S. (2004). Structural and optoelectronic properties of SnO_2 nanowires synthesized from ball-milled SnO_2 powders. *Journal of Crystal Growth, 267*, 145–149.

Leite, E. R., Weber, I. T., Longo, E., & Varela, J.A. (2000). A new method to control particle size and particle size distribution of SnO_2 nanoparticles for gas sensor applications. *Advanced Materials, 12*, 965–968.

Leung, S. F., Ho, K. T., Kung, P. K., Hsiao, V. K. S., Alshareef, H. N., Wang, Z. L., & He, J.-H. (2018). A self-powered and flexible organometallic halide perovskite photodetector with very high detectivity. *Advanced Materials, 30*, 1704611.

Li, C., Zhang, D., Liu, X., Han, S., Tang, T., Han, J., & Zhou, C. (2003). In_2O_3 nanowires as chemical sensors. *Applied Physics Letters, 82*, 1613–1615.

Li, J. T., Cushing, S. K., Zheng, P., Senty, T., Meng, F., Bristow, A. D., … Wu, N. (2014). Solar hydrogen generation by a CdS-Au-TiO_2 sandwich nanorod array enhanced with Au nanoparticle as electron relay and plasmonic photosentitizer. *Journal of the American Chemical Society, 136*(23), 8438–8449.

Li, N., Liu, G., Zhen, C., Li, F., Zhang, L., & Cheng, H-M. (2011). Battery performance and photocatalytic activity of mesoporous anatase TiO_2 nanospheres/graphene composites by template-free self-assembly. *Advanced. Functional Materials, 21*(9), 1717–1722.

Li, Y. M., Lv, X. J., Lu, J., & Li, J. (2010). Preparation of SnO_2-nanocrystal/graphene-nanosheets composites and their lithium storage ability. *Journal of Physical Chemistry C, 114*, 21770–21774.

Li, Z. D., Zhou, Y., Yu, T., Liu, J., & Zou, Z. (2012). Unique Zn-doped SnO_2 nano-echinus with excellent electron transport and light harvesting properties as photoanode materials for high performance dye-sensitized solar cell. *Crys Eng Comm, 14*, 6462–6468.

Lian, P. C., Zhu, X. F., Liang, S. Z., Li, Z., Yang, W., & Wang, H. (2011). High reversible capacity of SnO_2/graphene nanocomposite as an anode material for lithium-ion batteries. *Electrochimica Acta, 56*, 4532–4539.

Liang, X., Kim, T.-H., Yoon, J.-W., Kwak, C-H., & Lee, J-H. (2015). Ultrasensitive and ultraselective detection of H_2S using electrospun CuO-loaded In_2O_3 nanofiber sensors assisted by pulse heating. *Sensors and Actuators B, 209*, 934–942.

Liang, Y., Tsubota, T., Mooij, L. P. A., & van de Krol, R. (2011). Highly improved quantum efficiencies for thin film $BiVO_4$ photoanodes. *The Journal of Physical Chemistry C, 115*, 17594–17598.

Liberal, I., & Engheta, N. (2017). Near-zero refractive index photonics. *Nature Photonics, 11*, 149–158.

Lin, Z., Li, N., Chen, Z., & Fu, P. (2017). The effect of Ni doping concentration on the gas sensing properties of Ni doped SnO_2. *Sensors and Actuators B: Chemical, 239*, 501–510.

Liu, D., Li, S., Zhang, P., Wang, Y., Zhang, R., Sarvari, H., … Chen, Z. D. (2017). Efficient planar heterojunction perovskite solar cells with Li-doped compact TiO_2 layer. *Nano Energy, 31*, 462–468.

Liu, Q., Aroonyadet, N., Song, Y., Wang, X., Cao, X., Liu, Y., … Zhou, C. (2016). Highly sensitive and quick detection of acute myocardial infarction biomarkers using In_2O_3 nanoribbon biosensors fabricated using shadow masks. *ACS Nano, 10*, 10117–10125.

Loh, K. P., Bao, Q. L., Eda, G., & Chhowalla, M. (2010). Graphene oxide as a chemically tunable platform for optical applications. *Nature Chemistry, 2*, 1015–1024.

Longoni, G., Pena Cabrera, R. L., Polizzi, S., D'Arienzo, M., Mari, C. M., Cui, Y., & Ruffo, R. (2017). Shape-controlled TiO_2 nanocrystals for Na-Ion battery electrodes: The role of different exposed crystal facets on the electrochemical properties. *Nano Letters, 17*(2), 992–1000.

Lopez, I., Nogales, E., Mendez, B., & Piqueras, J. (2012). Resonant cavity modes in gallium oxide microwires. *Applied Physics Letters, 100*, 261910.

Luo, Z. T., Vora, P. M., Mele, E. J., Johnson, A. T. C., & Kikkawa, J. M. (2009). Photoluminescence and band gap modulation in graphene oxide. *Applied Physics Letters, 94*, 111909.

Maestre, D., Cremades, A., & Piqueras, J. (2005). Growth and luminescence properties of micro- and nanotubes in sintered tin oxide. *Journal of Applied Physics, 97*, 044316.

Maestre, D., Haussler, D., Cremades, A., Jäger, W., & Piqueras, J. (2011). Nanopipes in In_2O_3 nanorods grown by a thermal treatment. *Crystal Growth and Design, 11*(4), 1117–1121.

Mali, J. M., Arbuj, S. S., Ambekar, J. D., Jalindar. D., Rane, S. B., Mulik, U. P., & Amalnerkar, D. P. (2013). Synthesis of SnO_2 nano rods and their photocatalytic properties. *Journal of Nanoengineering and Nanomanufacturing*, *3*, 121–125.

Marchesan, S., Kostarelos, K., Bianco, A., & Prato, M. (2015). The winding road for carbon nanotubes in nanomedicine. *Materials Today*, *18*(1), 12–19.

Marom, R., Amalraj, S. F., Leifer, N., Jacob, D., & Aurbach, D. (2011). A review of advanced and practical lithium battery materials. *Journal of Materials Chemistry*, *21*, 9938–9954.

Mathkar, A., Tozier, D., Cox, P., Ong, P., Galande, C., Balakrishnan, K., ... Ajayan, P. M. (2012). Controlled, stepwise reduction and band gap manipulation of graphene oxide. *The Journal of Physical Chemistry Letters*, *3*, 986–991.

Mei, Q., Zhang, K., Guan, G., Liu, B., Wang, S., & Zhang, Z. (2010). Highly efficient photoluminescent graphene oxide with tunable surface properties. *Chemical Communications*, *46*, 7319–7321.

Momma, K., & Izumi, F. (2011). VESTA 3 for three-dimensional visualization of crystal, volumetric and morphology data. *Journal of Applied Crystallography*, *44*, 1272–1276.

Moseley, P. T. (2017). Progress in the development of semiconducting metal oxide gas sensors: A review. *Measurement Science and Technology*, *28*, 082001.

Neri, G., Bonavita, A., Micali, G., Rizzo, G., Galvagno, S., Niederberger, M., & Pinna, N. (2005). A highly sensitive oxygen sensor operating at room temperature based on platinum-doped In_2O_3 nanocrystals. *Chemical Communications*, *48*, 6032–6034.

Neri, G., Bonavita, A., Micali, G., Rizzo, G., Pinna, N., & Niederberger, M. (2007). In_2O_3 and Pt-In_2O_3 nanopowders for low temperature oxygen sensors. *Sensors and Actuators B: Chemical*, *127*, 455–462.

Nitta, N., & Yushin, G. (2014). High-capacity anode materials for Lithium-Ion batteries: Choice of elements and structures for active particles. *Particle and Particle Systems Characterization*, *31*, 317–336.

Park, P. W., Kung, H. H., Kim, D. W., & Kung, M. C. (1999). Characterization of SnO_2/Al_2O_3 lean NO_x catalysts. *Journal of Catalysis*, *184*, 440–454.

Park, S., Kim, S., Sun, G.-J., & Lee, C. (2015). Synthesis, structure, and ethanol gas sensing properties of In_2O_3 nanorods decorated with Bi_2O_3 nanoparticles. *ACS Applied Materials and Interfaces.*, *7*, 8138–8146.

Patra, D., Sengupta, S., Duan, W., Zhang, H., Pavlick, R., & Sen. A. (2013). Intelligent, self-powered, drug delivery systems. *Nanoscale*, *5*(4), 1273–1283.

Rai, P., Oh, S., Shyamkumar, P., Ramasamy, M., Harbaugh, R. E., & Varadan, V. K. (2014). Nano- bio-textile sensors with mobile wireless platform for wearable health monitoring of neurological and cardiovascular disorders. *Journal of the Electrochemical Society*, *161*(2), B3116–B3150.

Ramgir, N. S., Yang, Y., & Zacharias, M. (2010). Nanowire-based sensors. *Small*, *6*, 1705–1722.

Ramos Ramón, J. A., Cremades, A., Maestre, D., Silva González, R., & Pal, U. (2017). Fabricating necklace-, tower-, and rod-shaped In_2O_3 nanostructures by controlling saturation kinetics of catalyst droplets in a vapor–liquid–solid process. *Crystal Growth and Design*, *17*(9), 4596–4602.

Rombach, J., Papadogianni, A., Mischo, M., Cimalla, V., Kirste, L., Ambacher, O., ... Bierwagen, O. (2016). The role of surface electron accumulation and bulk doping for gas-sensing explored with single-crystalline In_2O_3 thin films. *Sensors and Actuators B: Chemical*, *236*, 909–916.

Rong, G., Tuttle, E. E., Reilly, A. N., & Clark, H. A. (2019). Recent developments in nanosensors for imaging applications in biological systems. *Annual Review of Analytical Chemistry*, *12*, 109–128.

Rout, C. S., Ganesh, K., Govindaraj, A., & Rao, C. N. R. (2006). Sensors for the nitrogen oxides, NO_2, NO and N_2O, based on In_2O_3 and WO_3 nanowires. *Applied Physics A*, *85*, 241–246.

Sagadevan, S., Chowdhury, Z. Z., Johan, M. R. B., Aziz, F. A., Roselin, L. S., Podder, J., ... Selvin, R. (2019). Cu-doped SnO_2 nanoparticles: Synthesis and properties. *Journal of Nanoscience and Nanotechnology*, *19*(11), 7139–7148.

Sberveglieri, G., Faglia, G., Groppelli, S., Nelli, P. & Perego, C. (1993). Oxygen gas-sensing properties of undoped and Li-doped SnO_2 thin-films. *Sensors and Actuators B:Chemical*, *13*, 117–120.

Scrosati, B., & Garche, J. (2010). Lithium batteries: Status, prospects and future. *Journal of Power Sources*, *195*, 2419–2430.

Shen, J., Shi, M., Yan, B., Ma, H., Li, N. & Ye, M. (2011). Ionic liquid-assisted one-step hydrothermal synthesis of TiO_2-reduced graphene oxide composites. *Nano Research*, *4*, 795.

Shogh, S., Mohammadpour, R., Zad, A. I., & Taghavinia, N. (2015). Improved photovoltaic performance of nanostructured solar cells by neodymium-doped TiO_2 photoelectrode. *Materials Letters*, *159*, 273–275.

Singh, N., Yan, C., & Lee, P. S. (2010). Room temperature CO gas sensing using Zn-doped In_2O_3 single nanowire field effect transistors. *Sensors and Actuators B: Chemical, 150*, 19–24.

Smith, A. M., & Nie, S. (2010). Semiconductor nanocrystals: Structure, properties, and band gap engineering. *Accounts of Chemical Research, 43*(2), 190–200.

Song, J., Kim, J., Kang, T., & Kim, D. (2017). Design of a porous cathode for ultrahigh performance of a Li-ion battery: An overlooked pore distribution. *Scientific Reports, 7*, 42521.

Song, Z., Wei, Z., Wang, B., Luo, Z., Xu, S., Zhang, W., … Liu, H. (2016). Sensitive room-temperature H_2S gas sensors employing SnO_2 quantum wire/reduced graphene oxide nanocomposites. *Chemistry of Materials, 28*, 1205–1212.

Stashans, A., Lunell, S., Bergstrom, R., Hagfeldt, A., & Lindquist, S. E. (1996). Theoretical study of lithium intercalation in rutile and anatase. *Physical Review B Condensed Matter, 53*, 159–170.

Štengl, V., Bakardjieva, S., Henych, J., Lang, K., Kormunda, M. (2013). Blue and green luminescence of reduced graphene oxide quantum dots. *Carbon, 63*, 537–546.

Suematsu, K., Shin, Y., Hua, Z., Yoshida, K., Yuasa, M., Kida, T., & Shimanoe, K. (2014). Nanoparticle cluster gas sensor: Controlled clustering of SnO_2 nanoparticles for highly sensitive toluene detection. *ACS Applied Materials and Interfaces, 6*, 5319–5326.

Tran, V. A., Truong, T. T., Phan, T. A. P., Nguyen, T. N., Huynh, T. V., Agresti, A., … Nguyen, P. T. (2017). Application of nitrogen-doped TiO_2 nano-tubes in dye-sensitized solar cells. *Applied Surface Science, 399*, 515–522.

Tshabalala, Z. P., Motaung, D. E., Mhlongo, G. H., & Ntwaeaborwa, O. M. (2016). Facile synthesis of improved room temperature gas sensing properties of TiO_2 nanostructures: Effect of acid treatment. *Sensors and Actuators B: Chemical, 224*, 841–856.

Udayabhaskar, R., Mangalaraja, R. V., Pandiyarajan, T., Karthikeyan, B., Mansilla, H.D., & Contreras, D. (2017). Spectroscopic investigation on graphene-copper nanocomposites with strong UV emission and high catalytic activity. *Carbon, 124*, 256–262.

Usui, H., Yoshioka, S., Wasada, K., Shimizu, M., & Sakaguchi, H. (2015). Nb-doped rutile TiO_2: A potential anode material for Na-ion battery. *ACS Applied Materials and Interfaces, 7*(12), 6567–6573.

Vásquez, C., Peche-Herrero, A., Maestre, D., Cremades, A., Ramírez-Castellanos, J., González-Calbet, J. M., & Piqueras, J. (2013). Cr doped titania microtubes and microrods synthesized by a vapor-solid method. *CrystEngComm., 15*, 5490–5495.

Vásquez, G. C., Maestre, D., Cremades, A., Peche-Herrero, M.A., Ramírez-Castellanos, J., Piqueras, J. & González-Calbet, J.M. Dióxido de titanio nanocristalino con mezcla de fases anatasa y rutilo en proporción y/o distribución espacial controlada mediante irradiación láser. Patent. ES 2525737 B2.

Vásquez, G. C., Peche-Herrero, M. A., Maestre, D., Alemán, B., Ramírez-Castellanos, J., Cremades, A., … Piqueras, J. (2014). Influence of Fe and Al doping on the stabilization of the anatase phase in TiO_2 nanoparticles. *Journal of Materials Chemistry C, 2*(48), 10377–11185.

Vásquez, G. C., Peche-Herrero, M. A., Maestre, D., Gianoncelli, A., Ramírez-Castellanos, J., Cremades, A., … Piqueras, J. (2015). Laser-induced anatase-to-rutile transition in TiO_2 nanoparticles: Promotion and inhibition effects by Fe and Al doping and achievement of micropatterning. *The Journal of Physical Chemistry C, 119*(21), 11965–11974.

Verardo, D., Agnarsson, B., Zhdanov, V. P., Höök, F., & Linke, H. (2019). Single-molecule detection with lightguiding nanowires: Determination of protein concentration and diffusivity in supported lipid bilayer. *Nano Letters, 19*(9), 6182–6191.

Vomiero, A., Bianchi, S., Comini, E., Faglia, G., Ferroni, M., & Sberveglieri, G. (2007). Controlled growth and sensing properties of In_2O_3 nanowires. *Crystal Growth and Design, 7*, 2500–2504.

Walsh, A., Da Silva, J. L. F., & Wei, S. H. (2008). Origins of band-gap renormalization in degenerately doped semiconductors. *Physical Review B, 78*, 075211.

Walsh, A., Da Silva, J. L. F., Wei, S. H., Körber, C., Klein, A., Piper, L. F., … Egdell, R. G. (2008). Nature of the band gap of In_2O_3 revealed by first-principles calculations and X-ray spectroscopy. *Physical Review Letters, 100*, 167402.

Wang, C. Y., Bagchi, S., Bitterling, M., Becker, R. W., Köhler, K., Cimalla, V., … Chaumette, C. (2012). Photon stimulated ozone sensor based on indium oxide nanoparticles II: Ozone monitoring in humidity and water environments. *Sensors and Actuators B: Chemical, 164*, 37–42.

Wang, C. Y., Becker, R. W., Passow, T., Pletschen, W., Köhler, K., Cimalla, V., & Ambacher, O. (2011). Photon stimulated sensor based on indium oxide nanoparticles I: Wide-concentration-range ozone monitoring in air. *Sensors and Actuators B: Chemical*, *152*, 235–240.

Wang, Ch. Y., Cimalla, V., Kups, Th., Röhlig, C-C., Stauden, Th., & Ambacher, O. (2007). Integration of In_2O_3 nanoparticle based ozone sensors with GaInN/GaN light emitting diodes. *Applied Physics Letters*, *91*, 103509.

Wang, C. Y., Dai, Y., Pezoldt, J., Lu, B., Kups, Th., Cimalla, V., & Ambacher, O. (2008). Phase stabilization and phonon properties of single crystalline rhombohedral indium oxide. *Crystal Growth and Design*, *8*, 1257–1260.

Wang, D., Choi, D., Li, J., Yang, Z., Nie, Z., Kou, R., … Liu, J. (2009). Self-assembled TiO_2-graphene hybrid nanostructures for enhanced Li-Ion insertion. *ACS Nano*, *3*(4), 907–914.

Wang, W., Dong, J., Ye, X., Li, Y., Ma, Y., & Qi, L. (2016). Heterostructured TiO_2 Nanorod@Nanobowl arrays for efficient photoelectrochemical water splitting. *Small*, *12*(11), 1469–1478.

Wang, X., Lu, X., Liu, B., Chen, D., Tong, Y., & Shen, G. (2014). Flexible energy-storage devices: Design consideration and recent progress. *Advanced Materials*, *26*, 4763–4782.

Wang, Y., Lee, J. Y., & Zeng, H. C. (2005). Polycrystalline SnO_2 nanotubes prepared via infiltration casting of nanocrystallites and their electrochemical application. *Chemistry of Materials*, *17*, 3899–3903.

Wang, Z., Nayak, P. K., Caraveo-Frescas, J. A., & Alshareef, H.N. (2016). Recent developments in p-type oxide semiconductor materials and devices. *Advanced Materials.*, *28*, 3831–3892.

Wang, Z. L. (2003). Nanobelts, nanowires, and nanodiskettes of semiconducting oxides – From materials to nanodevices. *Advanced Materials*, *15*, 432–436.

Wang, Z. L. (2008). Towards self-powered nanosystems: From nanogenerators to nanopiezotronics. *Advanced Functional Materials*, *18*(22), 3553–3567.

Wen, J., Li, X., Liu, W., Fang, Y., Xie, J., & Xu, Y. (2015). Photocatalysis fundamentals and surface modification of TiO_2 nanomaterials. *Chinese Journal of Catalysis*, *36*(12), 2049–2070.

Wilkes, G. C., Deng, X., Choi, J. J., & Gupta, M. C. (2018). Laser annealing of TiO_2 electron-transporting layer in perovskite solar cells. *ACS Applied Materials and Interfaces*, *10*(48), 41312–41317.

Williams, G., & Kamat, P. V. (2009). Graphene-semiconductor nanocomposites: Excited-state interactions between ZnO nanoparticles and graphene oxide. *Langmuir*, *25*, 13869–13873.

Williams, G., Seger, B., & Kamat, P. V. (2008). TiO_2-graphene nanocomposites. UV-assisted photocatalytic reduction of graphene oxide. *ACS Nano*, *7*(2), 1487–1491.

Wu, C.-H., Chou, T.-L., & Wu, R.-J. (2018). Rapid detection of trace ozone in TiO_2–In_2O_3 materials by using the differential method. *Sensors and Actuators B: Chemical*, *255*, 117–124.

Wu, H. B., Chen, J. S., Hng, H. H., & Lou, X.W. D. (2012). Nanostructured metal oxide-based materials as advanced anodes for lithium-ion batteries. *Nanoscale*, *4*, 2526–2542.

Xing, R., Xu, L., Song, J., Zhou, C., Li, Q., Liu, D., & Song, H. W. (2015). Preparation and gas sensing properties of In_2O_3/Au nanorods for detection of volatile organic compounds in exhaled breath. *Scientific Reports*, *5*, 10717.

Xiong, L., Guo, Y., Wen, J., Liu, H., Yang, G., Qin, P., & Fang, G. (2018). Review on the application of SnO_2 in perovskite solar cells. *Advanced Functional Materials*, *28*, 1802757.

Xu, L., Zhang, X., Guo, D., Zhou, L., Pang, Q., Zhang, G., & Wang, S. (2018). Design and synthesis of p-n conversion indium-oxide-based gas sensor with high sensitivity to NO_x at room-temperature. *Chemistry Select*, *3*, 2298–2305.

Xu, S., Gao, J., Wang, L., Kan, K., Xie, Y., Shen, P., … Shi, K. (2015). Role of the heterojunctions in In_2O_3-composite SnO_2 nanorod sensors and their remarkable gas-sensing performance for NO_x at room temperature. *Nanoscale*, *7*, 14643–14651.

Yang, J., & Guo, L. J. (2006). Optical sensors based on active microcavities. *IEEE Journal of Selected Topics in Quantum Electronics*, *12*, 143–147.

Yang, W., Wan, P., Jia, M., Hu, Y., Guan, Y., & Feng, L. (2015). A novel electronic nose based on porous microtubes sensor arrays for the discrimination of VOCs. *Biosensors and Bioelectronics*, *64*, 547–553.

Yang, W., Wei, D., & Jin, X. (2016). Effect of the hydrogen bond in photoinduced water dissociation: A double-edge sword. *Journal of Physical Chemistry Letters*, *7*(4), 603–608.

Yin, P. T., Shah, S., Chhowalla, M., & Lee, K-B. (2015). Design, synthesis, and characterization of graphene–nanoparticle hybrid materials for bioapplications. *Chemical Reviews*, *115*, 2483–2531.

Yin, W., Cao, M., Luo, S., Hu, C., & Wei, B. (2009). Controllable synthesis of various In_2O_3 submicron nanostructures using chemical vapor deposition. *Crystal Growth and Design*, *9*, 2173–2178.

Ying, Z., Wan, Q., Song, Z. T., & Feng, S. L. (2004). SnO_2 nanowhiskers and their ethanol sensing characteristics. *Nanotechnology, 15*, 1682–1684.

Yoshida, M., & Lahann, J. (2008). Smart nanomaterials. *ACS Nano, 2*(6), 1101–1107.

Zachariasen, W. (1927). The crystal structure of the modification C of the sesquioxides of the rare earth metals, and of indium and thallium. *Norw. J. Geol., 9*, 310–316.

Zakutayev, A., Paudel, T. R., Ndione, P. F., Perkins, J. D., Lany, S., Zunger, A., & Ginley, D. S. (2012). Cation off-stoichiometry leads to high -type conductivity and enhanced transparency in CoZnO and CoNiO thin films. *Physical Review B, 85*(8), 085204.

Zhai, T., Fang, X., Liao, M., Xu, X., Zeng, H., Yoshio, B., & Golberg, D. (2009). A comprehensive review of one-dimensional metal-oxide nanostructure photodetectors. *Sensors (Basel), 9*(8), 6504–6529.

Zhan, Z., Lu, J., Song, W., Jiang, D., & Xu, J. (2007). Highly selective ethanol In_2O_3-based gas sensor. *Materials Research Bulletin, 42*, 228–235.

Zhang, D., Li, C., Liu, X., Han, S., Tang, T., & Zhou, C. (2003). Doping dependent NH_3 sensing of indium oxide nanowires. *Applied Physics Letters, 83*, 1845–1847.

Zhang, D., Liu, Z., Li, C., Tang, T., Liu, X., Han, S., … Zhou, C. (2004). Detection of NO_2 down to ppb levels using individual and multiple In_2O_3 nanowire devices. *Nano Letters, 4*, 1919–1924.

Zhang, J., Liu, X., Neri, G., & Pinna, N. (2016). Nanostructured materials for room-temperature gas sensors. *Advanced Materials, 28*, 795–831.

Zhang, S., Liang, B., Fan, Y., Wang, J., Liang, X., Huang, H., … Guo, J. (2019). Ferrocene as a novel additive to enhance the lithium-ion storage capability of SnO_2/graphene composite. *ACS Applied Materials and Interfaces, 11*, 31943–31953.

Zhang, W., Tian, J. L., Wang, Y. A., Fang, X., Huang, Y., Chen, W., … Zhang, D. (2014). Single porous SnO_2 microtubes templated from Papilio maacki bristles: New structure towards superior gas sensing. *Journal of Materials Chemistry A, 2*, 4543–4550.

Zhao, N. H., Wang, G. J., Huang, Y., Wang, B., Yao, B., & Wu, Y. (2008). Preparation of nanowire arrays of amorphous carbon nanotube-coated single crystal SnO_2. *Chemistry of Materials, 20*, 2612–2614.

Zhao, Y., Huang, Y., Wang, Q. F., Wang, X., & Zong, M. (2013). Carbon-doped Li_2SnO_3/graphene as an anode material for lithium-ion batteries. *Ceramics International, 39*, 1741–1747.

Zhou, K., Zhu, Y., Yang, X., & Li, C. (2011). Preparation and application of mediator-free H_2O_2 biosensors of graphene Fe_3O_4 composites. *Electroanalysis, 23*, 862–869.

Zhu, Z., Chang, J.-L., Wu, C.-H., Chou, T-L., & Wu, R-J. (2016). Promotion effect of silver on indium-(III) oxide for detecting trace amounts of ozone. *Sensors and Actuators B: Chemical, 232*, 442–447.

Zhu, Z., Bai, Y. Liu, X., Chueh, C-C., Yang, S., & Jen, A.K-Y. (2016). Enhanced efficiency and stability of inverted perovskite solar cells using highly crystalline SnO_2 nanocrystals as the robust electron-transporting layer. *Advanced Materials, 28*, 6478–6484.

Zou, W., Zhu, J., Sun, Y., & Wang, X. (2011). Depositing ZnO nanoparticles onto graphene in a polyol system. *Materials Chemistry and Physics, 125*, 617–620.

6 Transition Metal Oxides for Magnetic and Energy Applications

Rada Savkina and Aleksej Smirnov

6.1 INTRODUCTION

As it is known, a broad structural variety as well as the unusual properties of transition metal oxides is due to the unique nature of the transition metal cation outer *d*- and *f*-electrons. These are the elements of the three *d*-group transition series and the two *f*-group series in the periodic table consisted from a total of 60 chemical elements (see Figure 6.1, *d*-block and *f*-block). Although strictly speaking, the most common definition of a transition metal is accepted by the IUPAC (International Union of Pure and Applied Chemistry). These are the elements with a partially filled *d* subshell or have the capacity to produce cations with an incomplete *d* subshell. The *f*-block lanthanide and actinide series are called the "inner" transition metals.

Because valence electrons are present in more than one shell, transition metals often exhibit multiple stable oxidation states that can give rise to a large number of oxide types such as monoxide (AO), dioxide (AO_2), perovskite (ABO_3), and spinel (AB_2O_4). Most important structure types for transition metal oxides as well as the relationship between structural features and physical and chemical properties of these materials are described, for example, by Greedan (2017).

Transition metal oxides are considered to be very fascinating functional materials in which the nature of metal–oxygen bonding can vary between nearly ionic to highly covalent or metallic one reflected in phenomenal range of electronic and magnetic properties exhibited by these materials. Their unique features involve colossal dielectric constant (Lunkenheimer et al., 2009), efficient charge separation (Kaspar et al., 2016), enhanced surface reactivity, as well as magnetization and polarization properties. There are oxides with metallic properties (e.g., RuO_2, ReO_3, $LaNiO_3$) and oxides with highly insulating behavior (e.g. $BaTiO_3$, Cr_2O_3), oxides which are characterized with charge ordering (e.g. $La_{2-x}Sr_xNiO_4$, $La_{2-x}Sr_xCuO_4$, Fe_3O_4) (Mizokawa, & Fujimori, 2002) and defect ordering (e.g. $Ca_2Mn_2O_5$, $Ca_2Fe_2O_5$) (Rao, 1993). It is also necessary to mention oxides with different magnetic properties – from ferromagnetics (e.g. CrO_2, γ- Fe_2O_3, $Y_3Fe_5O_{12}$) to anti-ferromagnetics (e.g. α- Fe_2O_3, NiO, $LaCrO_3$), as well as superconductive cuprates (Waldram, 2017) and materials with a close coupling of magnetization and polarization via magnetoelectric and magnetodielectric effects – multiferroics (Hill, 2000; Lawes, & Srinivasan, 2011).

Unique properties of transition metal oxides develop when their spatial dimensions are reduced to the nanoscale (Rajesh Kumar, Raj, & Venimadhav, 2019). Besides, applications of composite structures including two or more transition metal oxides allow even wider diversity in their electronic properties and chemical behavior. For example, in single-phase multiferroic metal–oxide-based materials, the magnetoelectric coupling is very weak and the ordering temperature is too low. In contrast, multiferroic composites incorporated

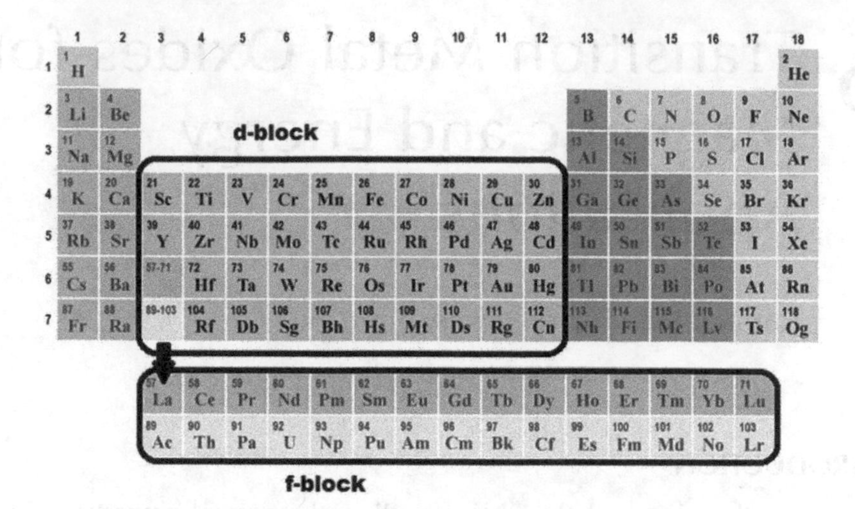

FIGURE 6.1 The transition metals (*d*-block) and the "inner" transition metals (*f*-block) on the periodic table. (www.technologyuk.net).

ferroelectric and ferromagnetic phases are characterized with giant magnetoelectric coupling response above room temperature.

Thus, studies of materials based on transition metal oxides and, especially, nanocomposite structures remain relevant in terms of practical application and are described in numerous scientific publications including monographs and reviews. The role of different ceramic oxide systems and their surface nano-architecture in governing the efficacy of a supercapacitor are presented in Balakrishnan and Subramanian (2017). Huge potential of metal oxide nanoparticles in technological field of current gas sensing tools is described by Eranna (2019). It focuses on the materials, devices, and techniques used for gas sensing applications, such as resistance and capacitance variations. Properties and applications of perovskite-type oxides are overviewed by Tejuca and Fierro (2019). This chapter is intended to provide recent results in the field of magnetic properties of transition metal oxides. In particular, we will consider such an interesting class of materials known as magnetoelectric multiferroics. In addition, we will look at recent advances in the use of transition metal oxides for solar fuel production.

6.2 MAGNETIC PHENOMENA IN OXIDE-BASED MATERIALS

Magnetic phenomena and materials are important and relevant in terms of practical application, since many decades they are the basis for mass storage devices as well as the subject of study in the field of spintronics. We are talking about sensors and memories based on giant magnetoresistive magnetic multilayers, in which magnetic fields cause order of magnitude changes in conductivity (Hartmann, 2000; Reig, Cubells-Beltrán, & Ramírez Muñoz, 2009). Other devices, for example, field-effect spin transistors using the ferromagnetic source and drain, are still under development (Sugahara, & Nitta, 2010; Sugahara, Takamura, Shuto, & Yamamoto, 2014).

It should be noted that the two very important factors that determine the progress in the field of research of magnetic materials and phenomena are advances in the characterization and growth techniques. In particular, in the surface probing methods such as the scanning tunneling microscopy and scanning tunneling spectroscopy (Wiesendanger, 2009), which not only provide spatial resolution on the atomic scale but also allow to investigate the atomic structure and the local density of states at the surface. Moreover, closely related methods such as the spin-polarized scanning tunneling microscopy (Bode, 2003) and the magnetic exchange force microscopy (Schwarz, Kaiser, Schmidt, & Wiesendanger, 2009) allow to obtain information about magnetic properties of materials. A flurry of activity in atomic and nanoscale growth techniques in the past decades has also led to the production of modern composite and hybrid magnetic materials that reveal a range of fascinating phenomena.

Among them, multiferroic magnetoelectrics are materials that are both ferromagnetic and ferroelectric in the same phase. Because of the combination of magnetic and electric properties, they are attractive materials for various electrically and magnetically cross-coupled devices in next generation of electronics and energy harvesting technologies, and at the same time they also represent a grand scientific challenge on understanding complex solid-state systems with strong correlations between multiple degrees of freedom (Lu, Hu, Tian, & Wu, 2015). Based on the type of ordering and coupling, multiferroic oxides have drawn increasing interest for a variety of device applications, such as magnetic field and electric current sensors (Palneedi, Annapureddy, Priya, & Ryu, 2016), ferroelectric photovoltaics, nanoelectronics (Ortega, Kumar, Scott, & Katiyar, 2015), and biomedicine (Kargol, Malkinski, & Caruntu, 2012). Such materials have the potential applications that include memory elements (Roy, Gupta, & Garg, 2012; Scott, 2007), in which data are stored both in the electric and the magnetic polarizations, or novel memory media, which might allow the writing of a ferroelectric data bit and the reading of the magnetic field generated by association. Since single-phase materials with strong cross-coupling properties exist rarely in nature, intensive research activity is directed toward the development of new multiferroic materials with strong magneto-electric coupling. In turn, the appearance of innovative materials with desired properties leads to the elaboration of new applications. Thus, the application-driven research in the field of multiferroics has focused on alternative materials such as composites from magnetic and ferroelectric materials coupled with mechanical strain, electric field effects, or exchange bias at the interfaces. It was recently published in the new book on multiferroics where the theory, materials, devices, design, and application of the ones are presented (Stojanovic, 2018).

6.3 WHAT IS A MULTIFERROIC?

"Crystals can be defined as multiferroic when two or more of the primary ferroic properties are united in the same phase." Hans Schmid (University of Geneva, Switzerland) in Fiebig, Eremenko, and Chupis (2004). We know three basic ferroic orders – ferromagnetism (FM) (spontaneous magnetization), ferroelasticity (spontaneous strain), and ferroelectricity (FE) (spontaneous polarization) (see Figure 6.2). Along with ferromagnets, ferroelectrics, and ferroelastics there is a fourth class of primary ferroics based on spontaneous magnetic vortex – ferrotoroidicity (Tolédano et al., 2015).

Multiferroicity is determined by a number of material parameters, including crystal symmetry, electronic, and chemical behavior. These materials can be divided into single-phase type with widely separated ferroelectric and magnetic ordering temperatures (type I multiferroics) and single-phase multiferroics having a magnetic transition with concurrent ferroelectric ordering (type II multiferroics) (Lawes, & Srinivasan, 2011; Lu et al., 2015).

If strong coupling between ferroic orders exists, such materials are named magneto-electrics (ME), which is the property that in certain materials a magnetic field induces an electric polarization and, conversely, an electric field induces a magnetization. Magneto-electric coupling may arise directly between the two-order parameters, or indirectly via strain. It is important to point out also that not every magnetic ferroelectric exhibits a linear magneto-electric effect and that not every material that exhibits a linear magneto-electric effect is

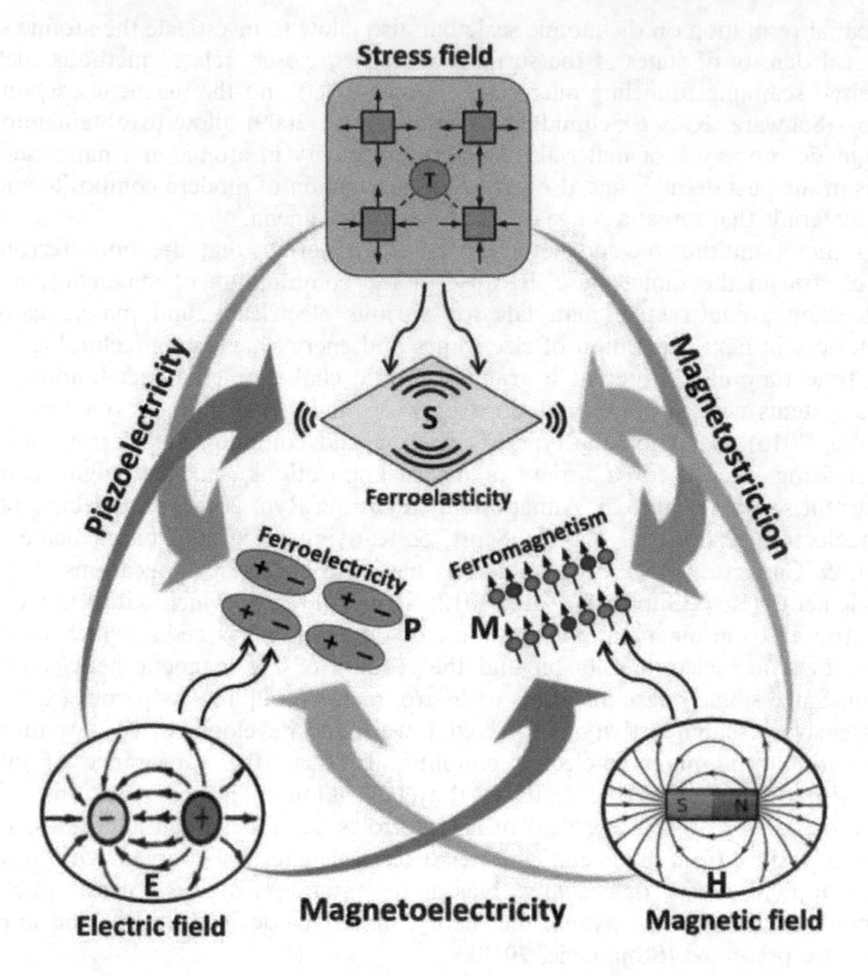

FIGURE 6.2 Schematic illustration of the magnetic–elastic–electric coupling in multiferroic materials: *M* is magnetization, *S* is mechanical strain, and *P* is dielectric polarization.

Reproduced from Palneedi et al. (2016) under the terms of the Creative Commons Attribution License.

also simultaneously multiferroic. Relationship between magnetic and electric materials shows that very small overlapping between these fields exists (Eerenstein, Mathur, & Scott, 2006). There is a fundamental reason behind the scarcity of ferroelectricity/ferromagnetism phenomena coexistence in the single-phase multiferroics. This is mutual exclusivity of the origins of magnetism and electric polarization – ferromagnetism needs unpaired 3d electrons and unfilled 3d orbitals, while ferroelectric polarization needs filled 3d orbitals of transition metals. Thus, only limited number of monolithic ME materials (in particular, transition metal oxides based) exhibit nonzero coupling at room temperature and an additional electronic or structural driving force must be present for FM and FE to occur simultaneously.

6.4 SINGLE-PHASE MULTIFERROIC COMPOUNDS

Single-phase multiferroic materials can be classified according to the multiferroicity origin. In type I multiferroics, the magnetic and electric orderings originate from different units (A-site driven, geometrically, and charge ordering-driven ferroelectricity). Ferroelectricity typically appears at higher temperatures than magnetism, and the spontaneous polarization P is often rather large (of order 100 μC/cm^2) while the ME coupling is weak. Examples are $BiFeO_3$ (T_N= 643 K, $P \sim 90$ μC/cm^2) and $YMnO_3$ (T_N= 76 K, $P \sim 6$ μC/cm^2, TN is Neel temperature) The multiferroic properties of bulk $BiFeO_3$ are fairly weak, but in thin-film form they are greatly enhanced.

In type II multiferroics, magnetic field can cause changes in symmetry through spin interactions, inducing ferroelectricity, ME coupling is much greater than that in type I, but the ferroelectric polarization in type II multiferroics is usually much smaller (~10 μC/cm^2). The important difference between multiferroics of types I and II is the nature of domain walls. FE and antiferromagnetic (AFM) domain walls may coincide in type I multiferroics. In multiferroics of type II, magnetoelectric coupling is originated from the interaction of magnetic and ferroelectric domain walls. One of the best-known examples of this behavior is $TbMnO_3$, an insulating perovskite that orders antiferromagnetically at T_N= 41 K and then undergoes a second magnetic transition at the Curie temperature T_C= 28 K (Lawes, & Srinivasan, 2011). Most of these materials are characterized with AFM spin configuration.

One of the most common ways of magnetic and ferroelectric phase coexistence in ABO_3 perovskites is hybridization between 2p orbitals of oxygen and 6p orbitals of A ions and localization of 6s^2 electrons on A-site ions. These are ferrites, manganites, and chromites of Bi, Pb and rare earth elements: $BiFeO_3$, $BiMnO_3$, $BiCrO_3$, and (RE)(Fe, Mn, Cr)O_3 as well as mixed perovskites with transitional metals such as (Bi, Pb, RE)$_2$(B B′)O_6. In the following sections, we will review the research progresses of some single-phase multiferroics and composite structures predominantly based on transition metal oxides.

6.5 Bi-Containing Multiferroic Oxides

Conventional ferroelectric, such as bismuth ferrite ($BiFeO_3$) BFO, is probably the most promising single-phase multiferroic to date and a hot favorite in sensorics and spintronics because of its large ferroelectric and magnetic ordering temperatures. Its Curie temperature is ~1123 K and the Néel temperature is ~640 K with a long periodic cycloidal spiral spin arrangement (Sosnowska, Loewenhaupt, David, & Ibberson, 1992). The basic physics and applications aspects of $BiFeO_3$ are summarized in Catalan and Scott (2009) and Sando, Barthélémy, and Bibes (2014). In 2016, Wu et al. have provided comprehensive review with unprecedented number of references subjected to the progress of BFO-based materials made in the past 15 years in the different forms of ceramic bulks, thin films, and nanostructures, focusing on the pathways to modify different structures and to achieve enhanced physical properties and new functional behavior.

The ferroelectricity in this material is driven by the stereochemically active Bi^{3+} cation, which exhibits a rhombohedrally distorted perovskite structure for the bulk phase, where all ionic sub-lattices are displaced relative to each other along the polar (111) direction, and the oxygen octahedral is rotated around the same (111) axis, alternately clockwise and counter-clockwise. The dielectric and switching behaviors of bismuth ferrite are greatly influenced by the domain structure (see Figure 6.3), which is determined by the interplay between the ferroelectric polarization energy and ferroelectric domain wall energy. Figure 6.4 summarizes the remnant polarization of $BiFeO_3$-based materials in the form of thin films (a) and bulks (b). Wu, Fan, Xiao, Zhu, and Wang (2016) have demonstrated that factors such as leakage currents, strain, film orientation, as well as interfacial coupling and construction of the phase boundaries play a critical role in the ferroelectric properties of $BiFeO_3$-based materials. A high Pr of >90 C/cm^2 (Wu et al., 2016) and fast switching behavior within about 125 ns (Ahn, Seo, Lim, & Son, 2015) were attained in the $BiFeO_3$ thin films.

The magnetization behavior of $BiFeO_3$ is attributed to both short- and long-range ordering. It originates from unpaired electrons in d orbitals of Fe^{3+} ions. The local short-range magnetic ordering of $BiFeO_3$ is G-type antiferromagnetic, that is, each Fe^{+3} spin is surrounded by

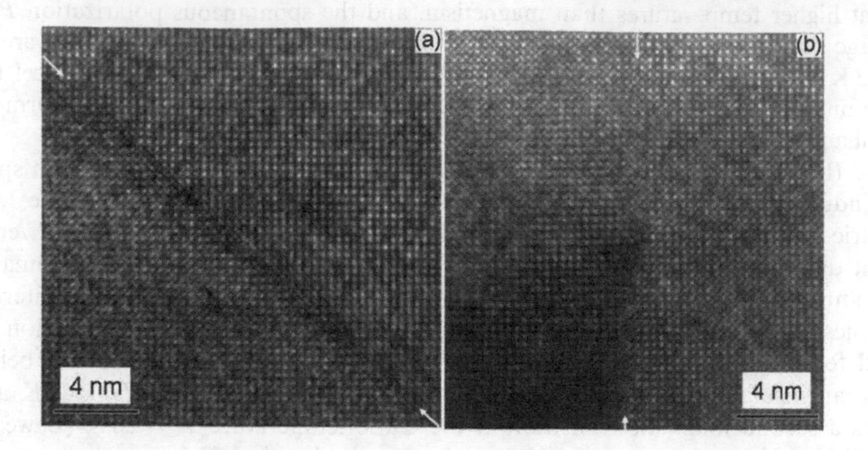

FIGURE 6.3 High-resolution TEM images of $BiFeO_3$ at (a) 71° and (b) 109° ferroelectric domain walls. **Reproduced with permission from Chen, Katz, and Pana (2007) Copyright 2007, AIP Publishing.**

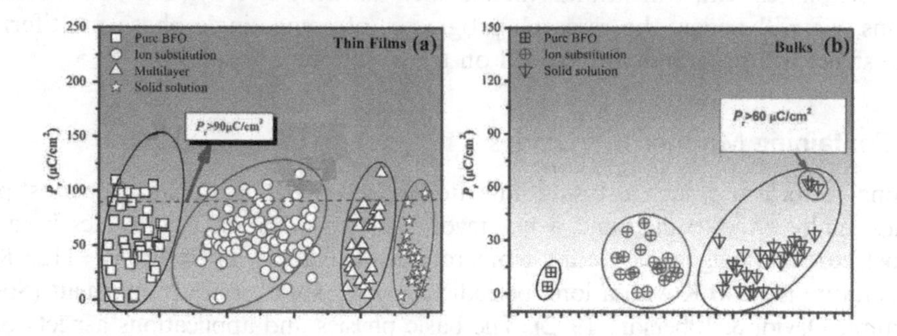

FIGURE 6.4 Ferroelectric properties of $BiFeO_3$-based materials: (a) thin films and (b) ceramic bulks. **(Reproduced with permission from Wu et al. (2016) Copyright 2016, Elsevier.)**

six antiparallel spins on the nearest Fe neighbors. The long-range magnetic ordering is an incommensurate spin cycloid of the antiferromagnetically ordered sublattice with a period of 62 nm (Lebeugle et al., 2008). Figure 6.5 shows magnetic properties of $BiFeO_3$-based materials in various forms such as thin films, ceramic bulks, and nanostructures (Wu et al., 2016).

A rather unique feature of $BiFeO_3$ is its ability to "morph" its ground state when an external mechanical constraint is imposed on it. The effect of epitaxial strain on the T_N and T_C of bismuth ferrite was found and systematically examined (Sando, Agbelele, et al., 2014). Figure 6.6 demonstrates tetragonality (c/a ratio) and the rotation angles of the oxygen octahedra about the *x*-, *y*-, and *z*-axes obtained from first-principles-based effective Hamiltonian calculations as well as T_C and T_N trends with misfit strain. The ferroelectric Curie temperature strongly decreases as strain increases in both the tensile and compressive ranges, which indicates that the strain-induced increase in the octahedral tilts is the main factor driving the ferroelectric response in $BiFeO_3$ films. Besides, weak dependence of the Néel temperature on strain is consistent with the variation of the Fe–O–Fe bond angles by a few degrees. At the same time, the magnetic Néel temperature of a T-type (highly elongated tetragonal-type) polymorph of $BiFeO_3$ is largely decreased to near room temperature

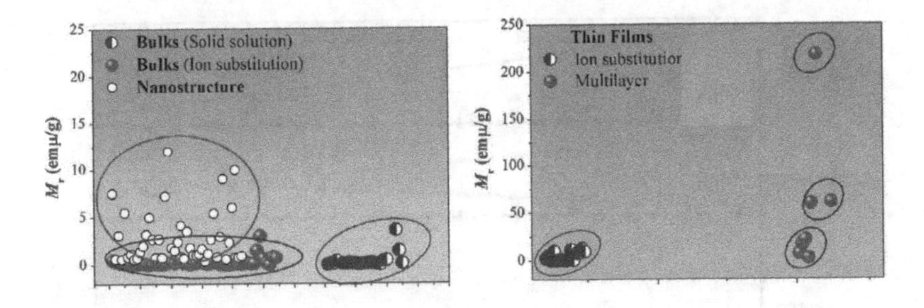

FIGURE 6.5 Magnetic properties of $BiFeO_3$-based materials in various forms. **(Reproduced with permission from Wu et al. (2016) Copyright 2016, Elsevier.)**

by heteroepitaxial misfit strain (Ko et al., 2011). Enhancement of polarization, magnetism, and morphotropic phase boundary-relevant piezoelectric response in heteroepitaxially constrained thin films of $BiFeO_3$ is reported in Huang and Chen (2014).

A particularly striking example is observed when a large (~4 to 5%) compressive strain is imposed on a $BiFeO_3$ thin film through the epitaxial constraint from the underlying substrate. Under these conditions, the ground state rhombohedral phase transforms into a tetragonal-like (or a derivative thereof) phase with a rather large unit cell (c/a ratio of ~1.26). When the epitaxial constraint is partially relaxed by increasing the film thickness, this tetragonal-like phase evolves into a "mixed-phase" state, consisting of a nanoscale admixture of the rhombohedral-like phase embedded in the tetragonal-like phase (Zhang, Zeches, He, Chu, & Ramesh, 2012).

It is known that the properties of unique materials develop during transition from macroscopic to nanoscale sizes. For example, temperature dependence of magnetization of single-crystal $BiFeO_3$ nanopowder synthesized with the hydrothermal method shows antiferromagnetic–paramagnetic phase transition at T_N= 220 K, while below this temperature, weak ferromagnetic ordering is detected (Čebela et al., 2017). The results of this study offer an overall conclusion that the local magnetic properties of pure $BiFeO_3$ nanoparticles mainly depend on the particle size and their diverse morphology due to the different preparation methods and annealing temperatures. The magnetic measurement indicated the existence of room-temperature weak ferromagnetism in $BiFeO_3$ nanoparticles and $BiFeO_3$ /SiO_2 core-shell structures prepared using a variation of the sol–gel procedure (Chauhan, Kumar, Chhoker, & Katyal, 2016).

According to the review by Lu et al. (2015), $BiMnO_3$ and $BiCrO_3$ are considered as another typical perovskite Bi-containing multiferroic oxides. $BiCoO_3$ also exhibits multiferroic properties with very large values of polarization – value about 180 μC/cm^2 is theoretically expected (Belik, 2012). The ferromagnetism of $BiMnO_3$was confirmed experimentally in both the film and bulk samples, and was generally explained in terms of super-exchange interaction between orbital ordering of Mn^{3+} ions. $BiMnO_3$ is the only compound among $BiBO_3$ (B = Cr, Fe, Co, and Ni) that shows true ferromagnetic ordering at T_C= 99 – 105K (Belik et al., 2007). The electronic, magnetic, and ferroelectric properties of $BiCrO_3$ in *C*2/*c* and *R*3*c* structure are investigated by first principles calculations by Ding, Kang, Wen, Li, and Zhang (2014) and Hill, Bättig, and Daul (2002). It was found that the easy magnetization axis in *C*2/*c* structure is along the *b* axis, the magnetic order in *R*3*c* structure is G-type antiferromagnetic, and the easy magnetization axis is along the rhombohedral [111] direction. Berry phase theory predicts that the *R*3c structure of $BiCrO_3$ has a large spontaneous polarization of 73.9 μC/cm^2 along the rhombohedral [111] direction.

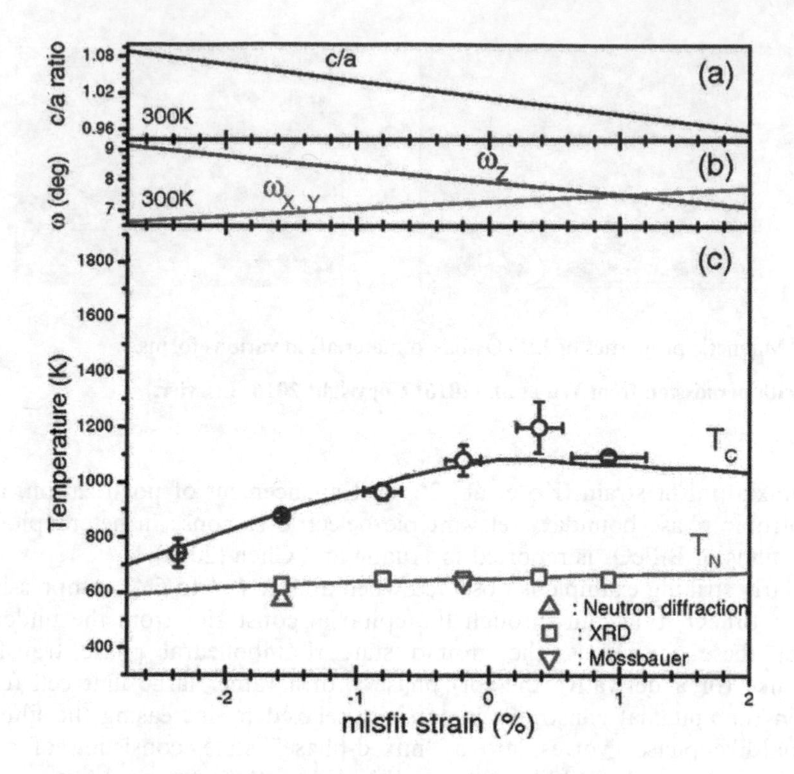

FIGURE 6.6 Effective Hamiltonian results evaluated in a $BiFeO_3$ film, at 300 K and as a function of misfit strain: (a) tetragonality (c/a ratio) and (b) antiferrodistortive angles ω along the *x*-, *y*-, and *z*-axes (ω_X and ω_Y – red line, ω_Z – blue line); (c) theoretical results on $BiFeO_3$ film transition temperatures as a function of misfit strain. Experimental T_C (circles) and T_N (squares and triangles) values are shown. **(Reproduced with permission from Sando, Agbelele, et al. (2014) Copyright 2014, ROYAL SOCIETY.)**

At the same time, ferroelectricity in $BiMnO_3$ is still the subject of many controversial research studies. Some authors did not confirm any ferroelectricity (Jung, Yang, & Jeong, 2013). The magnetic and multiferroic properties of $BiMnO_3$ have been studied by Zhai and Wang (2017). No ferroelectric order was observed. This finding is consistent with the centrosymmetric crystal structure recently suggested by theoretical calculations and structural studies on ceramic samples of stoichiometric $BiMnO_3$ (Lee, Yoo, Nam, Toreh, & Jung, 2015). Despite these claims, reports of ferroelectricity in thin films of $BiMnO_3$ remain, suggesting the presence of alternate extrinsic mechanisms. Some papers report about $BiMnO_3$ polarizations from $P = 0.004 - 0.03\ \mu C/cm^2$ at low temperatures of around 120 K (Grizalez, Martinez, Caicedo, Heiras, & Prieto, 2008; Moreira dos Santos, Parashar, et al., 2002) to $P = 9 - 16\ \mu C/cm^2$ at room temperature in thin films (Son, & Shin, 2008). Signatures of room temperature ferroelectricity were found in the pseudo-cubic ultra-thin films of $BiMnO_3$ (De Luca et al., 2013). A giant linear magnetoelectric coupling was found in strained $BiMnO_3$ thin films in which the disorder associated with an islanded morphology gives rise to extrinsic relaxor ferroelectricity that is not present in bulk centrosymmetric ferromagnetic crystalline $BiMnO_3$ (Mickel, Jeen, Kumar, Biswas, & Hebard 2016). In addition, the ME coupling coefficient was found to be quite large, reaching a value of −1.25 ns/m at T = 65 K. This value is approximately 35 times larger than the current record for single-phase linear ME coupling of 36.7 ps/m found inTbPO4.

The same controversial results were obtained for room-temperature crystal structure of $BiMnO_3$. Some reports defend noncentrosymmetric C2 structure with off-centered Bi $6s^2$ lone pairs breaking the antisymmetry (Atou, Chiba, Ohoyama, Yamaguchi, & Syono, 1999; Moreira dos Santos, Cheetham, et al., 2002), other papers propose centrosymmetric C2/c structure of $BiMnO_3$ crystal symmetry (Belik et al., 2007; Goian et al., 2012; Yokosawa et al., 2008), which rules out ferroelectricity in this material. At the same time, it was pointed out that BiMnO3 thin-film sample can have strain-sensitive structural and compositional modifications that are strongly correlated with the appearance of ferroelectricity (Belik, 2012). Thus, ferroelectricity in $BiMnO_3$ remains open to question.

6.6 DOUBLE PEROVSKITE COMPOUNDS

New properties are shown by $BiFeO_3$ with addition of 50% concentration of Mn or Cr ions and ordered double perovskite such as ($A_2BB'O_6$). Very large polarization ($P \sim 60$ μC/cm^2) was experimentally evidenced in $BiFe_{0.5}Cr_{0.5}O_3$ (Vinai et al., 2015). Another compound $BiFe_{0.5}Mn_{0.5}O_3$ demonstrates a large magnetization of ~90 emu/cc (Choi et al., 2011; Xu et al., 2015). It was found that magnetic properties can be well modified by introduction of even a small amount of foreign elements to Fe site. For example, $BiFe_{0.95}Co_{0.05}O_3$ (BFC) nanotube arrays showed a saturated M-H loop of ~0.021 emu/g, which was superior to undoped $BiFeO_3$ nanotubes, indicating that the Co doping enhanced the magnetic properties due to the possible breakage in space-modulated spin cycloid (see figure 6.7 and Wu et al., 2016).

In the ordered double perovskite ($A_2BB'O_6$) system, the polarization is induced predominantly by structural distortion arising from different valencies of ions occupying the B and B′ sites. It is also possible that ferroelectricity is caused by A-site localized $6s^2$ electronic pairs. Ferromagnetism of the mixed perovskites with randomly distributed various ratios of two or more transition metals such as (Bi, Pb)$_2$(BB′)O$_6$ occurs from B(d^n)-O-B′(d^n) superexchange (Kim et al., 2019; Moon et al., 2015). For example, ferromagnetic properties in the double perovskite Gd_2NiMnO_6 are caused by alternating Ni^{2+} and Mn^{4+} spins order and it

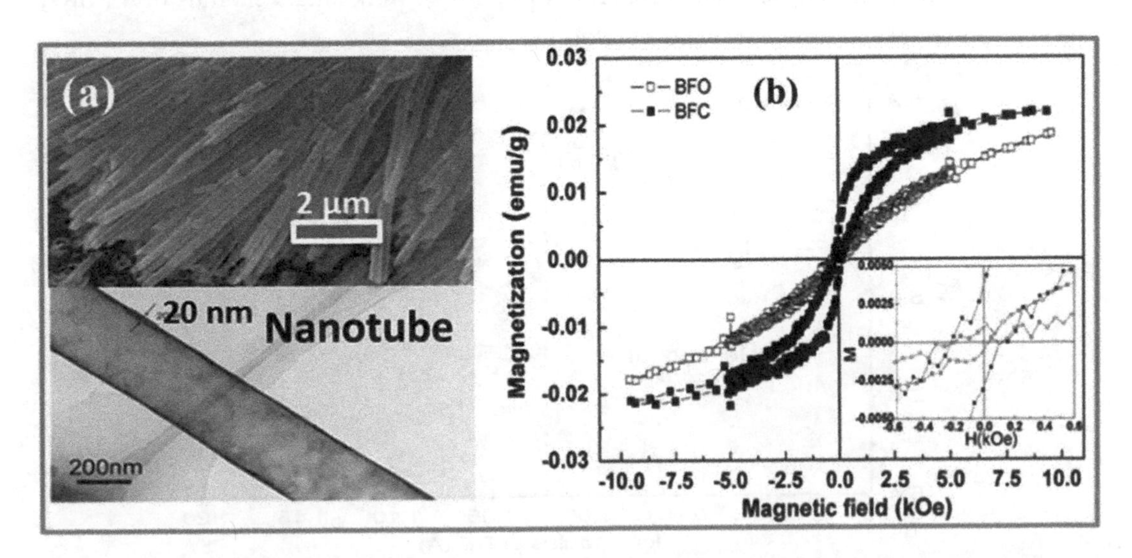

FIGURE 6.7 (a) SEM and TEM images of $BiFe_{0.95}Co_{0.05}O_3$ (BFC) nanotubes, and (b) M-H hysteresis loops of $BiFeO_3$ (BFO) and BFC nanotubes.

(Reproduced with permission from Wu et al. (2016) Copyright 2016, Elsevier.)

arises below T_C= 134K and an additional order of Gd^{3+} spins occurs at T= 33K (Oh et al., 2015). The formation of short-range ferromagnetic clusters occurs below T = 230 K. In Bi-based double perovskites, ferromagnetic to paramagnetic phase transitions occur at 140 K (Bi_2NiMnO_6), 440 K (Bi_2FeMnO_6), and even as high as 800 K (Bi_2CoMnO_6), which was explained by the monoclinic tilt and possible interaction between the local magnetization and $6s^2$ pairs of electrons (Stojanovic et al., 2018). This high transition temperature for magnetic ordering can be further tuned by varying the strain in the films.

Recently, the topic of single-phase multiferroics has been updated with a number of reports and reviews on the structural, magnetic, and ferroelectric properties of double perovskites (see, for example, Ding, Gao, et al., 2019; Ding, Khalyavin, et al., 2019; Gaikwad, Brahma, Borah, & Ravi, 2019; Maiti, Saxena, & Roy, 2019; Rahmani et al., 2019; Roknuzzaman et al., 2019; Gauvin-Ndiaye, Tremblay, & Nourafkan, 2019). It should be noted that this topic is worthy of consideration in a separate chapter or book.

6.7 THE RARE-EARTH PEROVSKITE MULTIFERROICS

Another class of multiferroics is materials with geometrically driven ferroelectricity caused of large mismatching in sizes of ions. First of all, these are a small size of A-site ions in a hexagonal structure (*P63cm*) such as $ReMnO_3$ (Re = Sc, Y, In, lanthanides from Ho to Lu). It should be noted that manganites having a large Re ionic radius (from Dy to La) are characterized with a perovskite orthorhombic structure *Pnma*. Figure 6.8 shows structural evolution of $ReMnO_3$ as a function of the radius of A-site rare-earth ions.

The most studied hexagonal manganite $ReMnO_3$ is $YMnO_3$, in which layers of MnO_5 trigonal bipyramids (different from the MnO_6 octahedron in orthorhombic RE manganites) and layers of Y^{3+} are alternatively packed along the *c* axis. Although yttrium is not a rare-earth element, it forms a stable trivalent cation with an ionic radius similar to those of the smaller rare-earth ions (Hill, 2000). Besides, $YMnO_3$ forms in both the hexagonal and orthorhombic phases, making it appealing for use in a comparative study. The long-range AFM order of Mn spins arises at T_N ~70 K with a 120° configuration, which is typically observed in triangular antiferromagnets (Lu et al., 2015). Bulk single crystals of $YMnO_3$

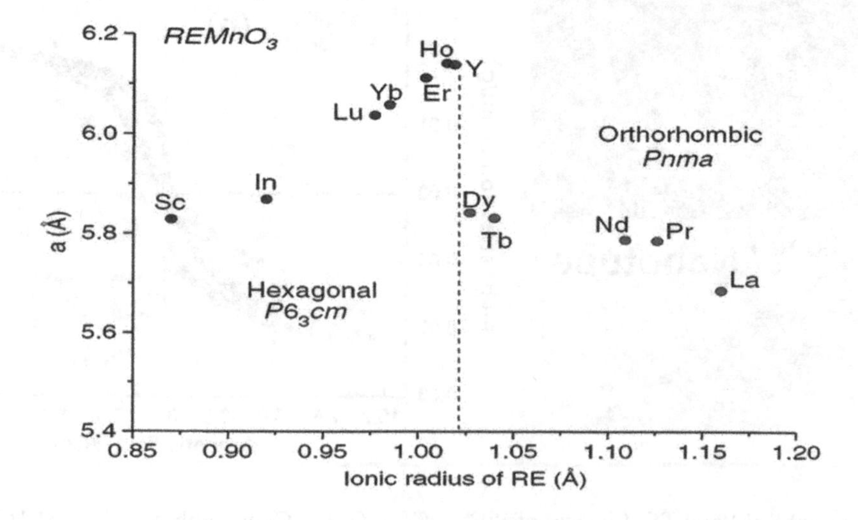

FIGURE 6.8 Structural evolution of $ReMnO_3$ as a function of the radius of A-site rare-earth ions. **Reproduced from Lu et al. (2015) under the terms of the Creative Commons Attribution (CC BY).**

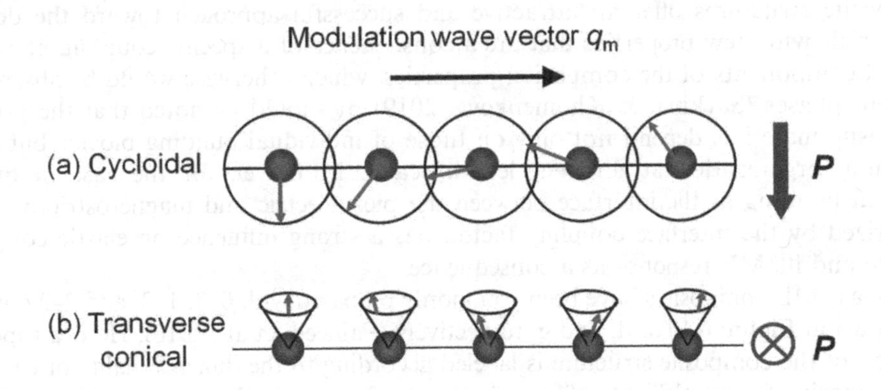

FIGURE 6.9 Schematic drawings of (a) cycloidal and (b) transverse conical spin ordering.

possess a remnant polarization of 5.5 μC/cm^2. In the orthorhombic – $YMnO_3$, the superexchange interaction-mediated long-range E-type AFM order can directly generate ferroelectricity (P ~ 0.8 μC/cm^2) due to the effect of exchange striction (Nakamura, Tokunaga, Kawasaki, & Tokura, 2011).

Orthorhombic RE manganites have centrosymmetric nonpolar structure with A-type AFM. Jahn-Teller distortion creates in these materials a deviation from perfect ordering of antiparallel magnetic moments, and the spontaneous polarization is induced by the spiral spin order, which exists in two types – cicloidal and transverse conical (see Figure 6.9 reproduced from Lu et al. (2015)).

In recent years, much research attention has been focused on rare-earth multiferroics with double-perovskite crystal structure since the origin of magnetic order and the polarize direction in these materials are still under debate. Jia, Zeng, and Lin (2017) have summarized the recent progress in multiferroics with an up-up-down-down (↑↑↓↓) magnetic structure, which causes an exchange striction and thus breaks inversion symmetry to get a polarization. These materials are characterized with E- and E*-type antiferromagnetic state. A magnetostrictive effect observed in $(Re)Fe_{0.5}Cr_{0.5}O_3$ perovskites (Re = Lu, Yb, Tm) was investigated by Pomiro et al. (2016). It associated with a negative thermal expansion attributed to a magnetoelastic effect produced by repulsion between the magnetic moments of neighboring transition metal ions. Effect of the spin phonon coupling is extensively investigated in rare-earth double perovskites due to its correlation with magnetodielectric response of Re_2BMnO_6 compounds (Filho et al., 2015; Silva et al., 2017). In comparison to the lack of a noticeable magnetodielectric effect in $YCr_{0.5}Fe_{0.5}O_3$, the strong correlation between magnetic and dielectric order parameters in $HoCr_{0.5}Fe_{0.5}O_3$ suggests the important role of magnetic rare-earth ions (Shin, Lee, & Choi, 2019).

6.8 MULTIFERROIC COMPOSITE SYSTEMS

Magnetoelectric composites occupy an important place in the field of multiferroic compounds due to their ease of fabrication and design flexibility (Andrew, Starr, & Budi, 2014). The disadvantages of low working temperatures and weak ME response of single-phase multiferroics are compensated by composite structures in which the ME response is enhanced via mechanical coupling between the ferroelectric and magnetic phases due to magnetostriction and electrostriction.

Composite structures offer an attractive and successful approach toward the design of new materials with new properties that are a consequence of a specific coupling between the individual components of the composite; properties, which otherwise would be absent in the constituent phases (Savkina, & Khomenkova, 2019). It should be noted that the properties of composite materials depend not only on those of individual building blocks, but also on their spatial organization at different length scales. Moreover, for the case in question, mechanical bonding at the interface between the piezoelectric and magnetostrictive phases, characterized by the interface coupling factor, has a strong influence on elastic coupling of composite and its ME response as a consequence.

Two-phase ME composites have been commonly prepared with 0–3, 1–3, and 2–2 connectivity, as shown in Figure 6.10 a, d, and g, respectively (Palneedi et al., 2016). Here, a topological description of the composite structure is labeled according to the dimensionality of each phase of the composite: 0 – particles, 1 – fibers, 2 – layers, 3 – matrix (Newnham, Skinner, & Cross, 1978). Namely, in the 0–3 particle–matrix composites, magnetic particles are embedded in the piezoelectric matrix. Particles in the 0–3 structures or fibers in the 1–3 composites can be either randomly dispersed or periodically aligned. A 2–2 laminate composite consists of alternating magnetic and ferroelectric layers. Last type of structures can be poled to a higher degree since some of the difficulties associated with the particulate composites can be overcome by forming 2–2 type composite structures (Vaz, Hoffman, Ahn, & Ramesh, 2010). Primarily, the strong directionality of the piezoelectric and magnetostrictive effects imply that the magnetoelectric coupling is expected to depend strongly on the crystalline structure and relative orientation, which may be difficult to control in particulate composites. Besides, the 0–3 type composites are characterized with high sintering temperatures, which can lead to produce some unpredictable phases due to interdiffusion between the piezoelectric and magnetic phases. The low electrical resistivity of the magnetostrictive phases and the large leakage current of the particulate composites reduces the ME coefficient. The thermal expansion mismatch between two phases degrades their performance too.

These problems can be eliminated in the laminate 2–2 composite ceramics. It should be noted the $BaTiO_3$–$CoFe_2O_4$ system exhibits the highest ME effect among the multiferroic composite materials without using lead and rare-earth elements. Its ME voltage coefficient α_{ME} is about 135 mV (cm Oe)$^{-1}$ at a bias magnetic field of 2600 Oe and a frequency of 1 kHz (Yang, Zhang, & Lin, 2015). This corresponds to a linear ME coefficient $\alpha \sim 720$ ps/m. ME voltage coefficient for laminate composite exceeds the typical value for the composite material in the form of particles, 102 mV (cm Oe)$^{-1}$. At the same time, direct magnetoelectric coupling coefficient of multiferroic fluids composed of $BaTiO_3$–$CoFe_2O_4$ composite nanoparticles dispersed in a highly insulating nonpolar oleic acid/silicone oil mixture is estimated to be $\alpha_{ME} = 1.58 \times 10^4$V (cm Oe)$^{-1}$ at a bias magnetic field of 1000 Oe (Gao et al., 2018). These results imply that besides magnetoelectric fluids that consist of core/shell-structured nanoparticles, conventional multiferroic fluids based on composite particles may provide an opportunity to gain electrical control of magnetization and *vice versa*, which implies potential application.

The fabrication and application of multiferroic composite structures as a new functional materials were discussed in the remarkable reviews (Chu, PourhosseiniAsl, & Dong, 2018; Palneedi et al., 2016; Spaldin, & Ramesh, 2019; Vaz et al., 2010) and books (Stojanovic, 2018). The theoretical modeling of ME effects in layered and bulk composites based on magnetostrictive and piezoelectric materials is presented in Bichurin and Petrov (2014). Scientific interest in this field has considerably increased and focused on various topics such as nanostructured composites fabrication, electric-field control of magnetism, radio- and high-frequency devices, and ultralow power logic-memory devices. This review summarizes pioneering, advanced, and prospected work on multiferroic and magnetoelectric composites, which are at the breakthrough toward technological applications. Advances and prospects

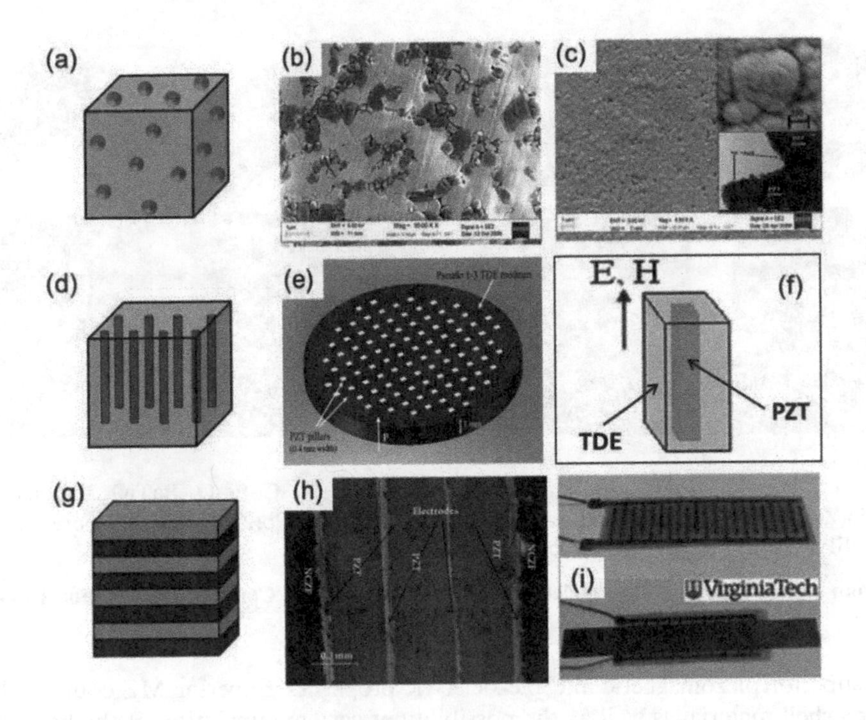

FIGURE 6.10 Bulk ME composites with different phase connectivity: (a)–(c) 0–3 connectivity; (d)–(f) 1–3 connectivity; and (g)–(i) 2–2 connectivity.

work on multiferroic and magnetoelectric composites, which are at the breakthrough toward technological applications of oxide-based multiferroics, were summarized by Kleemann (2017). In this review, three different multiferroic composites were presented in some detail and their use as novel media for data processing was proposed: (1) electric switching of magnetism in the 2–2 composite Cr_2O_3/(Pt/Co/Pt) via exchange bias. It was proposed to use magnetoelectric $Cr_2O_{2.97}B_{0.03}$ in order to achieve a sufficiently high antiferromagnetic ordering temperature, T_N = 400 K (Binek et al., 2015); (2) magnetic switching of electric polarization in the 2–1 composite $BaTiO_3/CoFe_2O_4$ via stress–strain coupling; (3) magnetic and/or electric switching of multiferroic nanodots in the 2–0 composite (K, Bi)TiO_3/Bi(Fe, Co)O_3 with giant ME effect enabling scanning probe control at ultrahigh packing density. A 2–2 laminated structure with a spin modulation element, which can sufficiently modulate the spin polarization rate of a ferromagnetic material through an electric field is presented in Suzuki, Nakada, and Yonemura (2018). A laminated structure includes ferromagnetic, multiferroic layer included any one selected from the group consisting of $BiFeO_3$, $BiMnO_3$, $GaFeO_3$, $AlFeO_3$, (Ga, Al)FeO_3, $YMnO_3$, $CuFeO_2$, Cr_2O_3, $Ni_3Bi_7O_{13}I$, $LiMnPO_4$, $Y_3Fe_5O_{12}$, $TbPO_4$, and $LiCoPO_4$, and a ferroelectric layer that can be selected from $La_xSr_{1-x}MnO_3$ ($0 \leq x \leq 1$), $Ba_xSr_{1-x}TiO_3$ ($0 \leq x \leq 1$), or $PbZr_xTi_{1-x}O_3$ ($0 \leq x \leq 1$).

In recent years, ME composites with new connectivity designs are being developed through the nanotechnology possibilities (Palneedi et al., 2016). ME composites with different types of core/shell designs are presented in Figure 6.11. Among them, $BaTiO_3$–$CoFe_2O_4$ composite is a highly promising material for the design of devices based on ME multiferroics, especially for biomedical application. A brief literature survey on various magnetic and ferroelectric ceramics indicated that components of this composite can be easily synthesized, no toxicity,

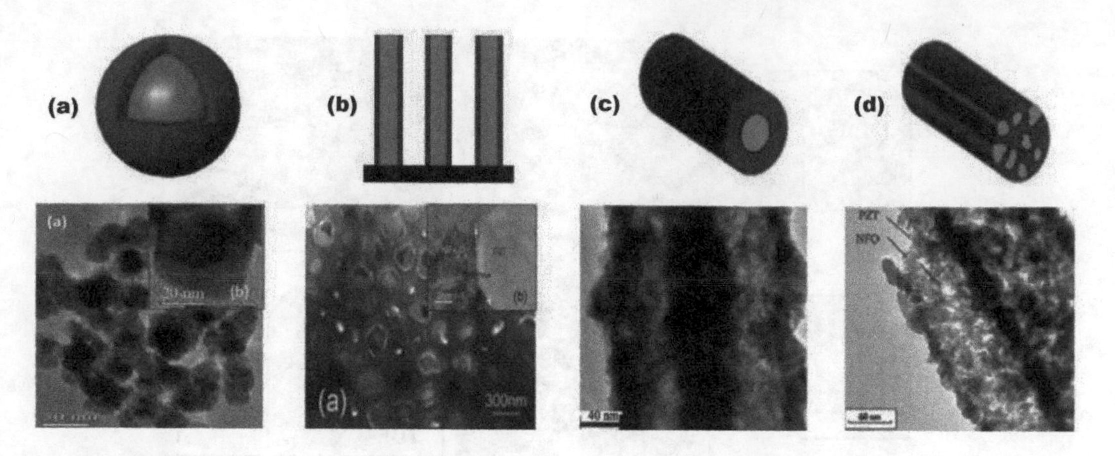

FIGURE 6.11 Examples of the core–shell ME composites: (a) – $CoFe_2O_4$-$BaTiO_3$ nanoparticles, (b) $NiFe_2O_4$-$Pb(Zr_{0.52}Ti_{0.48})O_3$ core–shell nanowires, (c) – $CoFe_2O_4$-$Pb(Zr_{0.52}Ti_{0.48})O_3$ core–shell nanofibers, (d) – $NiFe_2O_4$- $Pb(Zr_{0.52}Ti_{0.48})O_3$ core–shell nanofibers.

Adapted from Palneedi et al. (2016) under the terms of the Creative Commons Attribution License.

and have superior piezomagnetic and piezoelectric properties. Superior ME coupling behavior in this core–shell material as well as the possibility of easy manipulation of the band gap over a range of energies by mere control of the molar ratio of the phases (Thankachan et al., 2018) establishes the potential for the use of $BaTiO_3$–$CoFe_2O_4$ core–shell structures for stimulation of vital functions of living cells under the influence of external magnetic field (Kargol et al., 2012; Rao et al., 2017).

6.9 IRON AND CHROMIUM OXIDE FOR MAGNETIC AND ENERGY APPLICATIONS

Isostructural and isovalent iron and chromium oxides are the subject of active experimental and theoretical investigations last time. Bulk α-Cr_2O_3 and α-Fe_2O_3 both are hexagonal corundum structure with space group of *R3c*. At the same time, their magnetic properties are determined by Néel temperature, $T_{N\alpha-Fe2O3} = 955K$; $T_{NCr2O3} = 307K$, which shows that these materials have different magnetic properties. Hematite exhibits weak ferromagnetism between 260 K and 955 K and has not manifested the linear magnetoelectric effect. Cr_2O_3 is antiferromagnetic up to $T_{NCr2O3} = 307K$ and magnetoelectric. The nature of magnetic ordering in non-van der Waals 2D metal oxides, hematene and chromene, was investigated (Bandyopadhyay, Frey, Jariwala, & Shenoy, 2019). Two-dimensional hematene is found to be fully oxygen-passivated and stable under ambient conditions. It exhibits a striped ferrimagnetic ground state with a small net magnetic moment, whereas chromene has a ferromagnetic ground state. Bandyopadhyay et al. (2019) also show that tuning the magnetic ordering in these materials controls the transport properties by modulating the band gap, which may be of use in spintronic or catalytic applications.

The solid solution or multilayer composite structures based on Fe/Cr oxides combine its different functional peculiarities in one system to achieve novel magnetic, electric, or photoelectric properties. For example, Fe_2O_3 can be applied as a sensitizer for wide band gap photocatalysts such as Cr_2O_3. Enhancing the antiferromagnetic spin correlation in the Cr_2O_3 film by forming a junction with the Fe_2O_3 layer was also demonstrated (Kota, Imamura, & Sasaki, 2014). Besides, the composite system of the alternate Fe_2O_3/Cr_2O_3 layers have demonstrated the positive magnetoresistance as well as the magnetic hysteresis and magnetoresistivity switching effect in the low magnetic fields (Smirnov et al., 2016).

Before considering the properties of structures and solid solutions of iron/chromium oxides system, we list the properties of these oxides.

6.10 IRON AND CHROMIUM OXIDES' PROPERTIES

Iron oxides have extensive and significant applications in semiconductor devices, magneto-optic memories, audio–video systems, computer chips, and in memory storage devices. Iron (III) oxide exists in four phases: α-Fe_2O_3, β-Fe_2O_3, γ-Fe_2O_3, and ε-Fe_2O_3. Some of their physical and magnetic properties are summarized in Table 6.1. Thermodynamically, hematite (α-Fe_2O_3) is the most stable in the family of iron oxides: α-Fe_2O_3, β-Fe_2O_3, γ-Fe_2O_3. The *d–d* transitions and metal charge transfer play important roles in tuning the *n*-type semiconducting band gap of hematite. This material exhibits weak ferromagnetism between 260 K and the Néel temperature. It finds application in wide varieties of uses such as gas sensors (Wang, Yin, Zhang, Xiang, & Gao, 2010), water splitting (Bouhjar, Bessaïs, & Marí, 2018), anode material for lithium-ion batteries (Lin, Abel, Heller, & Buddie Mullins, 2011), photocatalyst (band gap 2.2 eV) (Mishra, & Chun, 2015), electron transport layer in solar cell (Guo et al., 2017), and supercapacitor (Binitha et al., 2013).

Iron oxides have a huge potential in biomedicine. Fe_3O_4 nanoparticles are used as a contrast agent in MRI scanners. Therapeutic applications of superparamagnetic nanoparticle include targeted delivery of drugs (Duli´nska-Litewka et al., 2019; Mahmoudi, Sant, Wang, Laurent, & Sen, 2011) and radioactive isotopes for chemotherapy and radiotherapy and contrast enhancement in magnetic resonance imaging. In hyperthermia treatment,

TABLE 6.1
Physical and magnetic properties of iron oxides.[1]

Property	Hematite	Magnetite	Maghemite
	Oxides		
Molecular formula	α-Fe_2O_3	Fe_3O_4	γ-Fe_2O_3
Density(g/cm^3)	5.26	5.18	4.87
Melting point (°C)	1350	1583–1597	—
Hardness	6.5	5.5	5
Type of magnetism	Weakly ferromagnetic or antiferromagnetic	Ferromagnetic	Ferrimagnetic
Curie temperature (K)	956	850	820–986
M_S at 300 K(A-m^2/kg)	0.3	92–100	60–80
Standard free energy of formation ΔG_f (kJ/mol)	−742.7	−1012.6	−711.1
Crystallographic system	Rhombohedral, hexagonal	Cubic	Cubic or tetrahedral
Structural type	Corundum	Inverse spinel	Defect spinel
Space group	*R3c* (hexagonal)	*Fd3m*	*P4₃32* (cubic); *P4₁2₁2* (tetragonal)
Lattice parameter (nm)	a = 0.5034, c = 1.375 (hexagonal) a_{Rh} = 0.5427, α = 55.3° (rhombohedral)	a = 0.8396	a = 0.83474 (cubic); a = 0.8347, c = 2.501 (tetragonal)

1 Reproduced with permission from Teja and Koh (2009). Copyright 2009, Elsevier.

tumors can be killed by the local temperature increase that occurs when Fe_3O_4 nanoparticles are placed in a rapidly varying magnetic field.

Numerous studies suggest about the ability of iron oxide nanoparticles (Fe_3O_4 and γ-Fe_2O_3) to remove heavy metals from contaminated water (Hua et al., 2012). Besides, iron oxide (α-Fe_2O_3) with hematite crystalline structure has attracted much attention as a potentially convenient material to be used in photoelectrochemical (PEC) water splitting, performed via the reaction: $H_2O \rightarrow \frac{1}{2}O_2+H_2(E_0= 1.23V)$, which is a way to convert solar energy into chemical energy, and a major source of H_2.

In references, many crystalline modifications of chromium oxides such as rutile (CrO_2), CrO_3, CrO_4, corundum (Cr_2O_3), Cr_2O_5, and Cr_5O_{12} have been reported. Among these modifications, Cr_2O_3 is one of the most important wide band gap ($E_g \approx$ 3eV) *p*-type semiconductor transition metal-oxide material with eskolaite-like structure. This kind of *p*-type wide band gap oxide semiconductor may be a good candidate for UV-light emitter using nanolasers, and optical storage system. Moreover, chromium oxide (Cr_2O_3) has many applications including as a gas sensor (Suryawanshi, Patil, & Patil, 2008), cathode for lithium cells (Takeda, Kanno, Tsuji, Yamamoto, & Taguch, 1983), rechargeable sodium ion batteries (Feng et al., 2016), and as a catalyst for industrial processes (Abu-Zied, 2000). Cr_2O_3 films, due to their high hardness, low friction coefficient, and excellent chemical stability, are also used as protective films and anti-wear surface on mechanical components (Bagde, Sapate, Khatirkar, Vashishtha, & Tailor, 2018). At the same time, chromium oxide coatings usually exhibit great brittleness and high porosity and only composite coatings based on this material with metals, single ceramics and metal/ceramic inclusions are deprived of such disadvantage (Yang et al., 2019).

However, the magnetic properties of chromium oxide and composite structures based on it are of the greatest interest. Cr_2O_3 is antiferromagnetic up to $T_{NCr2O3} =$ 307K and magnetoelectric – material in which magnetic and electric order coexist. In contrast, chromium dioxide (CrO_2) is strongly ferromagnetic at room temperature (T_C= 393K), with a half-metallic band structure fully spin-polarized at the Fermi level (Dalakova et al., 2012; Ji et al., 2019). Two-phase system Cr_2O_3-CrO_2 shows remarkable magnetoresistance due to intergrain tunneling between high spin polarized crystals. The growing temperature of such composite structure is of great importance, since Cr_2O_3 is the most stable magnetic-dielectric oxide-material whereas CrO_2 easily decomposes into the insulating antiferromagnetic Cr_2O_3 phase (Sousa, Silvestre, Popovici, & Conde, 2005).

Up to now, Cr_2O_3 has been the most promising material for realistic applications close to room temperature in magnetoelectric-controlled spintronic elements like MERAM.

6.11 HEMATITE (A-Fe_2O_3) PHOTOANODES FOR SOLAR WATER SPLITTING

Important properties of iron oxide are its low toxicity, earth abundant, and suitable redox potential for photocatalytic water dissociation as a source of H_2 fuel (Hamann, 2012). But, one of the major limiting factors affecting the utilization of hematite as a photocatalyst includes fast electron–hole recombination, facilitated in part by slow carrier transport kinetics. Current research aims to improve the efficiency with which charge carriers can be separated in α- Fe_2O_3 via control of the morphology, and significant effort is made to reduce the required overpotential through surface engineering. To overcome recombination limitations, a variety of approaches have been suggested such as

- elemental doping to boost electrical conductivity (Malviya et al., 2016);
- designing heterojunctions for efficient charge carrier separation (Kaspar et al., 2016);
- increasing photocurrent with morphology control and surface modification (Pyeon et al., 2018); and

- a dual absorber approach [$TiO_2/\alpha\text{-}Fe_2O_3$ (Luan, Xie, Liu, Fu, & Jing, 2014), Fe_2O_3 /WO_3 (Jin et al., 2014; Müller et al., 2017; Sivula, Le Formal, & Grätzel, 2009, 2011)].

One route to improve excited carrier transport properties is reducing the path length that carriers must travel (Kay, Cesar, & Gratzel, 2006). Another method is to introduce material components to provide a dedicated charge-transport pathway. A general strategy of forming heteronanostructures to help meet the charge transport challenge is presented in Lin, Yuan, Sheehan, Zhoua, and Wang (2011). Investigation of the hematite-based heteronanostructures and their performance for water splitting by time-resolved photoconductivity measurements verifies the hypothesis that the integration of conductive components such as webbed nanonets or vertically aligned transparent conductive nanotubes with hematite indeed increases charge lifetimes in such structures (see Figure 6.12). The PEC performance of the resulting heteronanostructures was compared with that of planar films, and the difference was obvious, as shown in Figure 6.12 (b) and (b′). The incident photon-to-charge

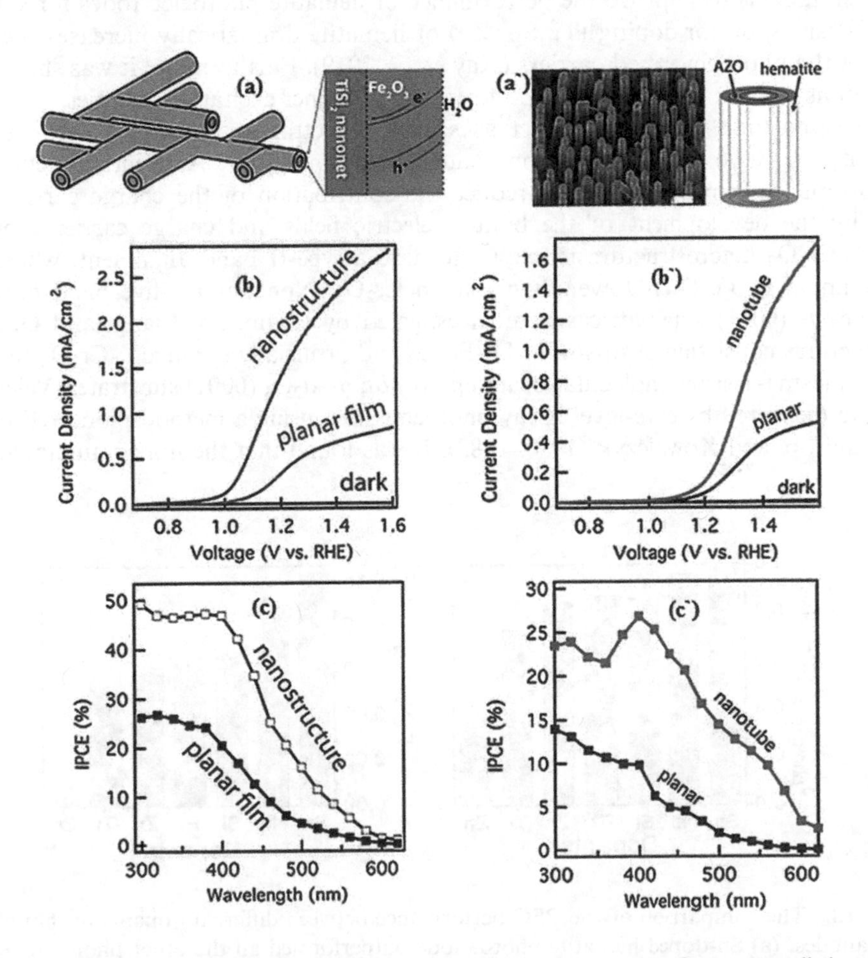

FIGURE 6.12 Hematite-based heteronanostructures and their performance for water splitting: (a) $TiSi_2$-nanonets-α-Fe_2O_3, (a′) – Al:ZnO-nanotubes-α-Fe_2O_3. The comparison of the PEC performance between (b) $TiSi_2$-nanonets-α-Fe_2O_3 and planar hematite, (b′) Al:ZnO-nanotubes-α-Fe_2O_3 and planar hematite. The external quantum efficiency comparison for (c) $TiSi_2$-nanonets-α-Fe_2O_3 and (c′) Al:ZnO-nanotubes-α-Fe_2O_3.

(Reproduced with permission from Lin, Yuan, et al. (2011) Copyright 2011, ROYAL SOCIETY.)

conversion efficiency (IPCE) measurements verified the improvement. The IPCE of $TiSi_2$ – hematite and Al:ZnO – hematite system reaches 25–50% (see Figure 6.12 (c) and (c′)).

Numerous studies have proved that doping α-Fe_2O_3 with other transition metal atoms can be used to modify its band gap so as to tune absorbance of sunlight and to improve water photo-oxidation. The effects of Sn, Nb, Si, Pt, Zr, Ti, Zn, Ni, and Mn dopants on the PEC properties of thin (~50 nm) film hematite photoanodes are presented in Figure 6.13 (Malviya et al., 2016). The results of systematic comparison of different dopants in thin-film hematite photoanodes for solar water splitting are not always consistent with other reports on doped hematite, suggesting that the PEC properties and performance of the (α-Fe_2O_3) photoanode depend not only on the identity of the dopant but also on its concentration, distribution, and on the morphology and microstructure of the photoanode in which it is incorporated. For example, small concentrations of Zr in α-Fe_2O_3 were found to enhance the photocurrent (0.33 mA/cm^2) by ~7.2 times with respect to that in undoped films of hematite (0.045 mA/cm^2), and the PEC activity was increased to ~14% (see Figure 6.14) (Shen et al., 2013). Kay, Grave, Ellis, Dotan, and Rothschild (2016) have demonstrated the potential of heterogeneous doping (Zn and Ti) to improve the performance of hematite photoelectrodes for solar water splitting. Besides, donor doping (Ti, Sn, Zn) of hematite dramatically increases the effective mobility of the photogenerated carriers (Kay et al., 2019). Furthermore, it was shown that all hematite films possess improved PEC performance at higher excitation energies.

At the same time, dopants can act as selective electron or hole traps, which enhance conductivity, or serve as recombination centers that suppress it. Alternative method of the PEC performance improvement is to reduce the contribution of the charge carrier's recombination by the development of the built-in electric fields and charge carriers' separation. The Fe_2O_3/Cr_2O_3 heterostructure is known to have a type-II band alignment, where valence band maximum in Fe_2O_3 is lower than that in Cr_2O_3. Noncommutative band offsets at α-Cr_2O_3/α-Fe_2O_3 (0001) superlattices were investigated by Chambers, Liang, and Gao (2000). These structures consisting of unstrained α-Fe_2O_3 and artificially strained α-Cr_2O_3 were grown by oxygen-plasma-assisted molecular-beam epitaxy on α-Al_2O_3 (0001) substrates. Valence-band offsets were measured by core-level X-ray photoemission, using a method pioneered by Kraut, Grant, Waldrop, and Kowalczyk (1980, 1983). It was found that the noncommutativity in the

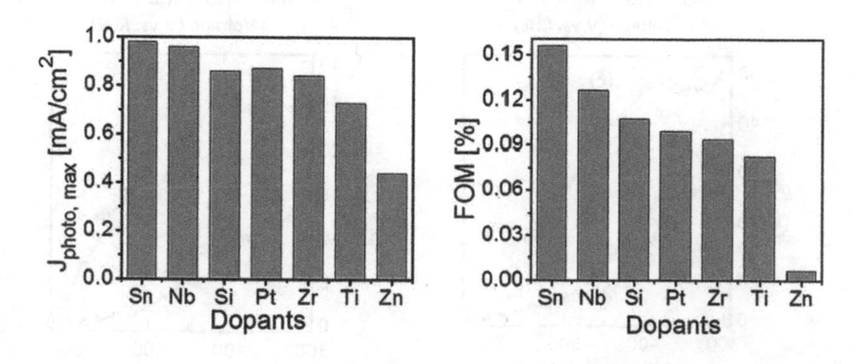

FIGURE 6.13 The comparison of the PEC performance between different dopants in thin-film hematite photoanodes: (a) Sn-doped hematite photoanode outperformed all the other photoanodes in both the photocurrent and photovoltage, achieving the highest photocurrent (~1 mA/cm^2) and lowest onset potential (~1.1V_{RHE}), (b) based on a figure of merit that accounts for the maximum photocurrent × photovoltage product (i.e., power) as well as the potential at which the maximum power is achieved, photoanodes were ranked in the following order: Sn > Nb > Si > Pt > Zr > Ti > Zn > Ni > Mn.

(Reproduced with permission from Malviya et al. (2016). Copyright 2016, ROYAL SOCIETY.)

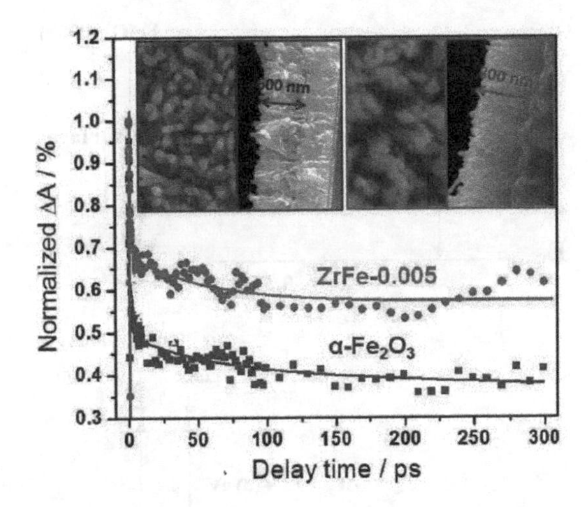

FIGURE 6.14 Normalized ultrafast transient absorption decay profiles of undoped and Zr-doped α-Fe_2O_3 (ZrFe-0.005) films. Zr doping into hematite has great effect on the thickness of nanorod films grown in an aqueous solution. Zr doping was also found to enhance the photocurrent for water splitting, by reducing the rate of electron–hole recombination.

(Reproduced with permission from Shen et al. (2013). Copyright 2013, Royal Society of Chemistry).

band offset is not due to anisotropic strain or quantum confinement, but rather appears to be due to a growth-sequence-dependent interface dipole. This feature results in an occurrence of the potential gradient over several periods of the multilayer structure (see Figure 6.15) that may be useful for enhanced effectiveness in spatially separating electrons and holes. Experimental confirmation that the noncommutative band offset property of the Fe_2O_3/Cr_2O_3 system is a consequence of the interface structure was found from the XPS spectra (Kaspar et al., 2016).

The largest difference in noncommutative band offset results when the naturally occurring Fe_2O_3 and Cr_2O_3 terminations are employed. To elucidate the effect of $(Fe_2O_3)_n$-$(Cr_2O_3)_n$ superlattice interfaces on the competition between separation and recombination of photo-generated electron–hole pairs, photocurrent spectra were investigated by Kaspar et al. (2016) on two thick (450 Å) epitaxial Fe_2O_3 films, one capped with a 54 Å thick $(Fe_{0.5}Cr_{0.5})_2O_3$ alloy layer and the other capped with a 54 Å thick $(Fe_2O_3)_3$-$(Cr_2O_3)_3$ structure. The superlattice-capped Fe_2O_3 film exhibits strong photocurrent, while a weaker photoinduced signal is observed for the alloy-capped film. Unlike a thin film alloy, in the supperlattice structure, efficient visible-light absorption occurs at the Fe_2O_3/Cr_2O_3 interfaces, where Fe–O–Cr bonding allows occupied Cr t_{2g} → unoccupied Fe t_{2g} low-energy transitions (Chamberlin et al., 2013) to occur. Both atom probe tomography and scanning transmission electron microscopy results confirm some degree of mixing at the Fe_2O_3/Cr_2O_3 interfaces in the superlattice structure (Kaspar et al., 2016). The photogenerated electrons and holes are efficiently separated into the Fe_2O_3 and Cr_2O_3 layers, resulting in the formation of the built-in potential across the supperlattice structure.

Thus, α-Fe_2O_3–α-Cr_2O_3 superlattices were shown to exhibit stronger PEC than α-$FeCrO_3$ mixed alloys. Coupling of the UV-activated photocatalysts, such as TiO_2, ZnO, or Cr_2O_3, with narrow band gap semiconductor, such as Fe_2O_3, results in the absorption band shifts to the visible region, which is more desirable in photocatalysis. Coupling of Fe_2O_3 with Cr_2O_3 not only reduces the photoexcited pairs recombination, but also activates such system in

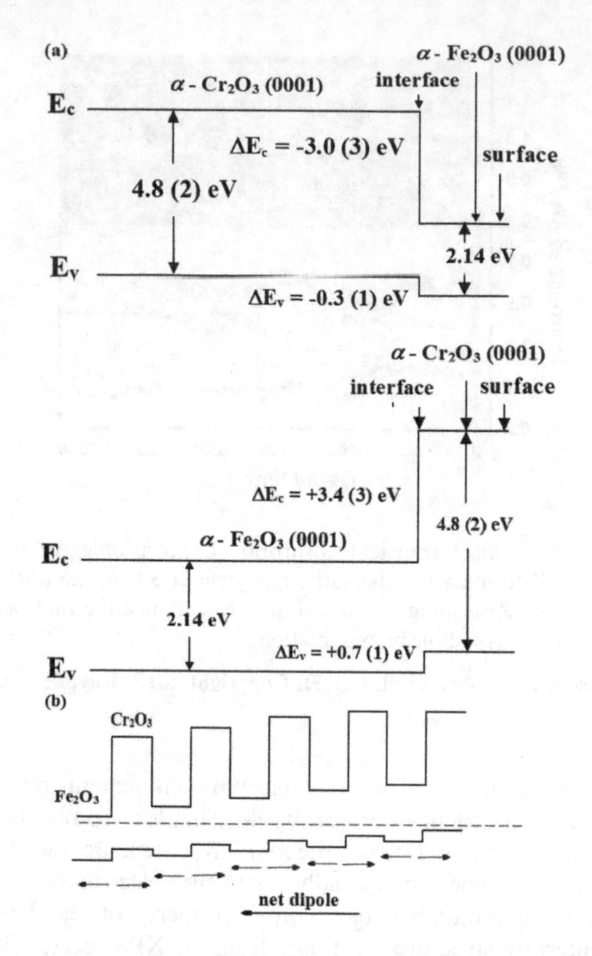

FIGURE 6.15 Energy-level diagrams showing (a) the noncommutative band offsets for α-Fe_2O_3 (0001)/α-Cr_2O_3(0001) and **α**-Cr_2O_3(0001)/**α**-Fe_2O_3 (0001) heterojunctions, and (b) the potential gradient that develops over several superlattice periods as a result of the noncommutative band offsets.

(Reproduced with permission from Chambers et al. (2000). Copyright 2000, American Physical Society).

visible light (Salari, 2019), and Fe_2O_3/Cr_2O_3 with narrow band gap and high surface area exhibited high photocatalytic activity.

One more iron oxide-based structure, $-\alpha-(Fe_{1-x}V_x)_2O_3$ at low (x = 0.04) and high (x = 0.5) doping levels, was investigated (see Chamberlin, Nayyar, Kaspar, Sushko, & Chambers, 2015; Nayyar et al., 2017). It was reported that both electrical resistivity and the band gap in $\alpha-(Fe_{1-x}V_x)_2O_3$ decrease with increasing x, which makes this material very promising for photovoltaic and PEC applications. This behavior contrasts that in $\alpha-(Fe_{1-x}Cr_x)_2O_3$, which becomes essentially insulating at $x > 0.5$. Nayyar et al. (2017) describe that at low V concentrations, the top of the valence band (VB) mainly comprises Vt_{2g} (Fe) and O_{2p} (Fe) states, which gives rise to metal-to-metal Vt_{2g}(Fe) $\rightarrow$ Fet_{2g} * low-energy excitations, reducing the onset of photo-absorption to ~0.5 eV from ~2.1 eV in pure α-Fe_2O_3. This is ~1.1 eV lower than what is observed for Cr-doped α-Fe_2O_3 thin films (Chamberlin et al., 2013). This difference is attributed to the occupied 3d3 orbitals of Cr^{3+} ions being lower in energy than the occupied 3d2 levels of V. For α-Fe_2O_3/V_2O_3 superlattices, the transitions from the top of the VB to these V–Fe hybridized states have stronger intensity and a lower energy than those in solid solutions: ~0.7 and ~1.2 eV, respectively. This demonstrates the sensitivity of spatial, atomic, and electronic arrangements of Fe and V atoms on their optical properties and suggests that layered α- Fe_2O_3/V_2O_3 superlattices are better photon absorbers than solid solutions of the same chemical composition (Nayyar et al., 2017).

REFERENCES

Abu-Zied, B. M. (2000). Structural and catalytic activity studies of silver/chromia catalysts. *Applied Catalysis A: General*, *198*(1–2), 139–153.

Ahn, Y., Seo, J., Lim, D., & Son, J. Y. (2015). Ferroelectric domain structures and polarization switching characteristics of polycrystalline $BiFeO_3$ thin films on glass substrates. *Current Applied Physics*, *15*, 584–587.

Andrew, J. S., Starr, J. D., & Budi, M. A. K. (2014). Prospects for nanostructured multiferroic composite materials. *Scripta Materialia*, *74*, 38–43.

Atou, T., Chiba, H., Ohoyama, K., Yamaguchi, Y., & Syono, Y. (1999). Structure determination of ferromagnetic perovskite $BiMnO_3$. *Journal of Solid State Chemistry*, *145*, 639–642.

Bagde, P., Sapate, S. G., Khatirkar, R. K., Vashishtha, N., & Tailor, S. (2018). Friction and wear behaviour of plasma sprayed Cr_2O_3-TiO_2 coating. *Mater Res Express*, *5*, 066424.

Balakrishnan, A., & Subramanian, K. R. V. (Ed.). (2017). *Nanostructured ceramic oxides for supercapacitor applications* (p. 209). Boca Raton, FL: CRC Press.

Bandyopadhyay, A., Frey, N. C., Jariwala, D., & Shenoy, V. B. (2019). Engineering magnetic phases in two-dimensional non-van der Waals transition-metal oxides. *Nano Letters*, *19*(11), 7793–7800.

Belik, A. A. (2012). Polar and nonpolar phases of $BiMO_3$: A review. *J Solid State Chem*, *195*, 32–40.

Belik, A. A., Iikubo, S., Yokosawa, T., Kodama, K., Igawa, N., Shamoto, S., … Takayama-Muromachi, E. (2007). Origin of the monoclinic-to-monoclinic phase transition and evidence for the centrosymmetric crystal structure of $BiMnO_3$. *J Am Chem Soc*, *129*, 971–977.

Bichurin, M., & Petrov, V. (2014). Modeling of magnetoelectric effects in composites. *Springer Series in Materials Science*, *201*. doi:10.1007/978-94-017-9156-4_1.

Binek, C., Dowben, P., Belashchenko, K., Wysocki, A., Mu, S., & Street, M. (2015) US Patent 20,150,243,414 A1.

Binitha, G., Soumya, M. S., Madhavan, A. A., Praveen, P., Balakrishnan, A., Subramanian, K. R. V., … Sivakumar, N. (2013). Electrospun α-Fe_2O_3 nanostructures for supercapacitor applications. *J Mater Chem A*, *1*, 11698–11704.

Bode, M. (2003). Spin-polarized scanning tunneling microscopy. *Rep Prog Phys*, *66*(4), 523.

Bouhjar, F., Bessaïs, B., & Marí, B. (2018). Ultrathin-layer α-Fe_2O_3 deposited under hematite for solar water splitting. *J Solid State Electrochem*, *22*, 2347–2356.

Catalan, G., & Scott, J. F. (2009). Physics and applications of bismuth ferrite. *Adv Mater*, *21*, 2463.

Čebela, M., Zagorac, D., Batalović, K., Radaković, J., Stojadinović, B., Spasojević, V., & Hercigonja, R. (2017). $BiFeO_3$ perovskites: A multidisciplinary approach to multiferroics. *Ceramics International*, *43*, 1256–1264.

Chamberlin, S. E., Nayyar, I. H., Kaspar, T. C., Sushko, P. V., & Chambers, S. A. (2015). Electronic structure and optical properties of α-$(Fe_{1-x}V_x)_2O_3$ solid-solution thin films. *Applied Physics Letters*, *106*(4), 041905–5.

Chamberlin, S. E., Wang, Y., Lopata, K., Kaspar, T. C., Cohn, A. W., Gamelin, D. R., … Chambers, S. A. (2013). Optical absorption and spectral photoconductivity in α-$(Fe_{1-x}Cr_x)_2O_3$ solid-solution thin films. *Journal of Physics: Condensed Matter*, *25*(39), 392002.

Chambers, S. A., Liang, Y., & Gao, Y. (2000). Noncummutative band offset at α−Cr_2O_3/α−Fe_2O_3 (0001) heterojunctions. *Phys Rev*, *B61*(19), 13223.

Chauhan, S., Kumar, M., Chhoker, S., & Katyal, S. C. (2016). A comparative study on structural, vibrational, dielectric and magnetic properties of microcrystalline $BiFeO_3$, nanocrystalline $BiFeO_3$ and core–Shell structured $BiFeO_3@SiO_2$ nanoparticles. *Journal of Alloys and Compound*, *686*, 454–467.

Chen, Y. B., Katz, M. B., & Pana, X. Q. (2007). Ferroelectric domain structures of epitaxial (001) $BiFeO_3$ thin films. *Appl Phys Lett*, *90*, 072907. doi:org/10.1063/1.2472092.

Choi, E. M., Patnaik, S., Weal, E., Sahonta, S. L., Wang, H., Bi, Z., … Macmanus-Driscoll, J. L. (2011). Strong room temperature magnetism in highly resistive strained thin films of $BiFe_{0.5}Mn_{0.5}O_3$. *Appl Phys Lett, 98*, 012509.

Chu, Z., PourhosseiniAsl, M., & Dong, S. (2018). Review of multi-layered magnetoelectric composite materials and devices applications. *Journal of Physics D: Applied Physics, 51*(24), 243001.

Dalakova, N. V., Belevtsev, B. I., Beliayev, E. Y., Bludov, O. M., Pashchenko, V. A., Osmolovsky, M. G., & Osmolovskaya, O. M. (2012). Resistive and magnetoresistive properties of CrO_2 pressed powders with different types of inter-granular dielectric layers. arxiv.org/ftp/arxiv/papers/1206/1206.1533.pdf.

De Luca, G. M., Preziosi, D., Chiarella, F., Di Capua, R., Gariglio, S., Lettieri, S., & Salluzzo, M. (2013). Ferromagnetism and ferroelectricity in epitaxial $BiMnO_3$ ultra-thin films. *Appl Phys Lett, 103*(6), 062902.

Ding, J., Kang, X.-B., Wen, L.-W., Li, H.-D., & Zhang, J.-M. (2014). Magnetic and ferroelectric properties of $BiCrO_3$ from first-principles calculations. *Chin Phys Lett, 31*, 107501–107504.

Ding, L., Khalyavin, D. D., Manuel, P., Blake, J., Orlandi, F., Yi, W., & Belik, A. A. (2019). Colossal magnetoresistance in the insulating ferromagnetic double perovskites Tl_2NiMnO_6: A neutron diffraction study. *Acta Materialia, 173*, 20–26.

Ding, X., Gao, B., Krenkel, E., Dawson, C., Eckert, J. C., Cheong, S. W., & Zapf, V. (2019). Magnetic properties of double perovskite Ln_2CoIrO_6 (Ln= Eu, Tb, Ho): Hetero-tri-spin 3 d− 5 d− 4 f systems. *Phys Rev B, 99*(1), 014438.

Duli´nska-Litewka, J., Łazarczyk, A., Hałubiec, P., Szafra´nski, O., Karnas, K., & Karewicz, A. (2019). Superparamagnetic iron oxide nanoparticles – Current and prospective medical applications. *Materials, 12*, 167–26.

Eerenstein, W., Mathur, N. D., & Scott, J. F. (2006). Multiferroic and magnetoelectric materials. *Nature, 442*, 759–765.

Eranna, G. (2019). *Metal oxide nanostructures as gas sensing devices.* Boca Raton, FL: CRC Press.

Feng, X., Chien, P.-H., Rose, A. M., Zheng, J., Hung, I., Gan, Z., & Hu, Y.Y. (2016). Chromium oxides as new cathodes for rechargeable sodium ion batteries. *Journal of Solid State Chemistry, 242*, 96–101.

Fiebig, M., Eremenko, V. V., & Chupis, I. E. (Ed.). (2004). *Magnetoelectric interaction phenomena in crystals.* Dordrecht: Kluwer.

Filho, R. B. M., Barbosa, D. A. B., Reichlova, H., Marti, X., Menezes, A. S. de, Ayala, A. P., & Paschoal, C. W. A. (2015). Role of rare-earth ionic radii on the spin–Phonon coupling in multiferroic ordered double perovskites. *Materials Research Express, 2*(7), 075201.

Gaikwad, V. M., Brahma, M., Borah, R., & Ravi, S. (2019). Structural, optical and magnetic properties of Pr2FeCrO6 nanoparticles. *Journal of Solid State Chemistry, 278*, 120903.

Gao, R., Zhang, Q., Xu, Z., Wang, Z., Cai, W., Chen, G., … Fu, C. (2018). Strong magnetoelectric coupling effect in $BaTiO_3$@$CoFe_2O_4$ magnetoelectric multiferroic fluids. *Nanoscale, 10*(25), 11750–11759.

Gauvin-Ndiaye, C., Tremblay, A. M., & Nourafkan, R. (2019). Electronic and magnetic properties of the double perovskites La_2MnRuO_6 and $LaAMnFeO_6$ (A= Ba, Sr, Ca) and their potential for magnetic refrigeration. *Phys Rev B, 99*(12), 125110.

Goian, V., Kamba, S., Savinov, M., Nuzhnyy, D., Borodavka, F., Vaněk, P., & Belik, A. A. (2012). Absence of ferroelectricity in $BiMnO_3$ ceramics. *J Appl Phys, 112*(7), 074112.

Greedan, J. E. (2017). Introduction to the crystal chemistry of transition metal oxides. In R. Dronskowski, S. Kikkawa, & A. Stein (Eds.), *Handbook of solid state chemistry* (1st ed.). Weinheim, Germany: Wiley-VCH Verlag GmbH & Co. KGaA.

Grizalez, M., Martinez, E., Caicedo, J., Heiras, J., & Prieto, P. (2008). Occurrence of ferroelectricity in epitaxial $BiMnO_3$ thin films. *Microelectronics Journal, 39*(11), 1308–1310.

Guo, Z., Hu, W., Liu, T., Yin, X., Zhao, X., Luo, S., … Liu, H. (2017). Hematite electron-transporting layer for environmentally stable planar perovskite solar cells with enhanced energy conversion and lower hysteresis. *J Mater Chem A, 5*, 1434–1441.

Hamann, T. W. (2012). Splitting water with rust: Hematite photoelectrochemistry. *Dalton Trans, 41*(26), 7830–7834.

Hartmann, U. (2000). *Magnetic multilayers and giant magnetoresistance: Fundamentals and industrial applications. Springer series in surface sciences.* Berlin: Springer-Verlag Berlin Heidelberg. doi:10.1007/978-3-662-04121-5.

Hill, N. A. (2000). Why are there so few magnetic ferroelectrics? *J. Phys. Chem. B, 104*(29), 6694–6709.

Hill, N. A., Bättig, P., & Daul, C. (2002). First principles search for multiferroism in $BiCrO_3$. *J Phys Chem B, 106*, 3383–3388.

Hua, M., Zhang, S., Pan, B., Zhang, W., Lv, L., & Zhang, Q. (2012). Heavy metal removal from water/wastewater by nanosized metal oxides: A review. *Journal of Hazardous Materials, 211–212*, 317–331.

Huang, C., & Chen, L. (2014). Effects of interfaces on the structure and novel physical properties in epitaxial multiferroic $BiFeO_3$ ultrathin films. *Materials (Basel)*, *7*(7), 5403–5426.

Ji, Y., Strijkers, G. J., Yang, F. Y., Chien, C. L., Byers, J. M., Anguelouch, A., … Gupta, A. (2001). Determination of the spin polarization of half-metallic CrO_2 by point contact Andreev reflection. *Phys Rev Lett*, *86*, 5585.

Jia, T., Zeng, Z., & Lin, H. Q. (2017). The collinear ↑↑↓↓ magnetism driven ferroelectricity in double-perovskite multiferroics. *IOP Conf Series: Journal of Physics: Conf Series*, *827*, 012005.

Jin, T., Diao, P., Wu, Q., Xu, D., Hu, D., Xie, Y., & Zhang, M. (2014). WO_3 nanoneedles/a-Fe_2O_3/cobalt phosphate composite photoanode for efficient photoelectrochemical water splitting. *Appl Catal B*, 148–149, 304–310.

Jung, M.-H., Yang, I. K., & Jeong, Y. H. (2013). Investigation of the magnetic and the ferroelectric properties of $BiMnO_3$ thin films. *Journal of the Korean Physical Society*, *63*(3), 624–626.

Kargol, A., Malkinski, L., & Caruntu, G. (2012). Biomedical applications of multiferroic nanoparticles. *Advanced Magnetic Materials*, 89–118. doi:10.5772/39100.

Kaspar, T. C., Schreiber, D. K., Spurgeon, S. R., Mc Briarty, M. E., Carroll, G. M., Gamelin, D. R., & Chambers, S. A. (2016). Built-in potential in Fe2O3-Cr2O3 superlattices for improved photoexcited carrier separation. *Adv Mater*, *28*(8), 1616–1622.

Kay, A., Cesar, I., & Gratzel, M. (2006). New benchmark for water photooxidation by nanostructured α-Fe_2O_3 films. *J Am Chem Soc*, *128*, 15714.

Kay, A., Fiegenbaum-Raz, M., Müller, S., Eichberger, R., Dotan, H., van de Krol, R., … Grave, D. A. (2019). Effect of Doping and Excitation Wavelength on Charge Carrier Dynamics in Hematite by Time-Resolved Microwave and Terahertz Photoconductivity. *Advanced Functional Materials*, 1901590.

Kay, A., Grave, D. A., Ellis, D. S., Dotan, H., & Rothschild, A. (2016). Heterogeneous doping to improve the performance of thin film hematite photoanodes for solar water splitting. *ACS Energy Lett*, *1*(4), 827–833.

Kim, M. K., Moon, J. Y., Oh, S. H., Oh, D. G., Choi, Y. J., & Lee, N. (2019). Strong magnetoelectric coupling in mixed ferrimagnetic-multiferroic phases of a double perovskite. *Scientific Reports*, *9* (1), 5456–10.

Kleemann, W. (2017). Multiferroic and magnetoelectric nanocomposites for data processing. *Journal of Physics D: Applied Physics*, *50*(22), 223001.

Ko, K. T., Jung, M. H., He, Q., Lee, J. H., Woo, C. S., Chu, K., … Liang, W. I. (2011). Concurrent transition of ferroelectric and magnetic ordering near room temperature. *Nature Communications*, *2*, 567.

Kota, Y., Imamura, H., & Sasaki, M. (2014). Enhancement of spin correlation in Cr_2O_3 film above Néel temperature induced by forming a junction with Fe_2O_3 layer: First-principles and Monte-Carlo study. *IEEE Transactions on Magnetics*, *50*(11), 1–4.

Kraut, E. A., Grant, R. W., Waldrop, J. R., & Kowalczyk, S. P. (1980). Precise determination of the valence-band edge in X-ray photoemission spectra: Application to measurement of semiconductor interface potentials. *Phys Rev Lett*, *44*, 1620.

Kraut, E. A., Grant, R. W., Waldrop, J. R., & Kowalczyk, S. P. (1983). Semiconductor core-level to valence-band maximum binding-energy differences: Precise determination by x-ray photoelectron spectroscopy. *Phys Rev B*, *28*, 1965.

Lawes, G., & Srinivasan, G. (2011). Introduction to magnetoelectric coupling and multiferroic films. *Journal of Physics D Applied Physics*, *44*(24), 243001. doi:10.1088/0022-3727/44/24/243001.

Lebeugle, D., Colson, D., Forget, A., Viret, M., Bataille, A. M., & Gukasov, A. (2008). Electric-field-induced spin flop in $BiFeO_3$ single crystals at room temperature. *Phys Rev Letters*, *100*(22), 227602.

Lee, B. W., Yoo, P. S., Nam, V. B., Toreh, K. R. N., & Jung, C. U. (2015). Magnetic and electric properties of stoichiometric BiMnO3 thin films. *Nanoscale Res Lett*, *10*, 47.

Lin, Y., Yuan, G., Sheehan, S., Zhoua, S., & Wang, D. (2011). Hematite-based solar water splitting: Challenges and opportunities. *Energy Environ Sci*, *4*, 4862–4869.

Lin, Y.-M., Abel, P. R., Heller, A., & Buddie Mullins, C. (2011). α-Fe2O3 nanorods as anode material for lithium ion batteries. *J Phys Chem Lett*, *2*(22), 2885–2891.

Lu, C., Hu, W., Tian, Y., & Wu, T. (2015). Multiferroic oxide thin films and heterostructures. *Appl Phys Rev, 2*, 02130–18.

Luan, P., Xie, M., Liu, D., Fu, X., & Jing, L. (2014). Effective charge separation in the rutile TiO_2 nanorod-coupled α-Fe_2O_3 with exceptionally high visible activities. *Sci Rep, 4*, 6180.

Lunkenheimer, P., Krohns, S., Riegg, S., Ebbinghaus, S. G., Reller, A., & Loid, A. (2009). Colossal dielectric constants in transition-metal oxides. *Eur Phys J Spec Top, 180*, 61–89.

Mahmoudi, M., Sant, S., Wang, B., Laurent, S., & Sen, T. (2011). Superparamagnetic iron oxide nanoparticles (SPIONs): Development, surface modification and applications in chemotherapy. *Advanced Drug Delivery Reviews, 63*, 24–46.

Maiti, T., Saxena, M., & Roy, P. (2019). Double perovskite (Sr 2 B′ B ″O 6) oxides for high-temperature thermoelectric power generation – A review. *Journal of Materials Research, 34*(1), 107–125.

Malviya, K. D., Dotan, H., Shlenkevich, D., Tsyganok, A., Mor, H., & Rothschild, A. (2016). Systematic comparison of different dopants in thin film hematite (α-Fe2O3) photoanodes for solar water splitting. *Journal of Materials Chemistry A, 4*(8), 3091–3099.

Mickel, P. R., Jeen, H., Kumar, P., Biswas, A., & Hebard, A. F. (2016). Proximate transition temperatures amplify linear magnetoelectric coupling in strain-disordered multiferroic $BiMnO_3$. *Phys Rev B, 93*, 134205.

Mishra, M., & Chun, D.-M. (2015). α-Fe2O3 as a photocatalytic material: A review. *Appl Catal A, 498*, 126–141.

Mizokawa, T., & Fujimori, A. (2002). Spin, charge, and orbital ordering in *3d* transition-metal oxides studied by model Hartree-Fock calculation. In A. Bianconi & N. L. Saini (Eds.), *Stripes and related phenomena. Selected topics in superconductivity* (Vol. 8, p. 121–128). Boston, MA: Springer.

Moon, J. Y., Kim, M. K., Choi, H. Y., Oh, S. H., Jo, Y., Lee, N., & Choi, Y. J. (2015). Direct observation of magnetodielectric effect in type-I multiferroic PbFe0.5Ti0.25W0.25O3. *Curr Appl Phys, 15*, 1545–1548.

Moreira dos Santos, A., Cheetham, A. K., Atou, T., Syono, Y., Yamaguchi, Y., Ohoyama, K., … Rao, C. N. R. (2002). Orbital ordering as the determinant for ferromagnetism in biferroic $BiMnO_3$. *Phys Rev B, 66*, 64425.

Moreira dos Santos, A., Parashar, S., Raju, A. R., Zhao, Y. S., Cheetham, A. K., & Rao, C. N. R. (2002). Epitaxial growth and properties of metastable $BiMnO_3$ thin films. *Solid State Commun, 122*, 49.

Müller, A., Kondofersky, I., Folger, A., Fattakhova-Rohlfing, D., Bein, T., & Scheu, C. (2017). Dual absorber Fe_2O_3/WO_3 host-guest architectures for improved charge generation and transfer in photoelectrochemical applications. *Mater Res Express, 4*, 016409.

Nakamura, M., Tokunaga, Y., Kawasaki, M., & Tokura, Y. (2011). Multiferroicity in an orthorhombic YMnO 3 single-crystal film. *Appl Phys Lett, 98*, 082902.

Nayyar, I. H., Chamberlin, S. E., Kaspar, T. C., Govind, N., Chambers, S. A., & Sushko, P. V. (2017). Effect of doping and chemical ordering on the optoelectronic properties of complex oxides: Fe_2O_3 –V_2O_3 solid solutions and hetero-structures. *Physical Chemistry Chemical Physics, 19*(2), 1097–1107.

Newnham, R. E., Skinner, D. P., & Cross, L. E. (1978). Connectivity and piezoelectric-pyroelectric composites. *Mat Res Bull, 13*, 525–536.

Oh, S. H., Choi, H. Y., Moon, J. Y., Kim, M. K., Jo, Y., Lee, N., & Choi, Y. J. (2015). Nonlinear magnetodielectric effect in double-perovskite Gd_2NiMnO_6. *J Phys D: Appl Phys, 48*, 445001.

Ortega, N., Kumar, A., Scott, J. F., & Katiyar, R. S. (2015). Multifunctional magnetoelectric materials for device applications. *J Phys: Condens Matter, 27*, 504002.

Palneedi, H., Annapureddy, V., Priya, S., & Ryu, J. (2016). Status and perspectives of multiferroic magnetoelectric composite materials and applications. *Actuator, 5*, 9–40.

Pomiro, F., Sánchez, R. D., Cuello, G., Maignan, A., Martin, C., & Carbonio, R. E. (2016). Spin reorientation, magnetization reversal, and negative thermal expansion observed in $RFe_{0.5}Cr_{0.5}O_3$ perovskites(R=Lu,Yb,Tm). *Phys Rev B, 94*, 13.

Pyeon, M., Ruoko, T.-P., Leduc, J., Gönüllü, Y., Deo, M., Tkachenko, N. V., & Mathur, S. (2018). Critical role and modification of surface states in hematite films for enhancing oxygen evolution activity. *Journal of Materials Research, 33*(4), 455–466.

Rahmani, N., Ghazi, M. E., Izadifard, M., Wang, D., Shabani, A., & Sanyal, B. (2019). Density functional study of structural, electronic and magnetic properties of new half-metallic ferromagnetic double perovskite Sr2MnVO6. *Journal of Physics: Condensed Matter, 31*(47), 475501.

Rajesh Kumar, R., Raj, R., & Venimadhav, A. (2019). Weak ferromagnetism in band-gap engineered α-$(Fe_2O_3)_{1-x}(Cr_2O_3)_x$ nanoparticles. *Journal of Magnetism and Magnetic Materials, 473*, 119–124.

Rao, B. N., Kaviraj, P., Vaibavi, S. R., Kumar, A., Bajpai, S. K., & Arockiarajan, A. (2017). Investigation of magnetoelectric properties and biocompatibility of $CoFe_2O_4$-$BaTiO_3$ core-shell nanoparticles for biomedical applications. *J Appl Phys, 122*(16), 164102.

Rao, C. N. R. (1993). Chemical synthesis of sold inorganic materials. *Material Science and Engineering, B18*, 1–21.

Reig, C., Cubells-Beltrán, M.-D., & Ramírez Muñoz, D. (2009). Magnetic field sensors based on giant magnetoresistance (GMR) technology: Applications in electrical current sensing. *Sensors, 9*, 7919–7942.

Roknuzzaman, M., Zhang, C., Ostrikov, K. K., Du, A., Wang, H., Wang, L., & Tesfamichael, T. (2019). Electronic and optical properties of lead-free hybrid double perovskites for photovoltaic and optoelectronic applications. *Scientific Reports, 9*(1), 718.

Roy, A., Gupta, R., & Garg, A. (2012). Multiferroic memories. *Advances in Condensed Matter Physics.* Article ID 926290. doi: 10.1155/2012/926290.

Salari, H. (2019). Kinetics and mechanism of enhanced photocatalytic activity under visible light irradiation using Cr_2O_3/Fe_2O_3 nanostructure derived from bimetallic metal organic framework. *Journal of Environmental Chemical Engineering, 7*, 103092.

Sando, D., Agbelele, A., Daumont, C., Rahmedov, D., Ren, W., Infante, I. C., … Bibes, M. (2014). Control of ferroelectricity and magnetism in multi-ferroic $BiFeO_3$ by epitaxial strain. *Phil Trans R Soc A, 372*, 20120438.

Sando, D., Barthélémy, A., & Bibes, M. (2014). $BiFeO_3$ epitaxial thin films and devices: Past, present and future. *J Phys Condens Matter, 26*(47), 473201.

Savkina, R., & Khomenkova, L. (2019). *Solid state composites and hybrid systems.* Boca Raton, FL: CRC Press.

Schwarz, A., Kaiser, U., Schmidt, R., & Wiesendanger, R. (2009). Magnetic exchange force microscopy. In S. Morita, F. Giessibl, & R. Wiesendanger (Eds.), *Noncontact atomic force microscopy. NanoScience and technology* (pp. 275–286). Berlin, Heidelberg: Springer.

Scott, J. F. (2007). Multiferroic memories. *Nature Materials, 6*, 256–257.

Shen, S., Guo, P., Wheeler, D. A., Jiang, J., Lindley, S. A., Kronawitter, C. X., … Mao, S. S. (2013). Physical and photoelectrochemical properties of Zr-doped hematite nanorod arrays. *Nanoscale, 5* (20), 9867–9874.

Shin, H. J., Lee, N., & Choi, Y. J. (2019). Nonlinear magnetodielectric effect of disordered perovskite $HoCr_{0.5}Fe_{0.5}O_3$: Role of magnetic rare-earth ions. *Journal of Alloys and Compounds, 785*, 1166–1172.

Silva, R. X., Castro Júnior, M. C., Yáñez-Vilar, S., Andújar, M. S., Mira, J., Señarís-Rodríguez, M. A., & Paschoal, C. W. A. (2017). Spin-phonon coupling in multiferroic Y_2CoMnO_6. *Journal of Alloys and Compounds, 690*, 909–915.

Sivula, K., Le Formal, F., & Grätzel, M. (2009). WO_3–Fe_2O_3 photoanodes for water splitting: A host scaffold, guest absorber approach. *Chem Mater, 21*, 2862–2867.

Sivula, K., Le Formal, F., & Grätzel, M. (2011). Solar water splitting: Progress using hematite (α-Fe2O3) photoelectrodes. *Chem Sus Chem, 4*, 432–449.

Smirnov, A. B., Kryvyi, S. B., Mulenko, S. A., Sadovnikova, M. L., Savkina, R. K., & Stefan, N. (2016). Structural and magnetoresistive properties of nanometric films based on iron and chromium oxides on the Si substrate. *Nanoscale Res Lett, 11*, 467.

Son, J. Y., & Shin, Y. H. (2008). Multiferroic BiMnO3 thin films with double $SrTiO_3$ buffer layer. *Appl Phys Lett, 93*, 062902.

Sosnowska, I., Loewenhaupt, M., David, W. I. F., & Ibberson, R. M. (1992). Investigation of the unusual magnetic spiral arrangement in $BiFeO_3$. *Physica B, 180–181*, 117–118.

Sousa, P. M., Silvestre, A. J., Popovici, N., & Conde, O. (2005). Morphological and structural characterization of CrO $_2$/Cr $_2$O $_3$ films grown by laser-CVD. *Applied Surface Science, 247*, 423–428.

Spaldin, N. A., & Ramesh, R. (2019). Advances in magnetoelectric multiferroics. *Nature Materials, 18*(3), 203–212.

Stojanovic, B. D. (2018). *Magnetic, ferroelectric, and multiferroic metal oxides.* Amsterdam, The Netherlands: Elsevier Books. Metal Oxides Series.

Sugahara, S., & Nitta, J. (2010). Spin-transistor electronics: An overview and outlook. *Proc IEEE, 98* (12), 2124–2154.

Sugahara, S., Takamura, Y., Shuto, Y., & Yamamoto, S. (2014). Devices and applications: Spin transistors and spin logic devices. In Y. Xu, D. Awschalom, & J. Nitta (Eds.), *Handbook of spintronics* (pp. 1243–1279). Dordrecht: Springer.

Suryawanshi, D. N., Patil, D. R., & Patil, L. A. (2008). Fe2O3-activated Cr2O3 thick films as temperature dependent gas sensors. *Sens Actuat B, 134*(2), 579–584.

Suzuki, E., Nakada, K., & Yonemura, S. (2018). Laminated structure and spin modulation element. US Patent 20,180,351,079 A1.

Takeda, Y., Kanno, R., Tsuji, Y., Yamamoto, O., & Taguch, H. (1983). Chromium oxides as cathodes for lithium cells. *J Power Sources, 9*(3), 325–328.

Teja, A. S., & Koh, P.-Y. (2009). Synthesis, properties, and applications of magnetic iron oxide nanoparticle. *Progress in Crystal Growth and Characterization of Materials, 55*(1–2), 22–45.

Tejuca, L. G., & Fierro, J. L. G. (2019). *Properties and applications of perovskite-type oxides.* Boca Raton, FL: CRC Press.

Thankachan, R. M., Raneesh, B., Mayeen, A., Karthika, S., Vivek, S., Nair, S. S., … Kalarikkal, N. (2018). Room temperature magnetoelectric coupling effect in $CuFe_2O_4$ -$BaTiO_3$ core-shell and nanocomposites. *Journal of Alloys and Compounds, 731*, 288–296.

Tolédano, P., Ackermann, M., Bohatý, L., Becker, P., Lorenz, T., Leo, N., & Fiebig, M. (2015). Primary ferrotoroidicity in antiferromagnets. *Physical Review B, 92*(9), 094431.

Vaz, C. A. F., Hoffman, J., Ahn, C. H., & Ramesh, R. (2010). Magnetoelectric coupling effects in multiferroic complex oxide composite structures. *Adv Mater, 22*, 2900–2918.

Vinai, G., Khare, A., Rana, D. S., Di Gennaro, E., Gobaut, B., Moroni, P., Yu., A., … Torelli, P. (2015). Unraveling the magnetic properties of $BiFe_{0.5}Cr_{0.5}O_3$ thin films. *APL Materials, 3*, 116107-7.

Waldram, J. R. (1996). *Superconductivity of metals and cuprates* (p. 410). Bristol: Institute of Physics.

Wang, C., Yin, L., Zhang, L., Xiang, D., & Gao, R. (2010). Metal oxide gas sensors: Sensitivity and influencing factors. *Sensors, 10*(3), 2088–2106.

Wiesendanger, R. (2009). Spin mapping at the nanoscale and atomic scale. *Rev Mod Phys, 81*, 1495–1550.

Wu, J., Fan, Z., Xiao, D., Zhu, J., & Wang, J. (2016). Multiferroic bismuth ferrite-based materials for multifunctional applications: Ceramic bulks, thin films and nanostructures. *Progress in Materials Science, 84*, 335–402.

Xu, Q., Sheng, Y., Khalid, M., Cao, Y., Wang, Y., Qiu, X., … Du, J. (2015). Magnetic interactions in $BiFe_{0.5}Mn_{0.5}O_3$ films and $BiFeO_3$/$BiMnO_3$ superlattices. *Scientific Reports, 5*, 9093.

Yang, H., Zhang, G., & Lin, Y. (2015). Enhanced magnetoelectric properties of the laminated $BaTiO_3$/$CoFe_2O_4$ composites. *Journal of Alloys and Compounds, 644*, 390–397.

Yang, X., Dong, S., Zeng, J., Zhou, X., Jiang, J., Deng, L., & Cao, X. (2019). Sliding wear characteristics of plasma-sprayed Cr2O3 coatings with incorporation of metals and ceramics. *Ceramics International, 45*, 20243–20250.

Yokosawa, T., Belik, A. A., Asaka, T., Kimoto, K., Takayama-Muromachi, E., & Matsui, Y. (2008). Crystal symmetry of $BiMnO_3$: Electron diffraction study. *Phys Rev B, 77*, 024111.

Zhai, L.-J., & Wang, H.-Y. (2017). The magnetic and multiferroic properties in BiMnO. *Journal of Magnetism and Magnetic Materials, 426*, 188–194.

Zhang, J. X., Zeches, R. J., He, Q., Chu, Y. H., & Ramesh, R. (2012). Nanoscale phase boundaries: A new twist to novel functionalities. *Nanoscale, 4*(20), 6196–6204.

7 Phonon and Plasmon–Phonon Interactions in ZnO Single Crystals and Thin Films

Oleksandr Melnichuk, Lyudmyla Melnichuk and Evgen Venger

7.1 INTRODUCTION

Optics of the surface of solids is the field of materials science that attracts considerable attention. It had arisen in the past millennium and it still continues its development. This considerable interest is governed by optical phenomena appeared due to the interaction of electromagnetic waves with lattice vibrations and charge carriers in materials.

The investigation of elementary excitations (such as phonons, plasmons, and polaritons) and their interactions with different materials and structures is one of the interesting problems of modern surface physics. The first theoretical consideration of elementary excitations such as *polaritons* was performed by K.B. Tolpygo (Tolpygo, 1950) who together with G.I. Pekar called them as "light-excitons." Later, J. Hopfield proposed the term "polaritons" for such excitations (Hopfield, 1958). Another type of such excitations, known today as "phonon polaritons," and their dispersion relation was also reported by Tolpygo in 1950 (Tolpygo, 1950) and, independently, by K. Huang in 1951 (Huang, 1951) for ionic crystals.

Surface plasma waves, that is, collective charge carrier motion (also called as surface plasmons (SPs)), are transverse magnetic electromagnetic waves, traveling along interfaces of two different media. The guiding of such surface waves can be considered in similar manner as light guiding in an optical fiber. The highest amplitude of electric field is observed on the solid surface being exponentially decreased from the interface toward media volume.The first study on surface plasmon–polaritons (SPPs), which are the result of the coupling of SPs with light whose wavelength depends on material nature and its geometry, was performed by A.Z. Otto in 1968 (see, for instance, Otto, 1968). SPPs are characterized by shorter wavelength than that of incident light (photons). They can have tighter spatial confinement and higher local field intensity. In the direction perpendicular to the interface, they have subwavelength-scale confinement. SPPs will propagate along the interface until their energy will be lost due to either absorption in the material (metal, semiconductor) or scattering into other directions (such as into free space).

SPPs can be both radiative and nonradiative. Radiative ones, being coupled with plane electromagnetic waves, are involved in emission process or plasma-resonance absorption. Nonradiative SPPs are known as solutions of Maxwell's equations (their theory concerns the consideration of the complex dielectric function) and result in total reflection. Both types of SPPs have different applications, for instance, in subwavelength optic microscopy or lithography beyond the diffraction limit, in photonic data storage, light generation and biophotonics, in the determination of electrophysical parameters of materials such as carrier concentration, mobility, and conductivity, or even in the measurement of a fundamental property of light itself such as the momentum of a photon in a dielectric medium. The investigation of such excitations and their collective interactions allowed to study not only inorganic solids, but also the organic ones and to develop different sensors. Recently, it was reported that specific light form, which may involve polaritons, could be useful in the development of quantum computers (Liang et al., 2018).The SPs and SPPs in isotropic solids have been studied in detail (see for instance, Agranovich & Mills, 1985; Dmitruk, Litovchenko, & Stryshevskii, 1989; Tarkhanyan & Uzunoglu, 2006; Venger, Melnichuk, & Pasychnik, 2001; Vinogradov & Dorofeev, 2009). However, such excitations and their interaction in anisotropic crystals were less addressed due to the difficulties appeared in the theoretical modeling of these excitations (Agranovich & Dubovskii, 1965; Melnichuk, 1998; Melnichuk, Melnichuk, & Pasechnik, 1998; Melnichuk & Pasechnik, 1998; Venger et al., 1995). Indeed, contrary to the consideration of isotropic solids, the modeling of dispersion relations of anisotropic ones requires to involve additional characteristic parameters, that is, the anisotropy of the dispersion laws of surface plasmon–phonon polaritons (SPPPs), effect of free carrier concentration and optical axis orientation with respect to SPPP propagation direction on the number of dispersion branches, anisotropy of damping coefficients of ordinary and extraordinary SPPP, and so on.

There are several techniques used to investigate phonon and plasmon systems of materials as well as the plasmon–phonon coupling. Among them, the most addressed are high-resolution electron energy loss spectroscopy and inelastic neutron scattering. Coupled with theoretical calculations, these techniques provide relevant information on free-carrier concentrations in the bulk and in the subsurface regime, band bending, phonon frequency dispersions, electron mobility, and plasmon damping for homogeneously doped materials. However, both techniques require specific complex equipment. In this regard, optical methods such as Raman scattering and infrared spectroscopy become more attractive. These methods are nondestructive and sensitive to materials' nature.

Among different approaches of infrared spectroscopy used for the investigation of elementary excitations and their interactions, the most attractive are specular infrared reflection (IRR) and attenuated total reflection (ATR). Both of them are sensitive to the interaction of the plasmons with the phonons because their coupling affects significantly the shape of the IRR and ATR spectra in the range of residual rays when the plasmon frequency (ν_p) is approaching the frequency of longitudinal optical phonon (ν_L). Modeling of these spectra allows, for instance, the concentration and mobility of free carriers, as well as material conductivity to be determined in semiconductor materials.

For the modeling of the IRR and ATR spectra in the range of residual rays of corresponding materials, the dielectric permittivity of the whole system is represented by the sum of the "phonon" and "plasmon" terms taking into account plasmon–phonon interaction (Melnichuk, 1998; Venger, Melnichuk, Melnichuk, & Semikina, 2016). When $\nu_p < \nu_L$, such interaction is insignificant, thus for the modeling of IRR spectrum only the contribution of phonon subsystem in dielectric permittivity of the material can be used. By contrast, for $\nu_p \geq \nu_L$, both "phonon" and "plasmon" terms are important. Otherwise, significant discrepancy will appear between the theoretical simulation and experimental spectra in the range of "residual rays" of the investigated materials.

In this chapter, main attention will be paid to theoretical consideration of surface polaritons and SPPPs in anisotropic materials. Besides, the effect of magnetic field on such collective excitations will be considered. As an example of anisotropic materials, zinc oxide (both single crystals and thin films) will be addressed.

ZnO has wurtzite structure with the $C_{6v}^4(P6_3mc)$ space group and shows strong anisotropy of the properties of the phonon and plasmon–phonon subsystems (Dmitruk et al., 1989; Kuzmina & Nikitenko, 1984; Venger et al., 2001). It attracts much attention among different polar materials because of the combination of unique optical, mechanical, and piezoelectric properties. This wide-bandgap semiconductor $\left(E_g = 3{:}37\text{ eV}\right)$ possesses high chemical inertness and durability in ambient conditions (Ellmer, Klein, & Rech, 2008). The growth of ZnO single crystals is a complicated process that requires a specific high-temperature equipment. Due to some technological difficulties in the growth of ZnO crystals and, as a consequence, their high cost, the production of ZnO films becomes more attractive today. One of the advantages of ZnO films is their capability to keep typical properties of single crystals. Another one is the possibility to grow with high structural quality at low temperature (even at room or near-room temperature) (Fortunato et al., 2005); Gieraltowska, Wachinski, Witkowski, Godlewski, & Guziewicz, 2012). These films can be used in different optical devices such as light-emitting devices, optical detectors, as well as in solar-cell windows, ultraviolet detectors, and so on (Ellmer et al., 2008; Fortunato et al., 2005).

One of the specific feature of ZnO is high anisotropy of phonon system and a low anisotropy of plasmon one. Consequently, the interaction of long-wavelength optical vibration of the lattice with electron plasma in ZnO materials will be different for the cases $E \perp c$ and $E||c$. This will affect IRR spectra obtained for both the orientations and their analysis permits to reveal the anisotropy of optical and electrophysical properties of these materials. Such effect was earlier demonstrated for pure and doped ZnO single crystals (Venger et al., 2001), as well as for pure and rare-earth-doped ZnO films grown semiconductor and insulating substrates such as 6H-SiC, Si, SiO_2, and Al_2O_3 (Korsunska et al., 2019; Melnichuk, Melnichuk, Korsunska, Khomenkova, & Venger, 2019; Melnichuk, Melnichuk, Tsykaniuk, et al., 2019; Venger et al., 2001, 2016). Using nondestructive and fast enough IRR method, free carrier concentration and their mobility as well as damping coefficients were estimated for such textured films when the results of direct electrical measurements were controversial due to a specific, column-like structure of the films (Korsunska et al., 2019; Melnichuk, Melnichuk, Tsykaniuk, et al., 2019).

7.2 SPPPS IN ZINC OXIDE SINGLE CRYSTALS

The study of the influence of strong anisotropy of the crystal lattice and weak anisotropy of the effective mass of electrons on the excitation and propagation of nonradiative SPPPs in uniaxial semiconductors has attracted the attention of the researchers only in last decade. Hereafter, the theory of the propagation of electromagnetic waves in hexagonal ZnO single crystals will be considered. The relationship of surface electromagnetic waves with plasma oscillations of free charge carriers (electrons) and with long-wavelength optical lattice vibrations is reported. It should be noted that the developed theoretical models can be implemented for the determination of the optical and electrophysical parameters of not only undoped and doped ZnO single crystals and thin films, but also of other anisotropic materials.

7.2.1 General Theory of Surface Plasmon–Phonon Polaritons in Anisotropic Crystals

The theory of SPPPs in anisotropic crystals was described by Gurevich and Tarkhanyan (1975) and developed later by Tarkhanyan and Uzunoglu (2006). According to this theory, a uniaxial crystal occupies the region $z < 0$, and the optical axis of the crystal lies in the xz plane and has an angle θ with the z axis. Then, the tensor of dielectric constant $\varepsilon_{ij}(\nu)$ has the form presented by Eq. (7.1):

$$\varepsilon_{ij} = \begin{Bmatrix} \varepsilon_{\perp}\cos^2\theta + \varepsilon_{||}\sin^2\theta & 0 & \sin\theta\cos\theta(\varepsilon_{||} - \varepsilon_{\perp}) \\ 0 & \varepsilon_{\perp} & 0 \\ \sin\theta\cos\theta(\varepsilon_{||} - \varepsilon_{\perp}) & 0 & \varepsilon_{\perp} sin^2\theta + \varepsilon_{||}\cos^2\theta \end{Bmatrix}, \tag{7.1}$$

where $\varepsilon_{\perp}$ and $\varepsilon_{||}$ are the components of $\varepsilon_{ij}(\nu)$, which in the region of plasmon–phonon interaction (without decay) are given by the expressions (7.2) – (7.4):

$$\varepsilon_{\perp} = \varepsilon_{\infty\perp}\left[\left(v^2 - v_{L\perp}{}^2\right)/\left(v^2 - v_{T\perp}{}^2\right)\right] - \mu_{\perp}\left(v_{P\perp}/v^2\right), \tag{7.2}$$

$$\varepsilon_{||} = \varepsilon_{\infty||}\left[\left(v^2 - v_{L||}{}^2\right)/\left(v^2 - v_{T||}{}^2\right)\right] - \mu_{||}\left(v_{P||}/v^2\right), \tag{7.3}$$

$$v_{L\perp}{}^2 = v_{T\perp}{}^2\left(\varepsilon_{0\perp}/\varepsilon_{\infty\perp}\right), v_{L||}{}^2 = v_{T||}{}^2\left(\varepsilon_{0||}/\varepsilon_{\infty||}\right), \tag{7.4}$$

where v is the frequency (in cm^{-1}), $\varepsilon_{\infty\perp,||}$ and $\varepsilon_{0\perp,||}$ are the main values of the high-frequency and static permittivity tensors perpendicular and parallel to the *C*-axis; $v_{T\perp,||}$ and $v_{L\perp,||}$ are the frequencies of transverse and longitudinal optical phonons; $\mu_{\perp,||}$ are the main values of the dimensionless tensor of the inverse effective mass.

When the wave vector of surface vibration ***K*** is perpendicular to the *C*-axis (***K*⊥*C***) and *xy*||***C*** ($\theta = \pi/2$), the SPPP is characterized by the dispersion relation:

$$\chi_x^2 = \varepsilon_{\perp}(v)/(1 + \varepsilon_{\perp}(v)),\ \chi_x = Kc/\omega, \tag{7.5}$$

In this case, the dispersion of surface waves is determined by dielectric function $\varepsilon_{\perp}(v)$ only, when $\varepsilon_{||}(v) < -1$.

It will be considered only the solutions of Eq. (7.5) for which $\chi_x^2 > 1(K > \omega/c)]$, where ω is the angular frequency and c is light velocity), that is, non-radiation SPPPs. In this case, two branches of SPPP appear. The high-frequency branch v^+ begins at the point $v = v_{T\perp}$. With the K increasing ($K \gg \omega/c$), the frequency v approaches asymptotically the value:

$$\begin{cases} v^{\pm} = (1/2)^{1/2}\left\{\tilde{v}_{T\perp}{}^2 + \tilde{v}_{P\perp}{}^2 \pm \left[\left(\tilde{v}_{T\perp}{}^2 + \tilde{v}_{P\perp}{}^2\right)^2 - 4\tilde{v}_{P\perp}{}^2\tilde{v}_{T\perp}{}^2\right]^{1/2}\right\}^{1/2}, \\ \tilde{v}_{T\perp}{}^2 = \left[(1+\varepsilon_{0\perp})/(1+\varepsilon_{0||})\right]v_{T\perp}{}^2, \quad \tilde{v}_{P\perp}{}^2 = \varepsilon_{\infty\perp}v_{P\perp}{}^2/(1+\varepsilon_{\infty\perp}), \end{cases} \tag{7.6}$$

this corresponds to the solution of the dispersion equation in the case of no delay of $\varepsilon_{\perp} = -1$ (biquadratic equation). The low-frequency branch v^- exists for all K values, and the SPPP frequency v increases with K rising from zero to the solution of Eq. (7.6) taken with sign "−."

For ***K*⊥*C***, *xy*⊥***C*** ($\theta = 0$), the SPPPs are characterized by the dispersion relation:

$$\chi_x^2 = \left[\varepsilon_{||}(v) - \varepsilon_{\perp}(v)\varepsilon_{||}(v)\right]/\left[1 - \varepsilon_{\perp}(v)\varepsilon_{||}(v)\right] \tag{7.7}$$

and inequalities:

$$\varepsilon_{\perp} < 0, K > \omega/c,\ \varepsilon_{||}\left(\chi_x^2 - \varepsilon_{||}\right) < 0. \tag{7.8}$$

The condition (7.8) along with (7.7) is correct either for $\varepsilon_{||} > \chi_x^2$ or for $\varepsilon_{||} < -(1/|\varepsilon_{\perp}|)$, that is,

$$\begin{aligned} &1)\ \varepsilon_{\perp} < 0, \quad \varepsilon_{||} < 0; \\ &2)\ \varepsilon_{\perp} < 0, \quad \varepsilon_{||} > \chi_x^2. \end{aligned} \tag{7.9}$$

This case differs qualitatively from the isotropic one, that is, new SPPP branches appear. Their number and the range of existence depend on the concentration of the electrons in the conduction band as well as on the values of $v_{T\perp,||}$, $v_{L\perp,||}$, $v_{P\perp,||}$, $v_{\perp,||}{}^{+,-}$, and $\Omega_{\perp,||}{}^{+,-}$ that are determined by the relations:

$$\begin{cases} \varepsilon_\perp(\nu_\perp{}^{+,-}) = 0,\ \varepsilon_{||}(\nu_{||}{}^{+,-}) = 0,\ \varepsilon_\perp(\Omega_\perp{}^{+,-}) = 1,\ \varepsilon_{||}(\Omega_{||}{}^{+,-}) = 1, \\ \nu_{\perp,||}{}^{+,-} = (1/2)^{1/2}\left\{(\nu_{L\perp,||})^2 + (\nu_{P\perp,||})^2 \pm \left[\left((\nu_{L\perp,||})^2 + (\nu_{P\perp,||})^2\right)^2 - 4(\nu_{P\perp,||})(\nu_{T\perp,||})^2\right]^{1/2}\right\}^{1/2} \end{cases} \tag{7.10}$$

and

$$\begin{cases} \Omega_{\perp,||}{}^{+,-} = (1/2)^{1/2}\left\{(\tilde{\nu}_{T\perp,||})^2 + (\tilde{\nu}_{P\perp,||})^2 \pm \left[\left((\tilde{\nu}_{T\perp,||})^2 + (\tilde{\nu}_{P\perp,||})^2\right)^2 - 4(\tilde{\nu}_{P\perp,||})(\nu_{T\perp,||})^2\right]^{1/2}\right\}^{1/2}, \\ \nu_{T\perp,||}{}^2 = \left[(\varepsilon_{0\perp,||}{}^{-1})/(\varepsilon_{\infty\perp,||}{}^{-1})\right]\nu_{T\perp,||}{}^2,\ \nu_{P\perp,||}{}^2 = \left[\varepsilon_{\infty\perp,||}(\nu_{P\perp,||})^2\right]/(\varepsilon_{\infty\perp,||}{}^{-1}). \end{cases} \tag{7.11}$$

For a given K value, the maximal number of branches is 4. The curves $\nu = \nu_i(K)$ (i = 1–4), that are the solution of Eq. (7.7), increase with K rise and at $K \gg \omega/c$ that approach asymptotically the frequencies on the surface phonons corresponding to the solutions of the equation

$$\varepsilon_\perp \varepsilon_{||} = 1. \tag{7.12}$$

However, only the segments of the dispersion curves that lie in the (ν, K) plane and for those the condition (7.8) is satisfied, correspond to the surface waves. Equation (7.12) is a fourth-order one with respect to ν^2, but only one, two, or three solutions lie in the range where $\varepsilon_\perp < 0$. Consequently, three, two, or one of the dispersion curve(s) end at finite K values, whereas the number of such solutions depends on the concentration of electrons in the conduction band.

Hereafter, the effects mentioned above will be discussed, at first, for different ZnO single crystals grown by hydrothermal method. Their electrophysical parameters such as carrier concentration (n_0), their mobility (μ) and effective masses ($m_\perp$* and $m_{||}$*), as well as conductivity (σ_0) given for different directions of electric field ($\boldsymbol{E}\perp\boldsymbol{C}$ and $\boldsymbol{E}||\boldsymbol{C}$) are collected in Table 7.1. The concentration and mobility of carriers were estimated using Hall measurements. The modeling of the crystal properties will be performed using self-consistent parameters taken from Melnichuk (1998); Venger et al. (2001) and are collected in the Table 7.2.

7.2.2 The Effect of Anisotropy on the High-Frequency SPPP Dispersion in Zinc Oxide

In this section, the effect of the crystal anisotropy on the SPPP dispersion dependences is discussed for three different orientations of hexagonal ZnO. Figure 7.1a (symbols) shows three experimental dispersion curves of $\nu_s(\boldsymbol{K})$, obtained for ZnO crystal with electron concentration $n_0 = 9.3 \times 10^{16}$ cm^{-3} (sample ZO2-3, Table 7.1). They are plotted versus the dimensionless wave vector $q = Kc/\omega_{T||}$, where $\boldsymbol{K}$ is the SPPP wave vector, c is the light speed, and $\omega_{T||}$ is the angular frequency of the transverse optical ZnO phonon when the vector of electric field of the infrared radiation $\boldsymbol{E}$ is parallel to the C-axis ($\boldsymbol{E}||\boldsymbol{C}$).

The high-frequency ν^+ branches of SPPPs were obtained in accordance with Eqs. (7.5) and (7.7), taking into account the anisotropy of electron effective mass in ZnO (Melnichuk, 1998; Venger et al., 2001). Curve 1 corresponds to the case $\boldsymbol{K}||\boldsymbol{C}$ and $xy||\boldsymbol{C}$ (x and y axes lie on the surface of the sample), whereas curve 2 was obtained at $\boldsymbol{K}\perp\boldsymbol{C}$ and $xy\perp\boldsymbol{C}$. Both these dependences are the features of extraordinary SPPPs. The ordinary SPPPs are manifested in

TABLE 7.1
Electrophysical parameters of ZnO crystals grown by hydrothermal method

Sample	n_0, cm^{-3}	μ_L, cm^2 V^{-1} s^{-1}		σ_0, Ω^{-1}cm^{-1}		$m^*_{\parallel}$	$m_{\perp}^*/m_{\parallel}^*$
		$E{\perp}C$	$E \parallel C$	$E{\perp}C$	$E \parallel C$		
ZO2-3	9.3×10^{16}	150	167	22	25	0.21	1.23
ZO1-3	6.6×10^{17}	80	96	85	101	0.23	1.13
ZO6-B	2.0×10^{18}	74	50	237	243	0.22	1.18
ZC1M	4.2×10^{18}	2.2	—	148	—	—	—
ZO9	5.0×10^{19}	1.5	—	1250	—	—	—

TABLE 7.2
Self-consistent parameters for ZnO single crystals used for theoretical simulation

ZnO	ε_0	ε_∞	ν_T, cm^{-1}	ν_L, cm^{-1}
$E{\perp}C$	8.1	3.95	412	591
$E \parallel C$	9.0	4.05	380	570

the orientation $\boldsymbol{K}{\perp}\boldsymbol{C}$ and $\boldsymbol{xy}{\perp}\boldsymbol{C}$ (curve 3). As one can see, the experimental results and their theoretical simulation are in a good agreement (Figure 7.1).

The optical parameters of ZnO (such as $\varepsilon_{\perp}(\nu)$ and $\varepsilon_{\parallel}(\nu)$) (Table 7.2) were obtained based on the dispersion analysis of IRR spectra, taking into account the anisotropy of electron effective mass (Table 7.1). The plasma frequencies of free carriers $\nu_{p\perp}$ and $\nu_{p\parallel}$ are linked by the relation

$$\nu_{P\parallel} = \left[\left(m_{\perp}^{*}\varepsilon_{\infty\perp}\right)/\left(m_{\parallel}^{*}\varepsilon_{\infty\parallel}\right)\right]^{1/2}\nu_{P\perp}, \tag{7.13}$$

where $m^*_{\perp,\parallel}$ is the transverse and longitudinal effective mass of electrons. Since for ZnO $m^*_{\perp}/m^*_{\parallel} = 1.15$, hence $\nu_{p\parallel} = 1.1\nu_{p\perp}$. The frequencies of plasmon–phonon oscillations are $\nu_{pf} = 548$ cm^{-1} (curve 2) and 561 cm^{-1} (curve 3) when the plasma frequencies $\nu_{p\perp} = 90$ cm^{-1} and $\nu_{p\parallel} = 100$ cm^{-1} ($\boldsymbol{E}{\perp}\boldsymbol{C}$, $\boldsymbol{E}\|\boldsymbol{C}$). The anisotropy of SPPPs manifests itself at $Kc/\omega_{T_{\parallel}} > 2$. Thus, when $Kc/\omega_{T_{\parallel}} = 2$, the difference $\delta\nu = 18$ cm^{-1} is seen between curves 1 and 3. At the same time, when $K \rightarrow 0$, curves 1 and 2 tend to achieve the frequency $\nu_{pf} = 548$ cm^{-1}.

The inset (a) in Figure 7.1 shows in detail the segment of the $\nu_s(K)$ dependence in the 380–420 cm^{-1} range. It is seen that up to $\nu = 412$ cm^{-1}, the SPPPs of the second type are manifested. Their existence is limited by the conditions $\varepsilon_x(\nu) < 0$ and $\varepsilon_z(\nu) > K^2$. The symbols represent the experimental data that are consistent with calculated ones (solid curve). The SPPPs of type 1 start at $\nu = 412$ cm^{-1} when $\varepsilon_x(\nu) < 0$ and $\varepsilon_z(\nu) < 0$.

Since the experimental dependence $\nu_s(K)$ represented by curve 1 in Figure 7.1 is a continuous line, for its modeling, it was considered the extremum value at $\nu = 412$ cm^{-1} as "stop point."

Figure 7.1b represents the experimental ATR curves obtained for the same ZnO crystal (curves 1–5) at different air gaps between ATR crystal and ZnO surface and the angles of IR light incidence. The minima on these spectra corresponds to the frequencies $\nu_{min} = 408$, 450, 496, 518, and 527 cm^{-1}, whereas their width are $\Gamma_p = 32$, 27, 22, 17, and 15 cm^{-1}, respectively. The modeling of the ATR spectra was performed for the orientation $\boldsymbol{K}\|\boldsymbol{C}$ and $xy\|\boldsymbol{C}$ in the extremum vicinity, that is near $\nu = 412$ cm^{-1}. Consequently, for the SPPPs of types 1 and 2, it was performed for $\nu_{sp1} = 413$ cm^{-1} and $\nu_{p2} = 411$ cm^{-1} taking into account the damping

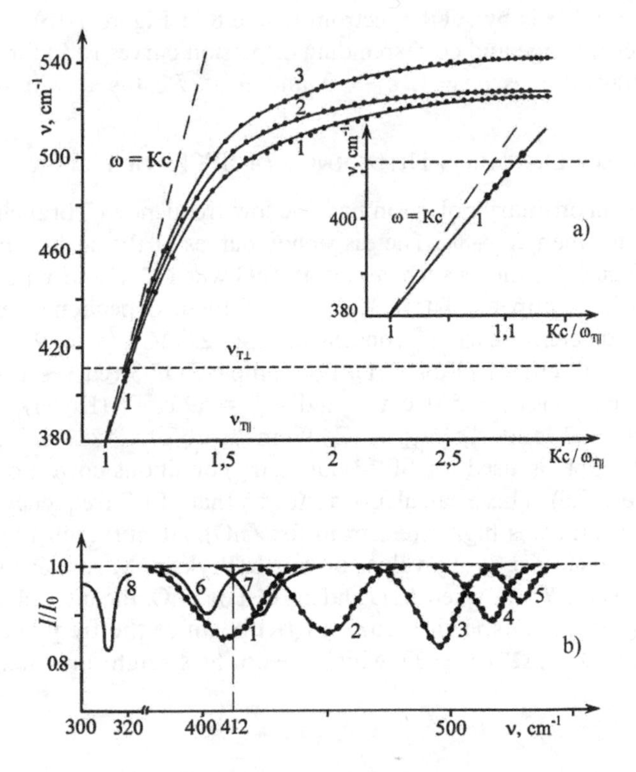

FIGURE 7.1 (**a**) Dispersion curves $\nu_s(K)$ for the SPPPs of ZnO crystal with electron concentration of $n_0 = 9.3 \times 10^{16}\text{cm}^{-3}$. The orientation of wave vector and C-axis is (1) $K||C$ and $xy||C$; (2) $K\perp C$ and $xy\perp C$; (3) $K\perp C$ and $xy||C$. Inset (**a**) shows curve 1 in detail in the range $\nu = 380 - 420\text{cm}^{-1}$; $K||C$ and $xy||C$. (**b**) ATR spectra of the same crystal recorded with an air gap 26 μm (1,2) and 3 mm (3–5), and the angle of incident light of $\alpha = 25.3°$ (1), 28° (2) 34° (3), 42° (4), and 52° (5). Curves 6 and 7 were calculated with $d = 72$ μm, $\alpha = 25.2°$ (6) and $d = 49$ μm, $\alpha = 25.7°$ (7) and demonstrate the presence of SPPPs of type 1 (7) and type 2 (6). Curve 8 is the modeling performed for the ZnO crystal with electron concentration $n_0 = 4.2 \times 10^{18}$ cm^{-3} that demonstrates the polaritons of type 3.

coefficient of transverse optical phonons $\gamma_f = 11$ cm^{-1} (Melnichuk, Melnichuk, & Pasechnik, 1994). Here, according to Eq. (7.5), the values of K_{sp1} and K_{sp2} were 1.11054 and 1.10415, respectively. Both theoretical spectra were found to have the same intensity in the minimum and the same width $\Gamma_p = 12$ cm^{-1}. The spectra practically overlap indicating that surface polaritons of both types can be generated simultaneously in the presence of damping at an emission frequency of 412 cm^{-1}. The experimental spectrum with a minimum at ν_{min} = 408 cm^{-1} and its width of $\Gamma_p = 32$ cm^{-1} (Figure 7.1b, curve 1) corresponds to the damping coefficient of SPPPs $\Gamma_{sp} = 4$ cm^{-1}. The cutoff frequencies are determined by Eq. (7.11).

Polaritons of type 2 can be observed only when $\varepsilon_\perp < 0$ and $\varepsilon_{||} > K_x^2$ (Gurevich & Tarkhanyan, 1975; Melnichuk et al., 1998). For ZnO crystals, the existence region of polaritons of type 2 is in the range 380–412 cm^{-1} (Figure 7.1a). They are exhibited when the orientation of the sample corresponds to $K||C$ and $xy||C$ as it was demonstrated experimentally by Melnichuk et al. (1998) for the first time.

One more interesting excitations is the polaritons of type 3 that can be present only in anisotropic crystals with the carrier concentration above a certain value (higher than) $n_0 = 2 \times 10^{18}\text{cm}^{-3}$ (Melnichuk et al., 1998). For instance, in ZnO crystals with

$n_0 = 4.2 \times 10^{18}$ cm^{-3}and it is by ATR spectrum (curve 8 in Figure 7.1b). The existence region lies in the low-frequency range and corresponding dispersion curves $v_s(K)$ for such polaritons are bounded by the straight lines $\varepsilon_{||} = 1$, $\varepsilon_{\perp} = 0$ and $\omega = Kc$ (as it will be discussed later).

7.2.3 Low-Frequency Dispersion Dependences of SPPPs of Hexagonal ZnO

In the isotropic case, an ordinary polariton has one low-frequency v^- branch. It begins at $v = 0$ and increases up to v^- when $K \to \infty$. The dispersion curves of the surface polaritons have two branches: $v^{+,-}(K)$. Figure 7.2 shows such curves of ZnO when $\boldsymbol{K} \perp \boldsymbol{C}$ and $\boldsymbol{xy} \| \boldsymbol{C}$. With this orientation, ordinary polaritons appear. The calculation of these dependences was performed for the samples with different carrier concentrations: ZO2–3 ($n_0 = 9.3 \times 10^{16}$ cm^{-3}) and ZO1 – 3 ($n_0 = 6.6 \times 10^{17}$ cm^{-3}) (Table 7.1). For sample ZO2-3 (curves 1 and 2), the parameters were $v_{p\perp} = 90$ cm^{-1}, $v_{pf}{}^+ = 561$ cm^{-1}, and $v_{pf}{}^- = 59$ cm^{-1} ($Kc/\omega_{T||} \to \infty$), whereas for sample ZO1-3 (curves 3 and 4): $v_{p\perp} = 240$ cm^{-1}, $v_{pf}{}^+ = 578$ cm^{-1}, $v_{pf}{}^- = 152$ cm^{-1} ($Kc/\omega_{T||} \to \infty$). Index "pf" is used for SPPP boundary conditions corresponding to the values $v^{+,-}$ obtained from Eq.(7.6). These calculations showed that SPPP frequency is higher when the concentration of free carriers is higher (electrons for ZnO). All curves in Figure 7.2 correspond to the SPPP of type 1 that exist at any values of $x_x^2 > 1$ (Bryksin, Mirlin, & Firsov, 1974).

Let us consider the case $\theta = 0$, when $K \perp C$ and $xy \perp C$. For ZnO, the interrelation of the frequencies is $v_{T||} < v_{T\perp} < v_{L||} < v_{L\perp}$. Dispersion curves v_s(K) begin at the frequencies $v = 0$; $v = v_{T\perp}$ ($\varepsilon_{\perp} = \infty$) and $v = \Omega_{||}^+$, $v = \Omega_{||}^-$ ($\varepsilon_{||} = 1$), which are on the straight line (light line) $\omega = Kc$. As

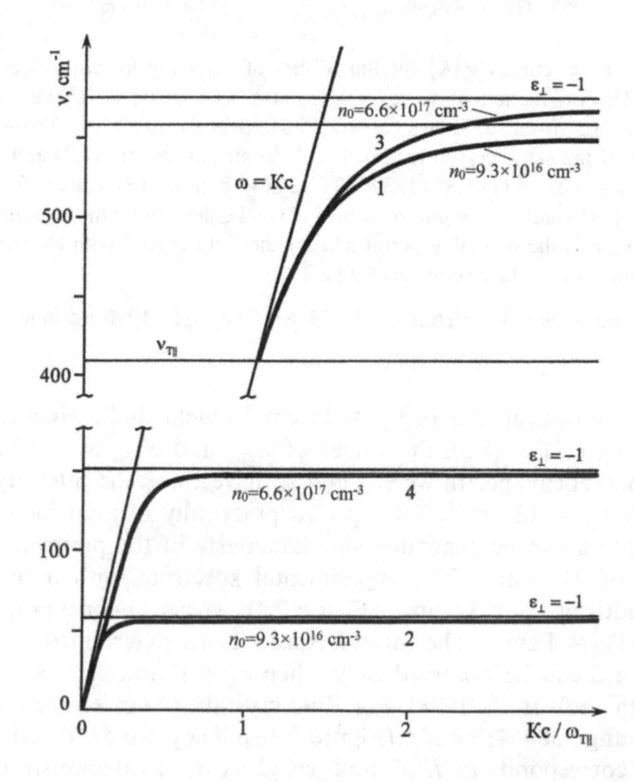

FIGURE 7.2 Dispersion curves v_s(K) for the SPPP of ZnO crystals with orientation $K \perp C$ and $xy \| C$; 1,2 – v_{p1} = 90 cm^{-1} (sample ZO2-3, $n_0 = 9.3 \times 10^{16}$ cm^{-3}); 3, 4 – v_{p1} = 240 cm^{-1} (sample ZO1-3, $n_0 = 6.6 \times 10^{17}$ cm^{-3}); 1 – $v_{pf}{}^+$ = 561 cm^{-1}; 2 – $v_{pf}{}^-$ = 59 cm^{-1}; 3 – $v_{pf}{}^+$ = 578 cm^{-1}; and 4 – $v_{pf}{}^-$ = 152 cm^{-1}

mentioned above, under certain conditions, up to four dispersion branches $\nu_s(K)$ are possible for an uniaxial semiconductor. At $K \gg \omega/c$, their number varies being depending on carrier concentration.

However, if for ZnO crystals the condition

$$\Omega_{||}^{+} < \nu < \nu_{\perp}^{+} \tag{7.14}$$

is failure then existence of only three dispersion branches $\nu_s(K)$ is possible.

Figures 7.3–7.6 show the dispersion dependences of ZnO with different carrier concentration for the orientation of $K \perp C$ and $xy \perp C$. It is seen from Figure 7.3 that for negligible carrier concentration ($n_0 = 4.0 \times 10^{15} \text{cm}^{-3}$, $\nu_p = 1\ \text{cm}^{-1}$), only the high-frequency $\nu_s(K)$ is observed being started at $\nu = \nu_{T\perp} = 412\ \text{cm}^{-1}$ and asymptotically approaches the value of $\nu_{pf}^{+} = 545.1\ \text{cm}^{-1}$ at $K_c/\omega_{T||} \to \infty$. This dispersion curve corresponds to extraordinary SPPPs of type 1 since they exist when $\varepsilon_{\perp} < 0$ and $\varepsilon_{||} < 0$, and the existence of $\nu_{pf}^{+,-}$ for $K_c/\omega_{T||} \to \infty$.

Figure 7.4 shows the dispersion dependence $\nu_s(K)$ of ZO2-3 sample with electron concentration $n_0 = 9.3 \times 10^{16}\ \text{cm}^{-3}$ when $\boldsymbol{K \perp C}$ and $\boldsymbol{xy \perp C}$, $\nu_{p\perp} = 90\ \text{cm}^{-1}$ and $\nu_{p||} = 100\ \text{cm}^{-1}$. In

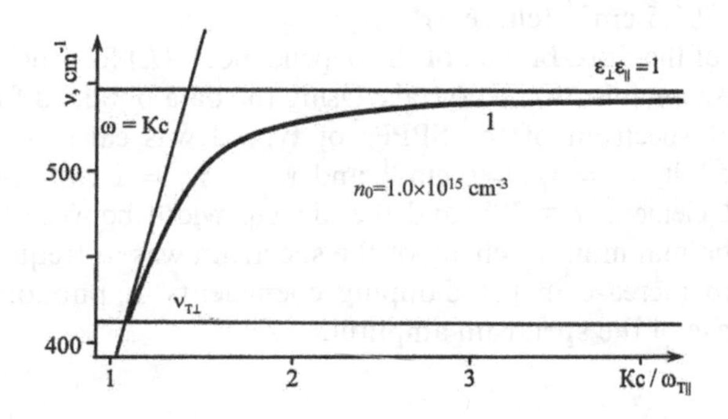

FIGURE 7.3 Dispersion curve $\nu_s(K)$ for SPPPs of ZnO ($n_0 = 4.0 \times 10^{15}\ \text{cm}^{-3}$) for orientation $\boldsymbol{K \perp C}$ and $xy \perp C$; $\nu_{p\perp} = 1\ \text{cm}^{-1}$; $\nu_{p||} = 1.1\ \text{cm}^{-1}$; 1. $\nu_{pf}^{+} = 545.1\ \text{cm}^{-1}$

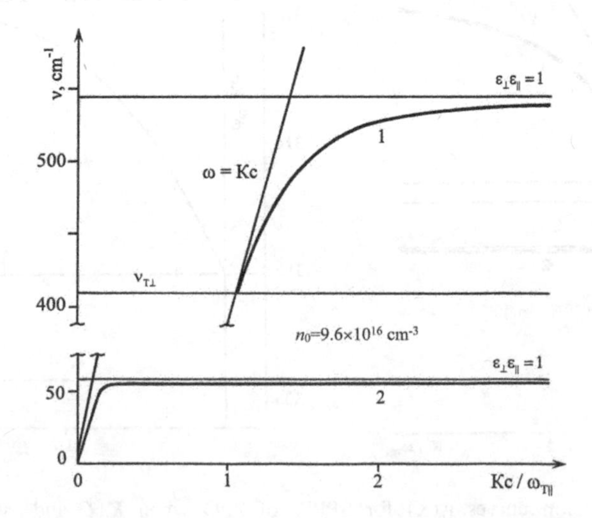

FIGURE 7.4 Dispersion curves $\nu_s(K)$ of the SPPPs in ZnO (sample ZO2-3, $n_0 = 9.6 \times 10^{16}\ \text{cm}^{-3}$) with orientation $\boldsymbol{K \perp C}$ and $xy \perp C$; $\nu_{p\perp} = 90\ \text{cm}^{-1}$; $\nu_{p||} = 100\ \text{cm}^{-1}$: 1 – $\nu_{pf}^{+} = 548\ \text{cm}^{-1}$; 2 – $\nu_{pf}^{-} = 60\ \text{cm}^{-1}$

this case, two branches $\nu_s(K)$ of the SPPPs of type 1 appear approaching the values of $\nu_{pf}^+ = 548$ cm^{-1} and $\nu_{pf}^- = 60$ cm^{-1} at $Kc/\omega_{T||} \to \infty$. Curve 2 has the features that are characteristic for plasma polaritons.

Figure 7.5 shows the $\nu_s(K)$ curves for ZnO crystal with higher electron concentration (sample ZC1M, $n_0 = 4.2 \times 10^{18}$ cm^{-3}) for the same orientation and $\nu_{p\perp} = 605$ cm^{-1} and $\nu_{p||} = 650$ cm^{-1}. The calculation shows the presence of three dispersion dependences (curves 1–3) with the next parameters: $\nu_{pf}^+ = 719$ cm^{-1} (curve 1) and $\nu_{pf}^- = 305$ cm^{-1} (curve 2). Curve 3 is shown in detail in Figure 7.5b. It is seen that it begins at a frequency of $\Omega_{||} = 309.9$ cm^{-1} at $Kc/\omega_{T||} = 0.815$ ($\varepsilon_{||} = 0$) and ends at a frequency of $\nu_{\perp}^- = 318.4$ cm^{-1} at $Kc/\omega_{T||} = 1.632$ ($\varepsilon_{\perp} = 0$).

Further increase of carrier concentration results in the transformation of the dispersion curves of the SPPPs of type 3. Figure 7.6a shows such dependences for ZnO crystal with $n_0 = 8.0 \times 10^{18}$ cm^{-3}, where $\nu_{p\perp} = 1300$ cm^{-1} and $\nu_{p||} = 1430$ cm^{-1}. In this case, $\nu_{pf}^+ = 1273.7$ cm^{-1}, and $\nu_{pf}^- = 363.58$ cm^{-1} at $Kc/\omega_{T||} \to \infty$. The curve $\nu_s(K)$ starts at a frequency of $\nu_3 = 363.8$ cm^{-1} and tends to the value $\nu_3 = 390$ cm^{-1} at $Kc/\omega_{T||} \to 0$ (Figure 7.6a, curve 3). The calculation of $\nu_s(K)$ for ZnO with higher electron concentration ($n_0 = 2.0 \times 10^{19}$ cm^{-3}, $\nu_{p\perp} = 1500$ cm^{-1} and $\nu_{p||} = 1650$ cm^{-1}) is presented in Figure 7.6b. Here, $\nu_{pf}^+ = 1450$ cm^{-1}, $\nu_{pf}^- = 367.7$ cm^{-1}. The cutoff frequency of the dispersion dependence is at $\nu = 395.5$ cm^{-1} (curve 3).

Thus, the behavior of the third branch of the dependence $\nu_s(K)$ for ZnO crystals differs from that observed previously in anisotropic crystals. Using the data obtained for the ZC1M sample (Figure 7.5), the ATR spectrum of the SPPPs of type 3 was calculated (Figure 7.7). This spectrum was obtained at $\gamma_{f\perp} = \gamma_{f||} = 6$ cm^{-1} and $\gamma_{p\perp} = \gamma_{p||} = 1$ cm^{-1}and an angle of light incidence in the ATR element $\alpha = 28°$, and the air gap width between the element and the sample $d = 26$ µm. The minimum intensity of the spectrum was at frequency $\nu = 312.5$ cm^{-1} at $Kc/\omega_{T||} = 0.92$. An increase in the damping coefficients of phonons and plasmons in ZnO leads to a decrease of the spectrum amplitude.

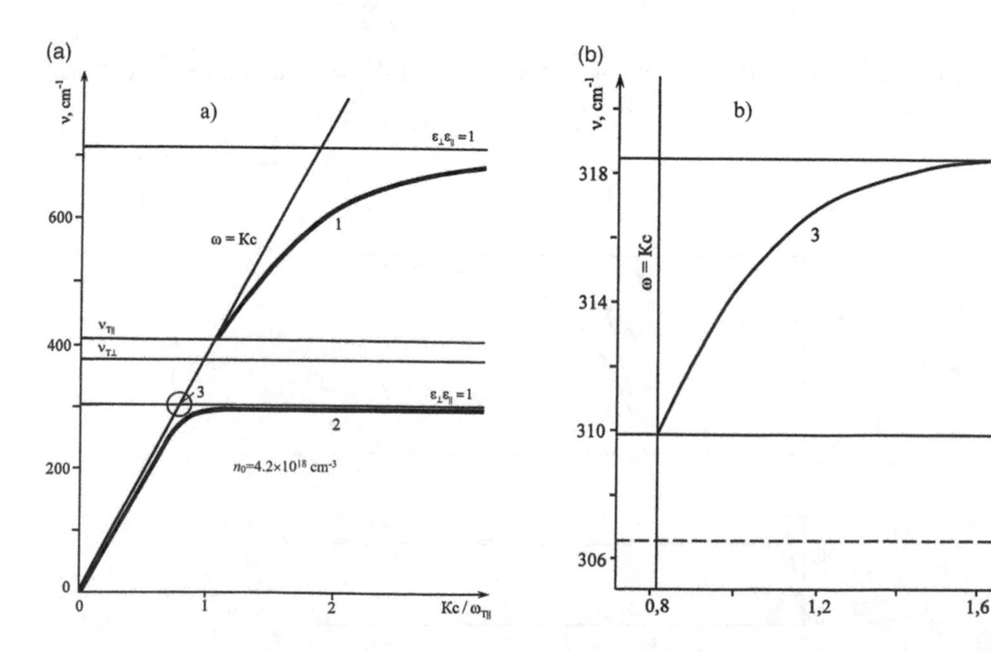

FIGURE 7.5 (**a**) Dispersion curves $\nu_s(K)$ for SPPPs of ZnO when $\boldsymbol{K}\perp\boldsymbol{C}$ and $xy\perp\boldsymbol{C}$; $\nu_{p\perp} = 605$ cm^{-1}; $\nu_{p||} = 650$ cm^{-1}(sample ZC1M, $n_0 = 4.2 \times 10^{18}$ cm^{-3}); (**b**) the detailed presentation of curve 3. $\nu_{\perp}^- = 318.4$ cm^{-1}; $Kc/\omega_{T||} = 1.632\,(\varepsilon_{\perp} = 0)$; $\Omega_{||} = 309.9\,\text{cm}^{-1}$; $Kc/\omega_{T||} = 0.815\,(\varepsilon_{||} = 1)$; $\nu_{||}^- = 306.5$ cm^{-1}; $Kc/\omega_{T||} = 0,017(\varepsilon_{||} = 0)$

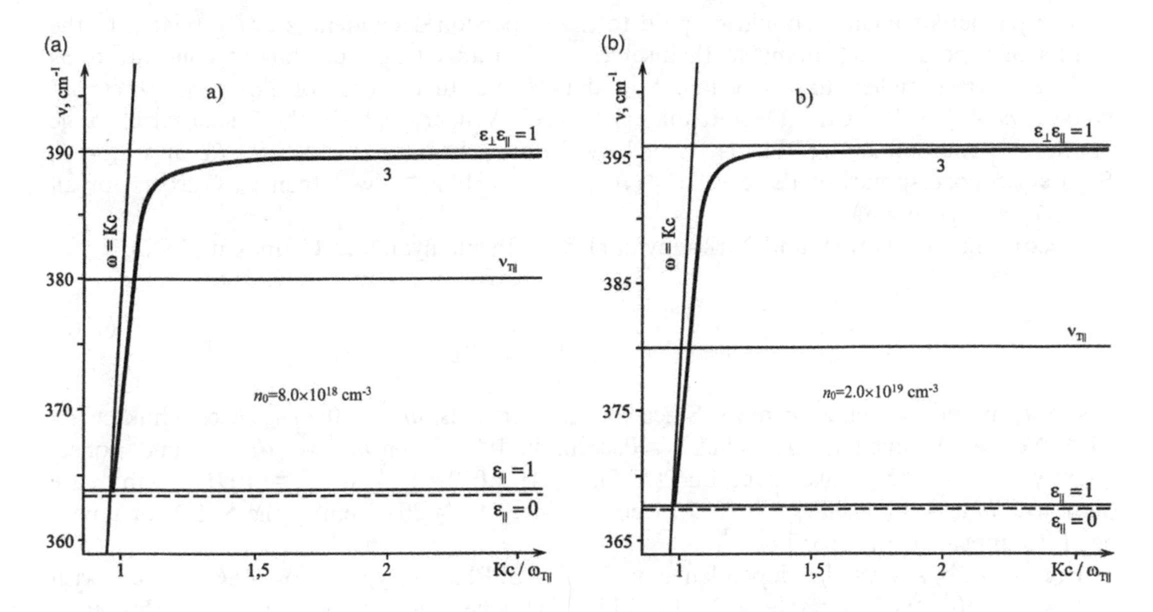

FIGURE 7.6 Dispersion curves $\nu_s(K)$ for SPPPs of type 3 in ZnO with $n_0 = 8.0 \times 10^{18}$ cm^{-3} (a) and $n_0 = 2.0 \times 10^{19}$ cm^{-3} (b) when $\boldsymbol{K} \perp \boldsymbol{C}$ and $xy \perp \boldsymbol{C}$

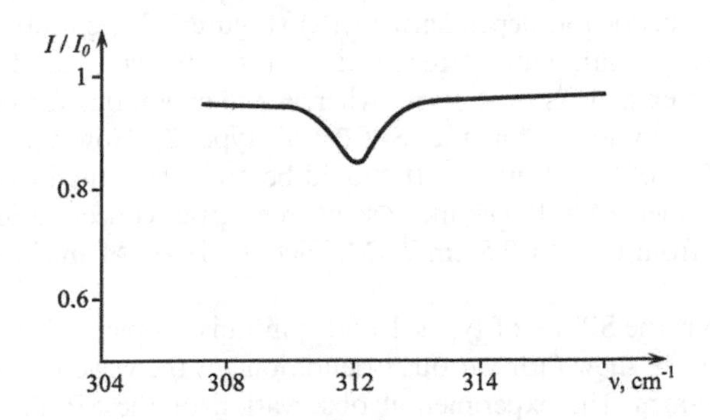

FIGURE 7.7 ATR spectrum of ZnO crystal (sample ZC1M, $n_0 = 4.2 \times 10^{18}$ cm^{-3}). Calculation parameters are: $\gamma_{f\perp} = \gamma_{f\|} = 6$ cm^{-1}, $\gamma_{p\perp} = \gamma_{p\|} = 1$ cm^{-1}, $\alpha = 28°$, $d = 26$ μm.

7.2.4 The Comparison of Experimental and Theoretical Data

The analysis of these results showed that the SPPPs of type 2 can occur only when $\varepsilon_\perp < 0$ and $\varepsilon_\| > \chi_x^2$ (Eq.(7.8)). In the case of ZnO, the range of the existence SPPPs of type 2 is limited by frequencies of 380–412 cm^{-1} (see Figure 7.1a). They appear when the orientation of the sample is $\boldsymbol{K} \| \boldsymbol{C}$ and $xy \| \boldsymbol{C}$. The dependence $\nu_s(K)$ practically represents a straight line passing towards the $\nu_s(K)$ of the SPPPs of type 1.

With orientations $\boldsymbol{K} \perp \boldsymbol{C}$, $\boldsymbol{K} \| \boldsymbol{C}$, and $xy \| \boldsymbol{C}$, the extraordinary SPPPs are excited, which is manifested in the two-component spatial structure of their fields. In Figures 7.3–7.5, in the $\nu, Kc/\omega_{T\|}$ coordinates, the regions, where condition (7.8) is fulfilled and the dispersion dependences of the SPPPs are observed at $\theta = 0$ and $\nu_{T\|} < \nu_{T\perp} < \nu_{L\|} < \nu_{L\perp}$ that occur for ZnO crystals. The dispersion curves $\nu_s(K)$ begin at the frequencies of $\nu = 0$, $\nu = \nu_{T\perp}$ $(\varepsilon_\perp = \infty)$ and $\nu = \Omega_\|^+$, $\nu = \Omega_\|^-$ $(\varepsilon_\| = 1)$, coincide with the straight line $\omega = Kc$.

The particular interest should be paid to the dispersion dependences $\nu_s(K)$, related to the SPPPs of type 3, which manifest themselves only in anisotropic crystals at concentrations of free carriers higher than a certain boundary value. In the case of ZnO, this threshold value is $n_0 = 2 \times 10^{18}$ cm^{-3} (Melnichuk et al., 1998; Venger et al., 2001). The spectral range of the existence of these SPPPs is framed by the straight lines $\varepsilon_{||} = 1$, $\omega = Kc$ and $\varepsilon_{\perp} = 0$. Figure 7.5 corresponds to the case $\nu_0 < \nu_1 (\nu_{\perp}^{-} < \nu_{T||})$. If $\nu_{\perp}^{-} > \nu_{T||}$, then ν_s(K) exists for all $K > \Omega_{||}^{-}/c$ (Figure 7.6).

According to Gurevich and Tarkhanyan (1975); Tarkhanyan and Uzunoglu (2006),

$$\begin{aligned} \nu_0 &= \nu_{P\perp}\left[\varepsilon_{\infty\perp}/\mu_{\perp}\right]^{1/2} \\ \nu_1^2 &= \nu_{T||}{}^2\left[\varepsilon_{\infty\perp}\nu_{T||}{}^2 - \varepsilon_{0\perp}\nu_{T\perp}{}^2\right]/\left[\mu_{\perp}\left(\nu_{T\perp}{}^2 - \nu_{T||}{}^2\right)\right], \end{aligned} \tag{7.15}$$

where $\mu_{\perp}$ is reduced electron mass. Since for ZnO crystals, $m_{\perp}^{*} = 0.26m_e$ (Melnichuk et al., 1994; Venger, Melnichuk, Melnichuk, & Pasechnik, 1995), then $m_{\perp}^{*} = (1/\mu_{\perp})m_e$ and, consequently, $\mu_{\perp} = 3.85$. Thus, from Eq. (7.15), $\nu_0 = 1.0129\nu_{p\perp}$ and $\nu_1 = 1091.17$ cm^{-1} are obtained. If $\nu_0 = \nu_1$, then $\nu_{p\perp} = 1077.27$ cm^{-1}. Since $\Omega_{||} \geq 301.1$ cm^{-1}, the SPPPs of type 3 begin to appear at $\nu_{p\perp} \geq 550$ cm^{-1}.

Figure 7.5b shows the dependence $\nu_s(K)$ for SPPPs of type 3 for ZnO crystal with $n_0 = 4.2 \times 10^{18}$ cm^{-3} (sample ZC1M, Table 7.1) where $\nu_{p\perp} = 605$ cm^{-1}. In this case, $\nu_{||}^{-} = 306.5$ cm^{-1}, $Kc/\omega = 0.017$ $(\varepsilon_{||} = 0)$; $\Omega_{||}^{-} = 309.9$ cm^{-1}, $Kc/\omega = 0.815$ $(\varepsilon_{||} = 1)$, $\nu_{\perp}^{-} = 318.4$ cm^{-1}, and $Kc/\omega = 1.632$ $(\varepsilon_{\perp} = 0)$.

Thus, the SPPPs of type 3 exist in the frequency range of $\nu = 309.9$–318.4 cm^{-1} for this sample. Corresponding dispersion dependences ν_s(K) (Figure 7.7) were obtained for $K \perp C$, $xy \perp C$ and $\nu_0 > \nu_1$ $(\nu_{\perp}^{-} > \nu_{T||})$ with cutoff frequencies $\nu_3 = 390$ cm^{-1} and 395.5 cm^{-1}. When $(\nu_{\perp}^{-} > \nu_{T||})$, the SPPPs exist at $Kc/\omega_{T||} \to \infty$, whereas the conditions for the SPPP existence at $\Omega_{||}^{-}$ $\nu < \nu_{T||}$ are similar to those for the SPPPs of type 2. However, in the range from $\nu_{T||} < \nu < \nu_{\perp}^{-}$, the SPPPs of type 1 appear. It should be noted that the low-frequency $\nu_s^{-}(K)$ for SPPPs begin at $\nu > 0$, when $\chi_x^2 > 1$. The increase of free carrier concentration results in the shift of the onset of $\nu_s(K)$ from 0.54 to 9.5 cm^{-1} (Melnichuk, 1998; Melnichuk et al., 1994, 1998; Venger et al., 2001).

Thus, in this section, the SPPPs of types 1 and 2 in anisotropic ZnO crystals with different carrier concentration were shown for various orientations of the wave vector toward the crystal surface and optical C-axis. The experimental observation of the SPPPs of type 2 is shown. Besides, new dispersion curves for polaritons of type 3 were presented. These excitations appear in the crystals with electron concentration higher than $n_o = 2 \times 10^{18}$ cm^{-3} and the orientation $\mathbf{K} \perp \mathbf{C}$ and $xy \perp \mathbf{C}$. The conditions for the existence of the SPPPs of type 3 are shown along with corresponding calculated dispersion curves and ATR spectrum.

7.3 EFFECT OF STRONG MAGNETIC FIELD ON SURFACE POLARITONS IN ZNO

It is known that when a semiconductor is situated in a magnetic field, a number of effects can appear, in particular, the Zeeman effect, diamagnetic shifting and transitions between the Landau levels, etc. (see, for instance, Haider, 2017; Landau & Lifhsitz, 1984; Sizov & Ukhanov, 1979; Sugano & Kojima, 2000; Zvezdin & Kotov, 1997). A comprehensive analysis of the latter was made for CdSe crystals by Kapustina, Petrov, Rodina, and Seisyan (2000). The spectral curves in a cubic magnetic semiconductor, the Cotton–Mouton effect quadratic in a magnetic field (the Voigt effect), nonreciprocal birefringence linear in a magnetic field, and the Faraday Effect were investigated by Krichevtsov and Weber (2004). It was shown that the Voigt effect is anisotropic in cubic magnetic semiconductors.

The propagation of electromagnetic waves in uniaxial semiconductors subjected to a magnetic field that is not parallel to the crystal axis was considered at first by Gurevich and Tarkhanyan (1975). It was shown, in particular, that the effect of cyclotron and plasma resonance shifting is related to their transformation to the combined cyclotron–plasma resonances. The Faraday Effect in cubic and hexagonal crystals was considered by Gurevich and Ipatova (1959); however, no comprehensive analyses of the expressions for the Faraday rotation were reported. Until end of the last century, the investigations of the effect of magnetic field on the specular IRR in uniaxial optically anisotropic single crystals were not addressed at all.

As it was mentioned in previous section, there are number of works devoted to the phonon and plasmon–phonon polaritons in polar optically isotropic and anisotropic semiconductors (Agranovich & Mills, 1985; Dmitruk et al., 1989; Venger, Ievtushenko, Melnichuk, & Melnichuk, 2008, 2010; Venger et al., 2001). In the previous section, it was shown that in the range of "residual rays" of semiconductor, the excitation of surface polaritons of the phonon type (i.e., SPP) and plasmon–phonon type (SPPP) is possible. Hereafter, the effect of magnetic field on the SPPs and SPPPs will be discussed.

One of the first papers on the effect of magnetic field on different excitations has appeared in 1973, when Brion et al. reported on the properties of surface magneto-plasmons in polar semiconductors under magnetic field of 100 kOe (see Brion, Wallis, Hartstein, & Burstein, 1973). Later, Palik et al. (1976) showed the formation of surface magneto-plasmons in InSb crystals (Palik et al., 1976). However, the first consideration of the effect of magnetic field on optically anisotropic crystals was performed by the authors of this chapter, especially for ZnO materials. Some main results will be presented below, that is, the effect of high uniform magnetic field on the surface phonon and plasmon–phonon polaritons will be discussed for ZnO crystals. The orientation was $C||y$, $\boldsymbol{K}\perp\boldsymbol{C}$, $xy||\boldsymbol{C}$, $\boldsymbol{H}\perp\boldsymbol{K}$, $\boldsymbol{H}||y$, $K_x = K$, $K_{y,z} = 0$ $\boldsymbol{H}||y$, $K_x = K$, $K_{y,z} = 0$ (Figure 7.8).

7.3.1 Theoretical Consideration of the Effect of Magnetic Field on the Surface Polaritons in ZnO

Let us consider a polar optically anisotropic ZnO single crystal in which the excitation and propagation of different types of surface polaritons occur along the crystal surface as shown in Figure 7.8. The x-axis is parallel to the direction of electromagnetic wave propagation at $C||y$.

The calculation of the reflection coefficient $R = I/I_0$, where I_0 is the intensity of incident infrared light at a given frequency and I is the intensity of reflected light at the same

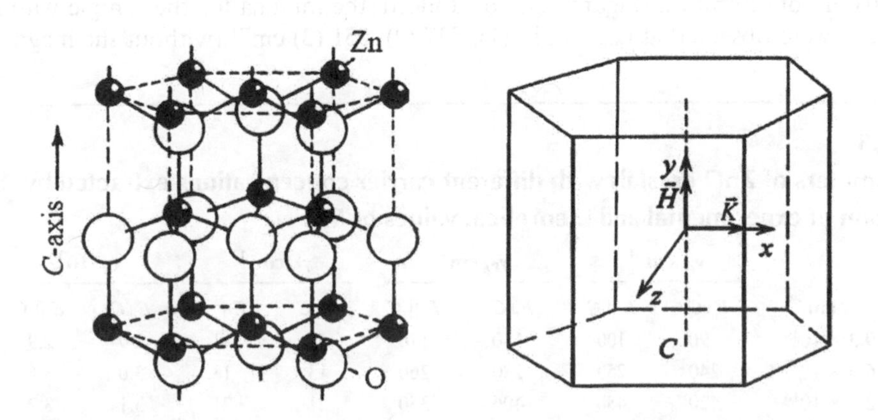

FIGURE 7.8 ZnO crystal structure and the configuration of vectors $\boldsymbol{H}$ and $\boldsymbol{K}$, $\boldsymbol{C}$-axis and x-, y-, and z-axes in the crystal

frequency, was performed taking into account an additive contribution of the phonon and plasmon subsystems in dielectric function of the semiconductor as well as the effect of magnetic field (Venger et al., 2001, 2010). The components of the dielectric constant in the magnetic field were considered as:

$$\varepsilon_1 = \varepsilon_{\infty\perp,\|}\left(1 + \frac{\nu_{L\perp,\|}^2 - \nu_{T\perp,\|}^2}{\nu_{T\perp,\|}^2 - \nu^2 - i\gamma_{f\perp,\|}\nu} + \frac{\nu_{p\perp,\|}^2\left(\nu + i\gamma_{p\perp,\|}\right)}{\nu\left(\Omega_e^2 - \left(\nu + i\gamma_{p\perp,\|}\right)^2\right)}\right),$$

$$\varepsilon_2 = \frac{\varepsilon_{\infty\perp,\|}\nu_{p\perp,\|}^2\Omega_e}{\nu\left(\left(\nu + i\gamma_{p\perp,\|}\right)^2 - \Omega_e{}^2\right)}, \tag{7.16}$$

$$\varepsilon_3 = \varepsilon_{\infty\perp,\|}\left(1 + \frac{\nu_{L\perp,\|}^2 - \nu_{T\perp,\|}^2}{\nu_{T\perp,\|}^2 - \nu^2 - i\gamma_{f\perp,\|}\nu} - \frac{\nu_{p\perp,\|}^2}{\nu\left(\nu + i\gamma_{p\perp,\|}\right)}\right),$$

where $\Omega_e = \frac{eH}{mc}$ is the cyclotron frequency; $\varepsilon_{\infty\perp,\|}$ – is high-frequency dielectric constant; $\nu_{L\perp,\|}$, $\nu_{T\perp,\|}$ are respectively the frequencies of the longitudinal and transverse optical phonons; $\nu_{p\perp,\|}$ is plasma resonance frequency; $\gamma_{p\perp,\|}$ is plasmon attenuation factor; and $\gamma_{f\perp,\|}$ is the damping coefficient of optical phonon.

The calculation was performed for the different ZnO crystals (Table 7.1) using self-consistent parameters of ZnO presented in Table 7.2. The obtained results are collected in Table 7.3.

7.3.2 Surface Polaritons in ZnO under Magnetic Field (Experimental Results)

Figure 7.9 (symbols) shows experimental ATR spectra of ZnO crystals with $n_0 = 9.3 \times 10^{16}$ cm^{-3} (Figure 7.9a), $n_0 = 6.6 \times 10^{17}$ cm^{-3} (Figure 7.9b), and $n_0 = 2.0 \times 10^{18}$ cm^{-3} (Figure 7.9c). The spectra were recorded at different air gaps (d = 3.4– 14.6 μm) and ATR prism angles (α = 30°, 35°, and 50°). The air gap between the ATR prism and the sample was varied until the absorbed wave intensity was reached about 20 % of minimum intensity of ATR spectrum.

The simulation of ATR spectra was performed according to the approach described by Venger et al. (2001) in the absence (solid curves 1–3) and the presence of magnetic field of H = 100 kOe (dashed curves 1′-3′). It was observed that the magnetic field caused the shift of the ATR minima toward higher frequencies. This shift is more pronounced for ZnO with higher carrier concentration (Figure 7.9a, b). Indeed, the minima for the sample with n_0 = 9.3 × 10^{16} cm^{-3} were observed at ν_{min} = 518 (1), 537 (2), 551 (3) cm^{-1} (without the magnetic field)

TABLE 7.3
The parameters of ZnO crystals with different carrier concentrations extracted by the comparison of experimental and theoretical values of $R(\nu)$

		ν_p, cm^{-1}		γ_P, cm^{-1}		γ_f, cm^{-1}		δ, ·10^3		
Sample	n_0, cm^{-3}	$E\perp C$	$E \| C$	$E\perp C$	$E \| C$	$E\perp C$	$E \| C$	$E\perp C$	$E \| C$	$\gamma_{p\perp}/\gamma_{p\|}$
ZO2-3	9.3×10^{16}	90	100	150	170	11	11	2.9	2.9	0.88
ZO1-3	6.6×10^{17}	240	250	280	260	13	13	3.0	3.6	1.08
ZO6-B	2.0×10^{18}	420	480	406	350	21	21	3.1	3.2	1.16
ZC1M	4.2×10^{18}	605	—	1020	—	14	—	2.0	—	—
ZO9	5.0×10^{19}	2115	—	1480	—	40	—	1.9	—	—

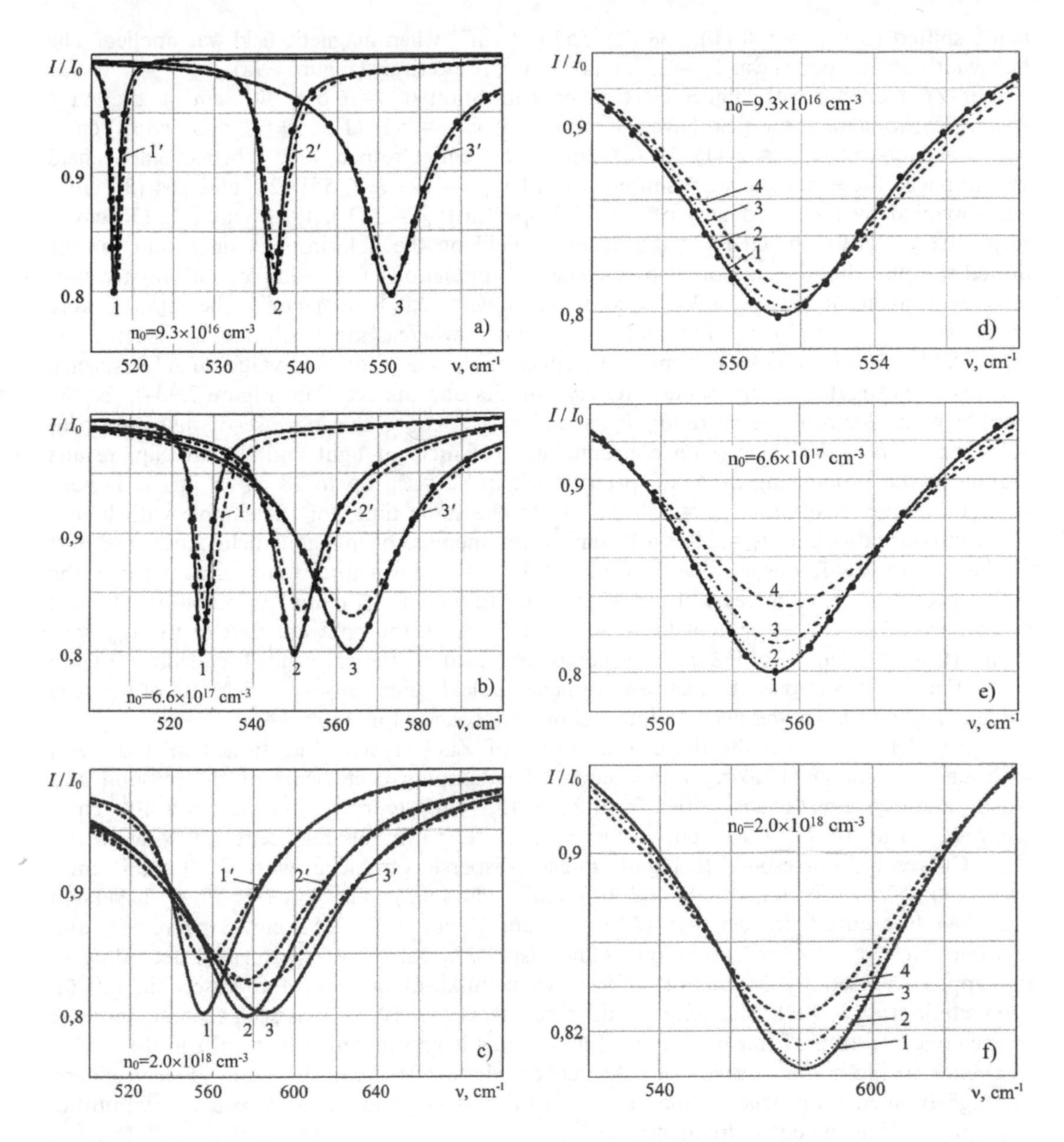

FIGURE 7.9 ATR spectra of ZnO crystals versus electron concentration and magnetic field value obtained for the samples: ZO2-3, $n_0 = 9.3 \times 10^{16}$ cm^{-3} (**a, d**); ZO1-3, $n_0 = 6.6 \times 10^{17}$ cm^{-3} (**b, e**), and ZO6-B, $n_0 = 2.0 \times 10^{18}$ cm^{-3} (**c, f**). The spectra recorded with the air gap of: (**a**) $d = 14.6$ μm (1,1′), 8.7 μm (2,2′), and $d = 4.1$ μm (3,3′); (**b**) $d = 13$ μm (1,1′), 7.4 μm (2,2′), and 3.4 μm (3,3′); c) $d = 10.2$ μm (1,1′), 5.6 μm (2,2′), and 2.6 μm (3,3′); (**d, f**) $d = 4.0$ μm (1–4); (**e**) $d = 5.0$ μm (1–4). The angle of ATR prism was: for (**a–c**), $\alpha = 30°$ (1,1′), 35° (2,2′), and 50° (3,3′); for (**d**) $\alpha = 50°$ (1–4); for (**e, f**) $\alpha = 40°$ (1–4). Magnetic field was: for (**a–c**) $H = 0$ (1–3) and 100 kOe (1′-3′); for (**d–f**) $H = 0$ (1), 30 (2), 65 (3), and 100 kOE (4). In all figures, the symbols correspond to experimental data and lines (solid and dashed) are the simulated ATR spectra

being shifted to ν_{min} = 518 (1′), 538 (2′), 552 (3′) cm^{-1} when magnetic field was applied. The half-width of the spectra was Γ_p = 2, 3, and 9 cm^{-1}, respectively (Figure 7.9a).

For ZnO crystal with higher electron concentration ($n_0 = 6.6 \times 10^{17}$ cm^{-3}), the ATR spectra without magnetic field have the minima at ν_{min} = 527 (1), 550 (2), and 563 (3) cm^{-1} with the half-width Γ_p = 5 (1), 17 (2), and 28 (3) cm^{-1}, respectively. When magnetic field was applied, these minima were shifted toward ν_{min} = 528 (1′), 551 (2′), and 564 (3′) cm^{-1} accompanied by the broadening of the ATR spectra (Γ_p = 7 (1′), 19 (2′), and 32 (3′) cm^{-1}, respectively). Thus, the effect of the magnetic field on the polaritons is significant for the doped samples that are supported by the spectra simulation of the samples with higher electron concentration (Figure 7.9c). Indeed, when $n_0 = 2.0 \times 10^{18}$ cm^{-3}, the minima were found at ν_{min} = 557 (1′), 574 (2′), and 582 (3′) cm^{-1}, whereas the width of ATR spectra was Γ_p = 58 (1′), 92 (2′), and 99 (3′) cm^{-1}.The effect of the magnetic field value on ATR spectra was also investigated for the same ZnO crystals. As one can see from Figure 7.9d–f, for the sample with electron concentration $n_0 = 6.6 \times 10^{17}$ cm^{-3}, the increase of magnetic field value up to H = 100 kOe (with constant angle of incident light and the air gap) results mainly in the broadening of ATR spectrum (from Γ_p = 20 up to 28 cm^{-1}) at nearly constant minimum position (ν_{min} = 558 cm^{-1}). At the same time, for the sample with higher electron concentration ($n_0 = 2.0 \times 10^{18}$ cm^{-3}), the increase of magnetic field causes the shift of the ν_{min} to low-frequency side from ν_{min} = 580 cm^{-1} down to 578 cm^{-1} and broaden the ATR spectrum, i.e. Γ_p increases from 83 cm^{-1} up to 98 cm^{-1} (Figure 7.9f). However, for the sample with low carrier content ($n_0 = 9.3 \times 10^{16}$ cm^{-3}), the opposite shift of the ν_{min} was found (from 551 cm^{-1} to 552 cm^{-1}) accompanied also by the spectrum broadening (Figure 7.9d). For all the samples, the increase of magnetic field value causes the decrease of the $R(\nu)$ value (Figure 7.9). The results mentioned above are collected in Table 7.4.

Figure 7.10a presents the dispersion curves of ZnO crystals (mentioned in Table 7.4) without magnetic field taking into account the damping coefficients of the phonon and plasmon subsystems. The n_0 varied from 9.3×10^{16} cm^{-3} (curves 1, 1′) via 6.6×10^{17} cm^{-3} (curves2, 2′) up to 2.0×10^{18} cm^{-3} (curves 3, 3′). The symbols represent the experimental data. Curves 1–3 correspond to high-frequency dispersion branches with $\nu^+(K)$ = 561 cm^{-1} (curve 1), 578 cm^{-1} (curve 2), and 627 cm^{-1} (curve 3). The low-frequency dispersion branches has cutoff frequency $\nu^-(K)$ = 59 cm^{-1} (curve 1′), 152 cm^{-1} (curve 2′), and 246 cm^{-1} (curve 3′). The simulation of the dispersion curves was performed according to the approach used by Melnichuk (1998); Melnichuk, Melnichuk, and Pasechnik (1996); Melnichuk et al. (1998); Melnichuk and Pasechnik (1996, 1998). It is seen that the increase in the concentration of free carriers in ZnO crystals is accompanied by a shift in the SPPPs frequency to the high-frequency side. Moreover, whatever the concentration of free carriers, the high-frequency branches of the SPPPs start at $\nu = \nu_{T\perp}$ and end at $K \gg \omega/c$, asymptotically approaching the cutoff frequency.

TABLE 7.4
Parameters ν^+, ν^-, and ν_f of surface polaritons in ZnO crystals versus magnetic field

		Magnetic field, H, kOe										
		0		30			65			100		
Sample	Electron Concentration, n_0, cm^{-3}	ν^-, cm^{-1}	ν^+, cm^{-1}	ν^-, cm^{-1}	ν_f, cm^{-1}	ν^+, cm^{-1}	ν^-, cm^{-1}	ν_f, cm^{-1}	ν^+, cm^{-1}	ν^-, cm^{-1}	ν_f, cm^{-1}	ν^+, cm^{-1}
ZO2-3	9.3×10^{16}	59	561	34	108	561	21	171	561	14	242	561
ZO1-3	6.6×10^{17}	152	578	124	201	576	98	245	575	80	293	574
ZO6-B	2.0×10^{18}	246	627	221	287	620	196	316	612	174	345	607

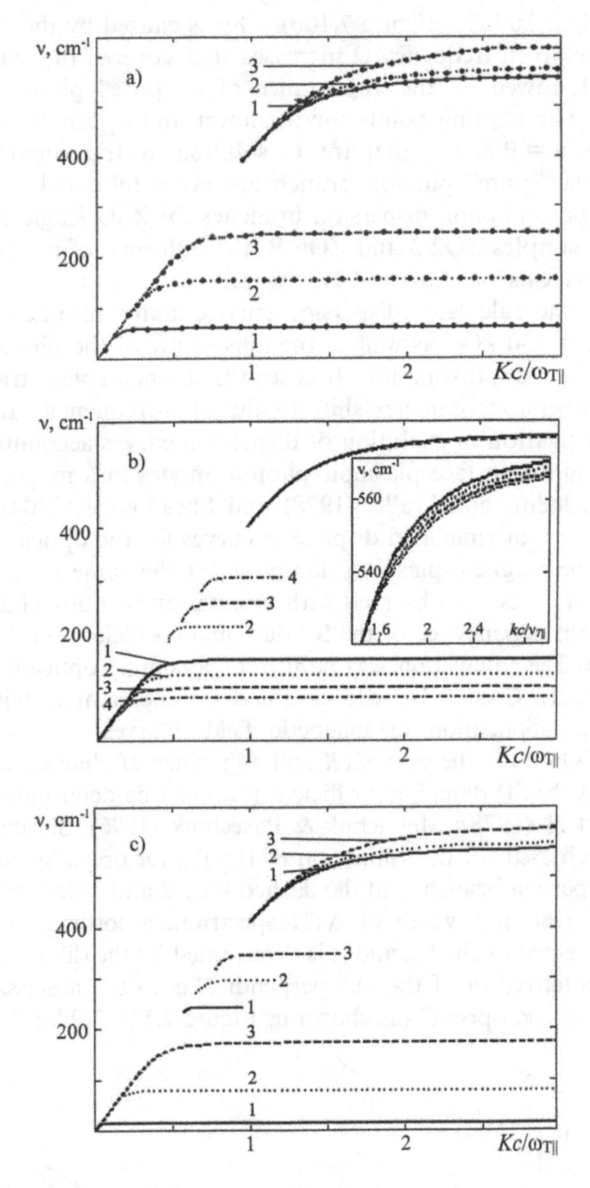

FIGURE 7.10 (**a**) Dispersion curves of ZnO crystals with electron concentration of 9.3×10^{16} cm^{-3} (1, 1′), 6.6×10^{17} cm^{-3} (2, 2′), and 2.0×10^{18} cm^{-3} (3, 3′) without (1–3) and with (1′–3′) application of magnetic field $H = 100$ kOe. The experimental data are shown by symbols and modeling is presented by the curves; (**b**) the modeling of the effect of magnetic field on dispersion curves of sample ZO1-3 ($n_0 = 6.6 \times 10^{17}$ cm^{-3}), $H = 0$ (1), 30 (2), 65 (3), and 100 (4) kOe; (**c**) Simulated dispersion dependences for the samples ZO2-3 (1), ZO1-3 (2), and ZO6-B (3) in the presence of magnetic field $H = 100$ kOe

Figure 7.10b shows the variation of dispersion curves of sample ZO1-3 ($n_0 = 6.6 \times 10^{17}$ cm^{-3}) with the increase of the magnitude of uniform magnetic field from $H = 0$ to 100 kOe. The inset in Figure 7.10b demonstrates negligible variation of high-frequency dispersion curves versus H value, whereas the stronger magnetic field causes the shift of low-frequency dispersion branch toward the lower frequencies. Besides, additional dispersion branch appears in the 190–350 cm^{-1} spectral range. This branch shifts to high-frequency side for the H values higher than 30 kOe (Figure 7.10b). This is caused by the fact that for stronger magnetic field, the cyclotron frequency Ω increases that governs the splitting between plasmons and phonons followed by the appearance of a "pure" phonon dispersion branch (Venger et al., 2001). The starting points for the lower and upper dispersion branches correspond to frequencies $\nu = 0$ and ν_T that are the solutions of the equations $\varepsilon_1 = 1$ or $\varepsilon_1 = 0$. The asymptotes for the "pure" phonon branch are given by equation $1+\varepsilon_1 \pm i\varepsilon_2 = 0$. The cutoff frequencies of pure phonon dispersion branches for ZnO single crystals are collected in Table 7.4. For the samples ZO2-3 and ZO6-B, the influence of magnetic field on corresponding dispersion relations was found to be similar.

Figure 7.10c shows the calculated dispersion curves, taking into account the presence of magnetic field with $H = 100$ kOe, as well as the anisotropy of the phonon and plasma subsystems in their harmonic approximation. It is seen that when concentration of free carriers increases, all three dispersion branches shift to the high-frequency range under magnetic field application. Investigation of evolution of dispersion curves accounting the damping for magnetic plasmons and for surface plasmon–phonon modes in a magnetic field was carried out by Martin, Maradudin, and Wallis (1978) and Shramkova (2004). The authors have observed a backward turn of simulated dispersion curves for the optically isotropic material, taking into account the magneto-plasmon damping. At the same time, for optically anisotropic crystals, such turn was not observed with orientation mentioned above.

Figure 7.11 shows the dependence of the SP damping coefficient on the frequency of $\Gamma_p(\nu)$ for ZnO single crystal. The calculation was performed assuming optically smooth crystal surface (Venger et al., 2001). Curves 1–3 are obtained for the samples with different electron concentrations without application of magnetic field. Curves 1′–3′ are the dependences obtained for sample ZO6-B for the case $H \perp K$ and $H \| y$ when H changes from 30 to 100 kOe.

It is interesting that the SP damping coefficient (Γ_{sp}) can be determined graphically as it is described in Martin et al. (1978); Melnichuk & Pasechnik (1996); Shramkova (2004). Figure 7.12 shows an approach used for determination of the Γ_{sp} for optically anisotropic ZnO crystals. Curve 1 is the dispersion branch, and the dashed lines 2 and 3 determine the half-width of ATR spectrum. The "real" half-width of ATR spectrum, according to Reshina, Mirlin, & Banshchikov (1976), is equal to the Γ_{sp} and it is determined by the difference of the frequencies corresponded to the intersection of the line perpendicular to the abscissa axis in the system "dashed lines – the dispersion point" (as shown in Figure 7.12). Tables 7.5 and 7.6 collect the

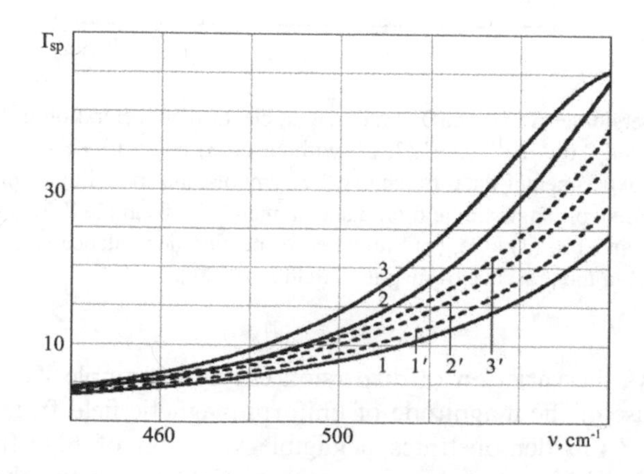

FIGURE 7.11 Spectral dependence of the SP damping coefficient (Γ_{sp}) for different ZnO crystals without application of magnetic field: 1 – ZO2-3; 2 – ZO1-3; 3 – ZO6-B. Curves 1′–3′ demonstrate the evolution of Γ_{sp} versus magnetic field for sample ZO6-B; $H = 30$ (1′), 65 (2′), and 100 (3′) kOe

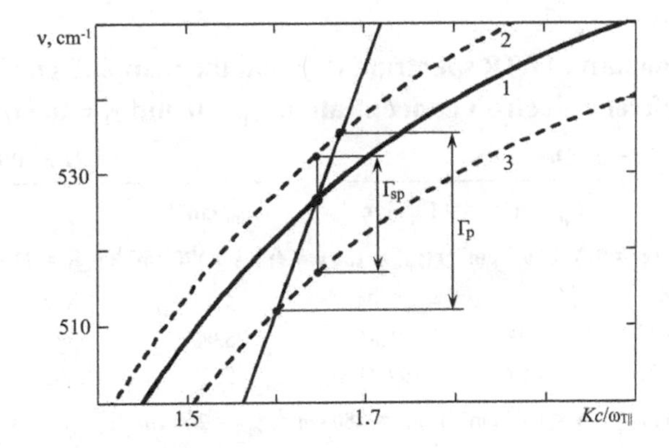

FIGURE 7.12 Dispersion curves for sample ZO1-3 ($n_0 = 6.6 \times 10^{17}$ cm^{-3}). Γ_p is the half-width of ATR spectrum and Γ_{sp} is the SP damping coefficient

results obtained for different ZnO crystals (without and with magnetic field (H = 100 kOe)) considering the case $\gamma_p = 0$ and $\gamma_f = 0$ (Table 7.5) and $\gamma_p \neq 0$ and $\gamma_f \neq 0$ (Table 7.6). The angle of ATR prism was α = 30°, 35°, and 50°.

As one can see from Tables 7.5 and 7.6, the increase of both the angles of ATR prism and free carrier concentration in ZnO causes the Γ_{sp} rise. The effect of plasmon–phonon damping and external magnetic field on Γ_p and Γ_{sp} is similar.

The results presented in this section showed the effect of strong uniform magnetic field on the plasmon and plasmon–phonon properties of optically anisotropic ZnO single crystals. This effect is the stronger as the higher concentration of free carriers. This can lead to the appearance of additional dispersion branch. This latter corresponds to the excitation of a "pure" phonon dispersion branch and demonstrates gradual shift toward higher frequencies caused by

TABLE 7.5
Half-width of the minimum of ATR spectrum (Γ_p) and the damping coefficient (Γ_{sp}) for ZnO crystals with different electron concentrations. γ_p = 0 and γ_f = 0, H = 0 and 100 kOe

	H = 0 kOe				H = 100 kOe			
α,°	ν_{min}, cm^{-1}	χ	Γ_p, cm^{-1}	Γ_{sp}, cm^{-1}	ν_{min}, cm^{-1}	χ	Γ_p, cm^{-1}	Γ_{sp}, cm^{-1}
				ZO2-3 ($n_0 = 9.3 \times 10^{16}$ cm^{-3})				
30	518	1.626	9.7	6.96	518	1.629	10.08	7.07
35	537	1.936	25.4	21.43	538	1.975	35.68	30.71
50	551	2.595	202.75	188.2	552	2.77	336.6	314.48
				ZO1-3 ($n_0 = 6.6 \times 10^{17}$ cm^{-3})				
30	527	1.646	22.54	15.68	528	1.68	36.13	25.16
35	550	1.98	114.58	91.89	551	2.11	243.48	191.03
50	563	2.44	—	—	564	—	—	—
				ZO6-B ($n_0 = 2.0 \times 10^{18}$ cm^{-3})				
30	555	1.735	248.57	121.53	557	1.83	—	268.45
35	576	1.95	547.11	325.65	574	—	—	—
50	585	—	—	—	582	—	—	—

TABLE 7.6
Half-width of the minimum of ATR spectrum (Γ_p) and the damping coefficient (Γ_{sp}) for ZnO crystals with different electron concentration. $\gamma_p = 0$ and $\gamma_f = 0$, $H = 0$ and 100 kOe

	$H = 0$ kOe				$H = 100$ kOe			
α,°	ν_{min}, cm^{-1}	χ	Γ_p, cm^{-1}	Γ_{sp}, cm^{-1}	ν_{min}, cm^{-1}	χ	Γ_p, cm^{-1}	Γ_{sp}, cm^{-1}
ZO2-3 ($n_0 = 9.3 \times 10^{16}$ cm^{-3}) ($\gamma_{p\perp} = 150$ cm^{-1}; $\gamma_{p\parallel} = 170$ cm^{-1}; $\gamma_{f\perp,\parallel} = 11$ cm^{-1})								
30	518	1.63	10.59	7.04	518	1.632	11.13	7.12
35	537	1.945	24.93	21.86	538	1.985	35.22	31.23
50	551	2.63	209.67	197.41	552	2.805	353.45	328.61
ZO1-3 ($n_0 = 6.6 \times 10^{17}$ cm^{-3}) ($\gamma_{p\perp} = 280$ cm^{-1}; $\gamma_{p\parallel} = 260$ cm^{-1}; $\gamma_{f\perp,\parallel} = 13$ cm^{-1})								
30	527	1.67	23.73	17.11	528	1.7	35.69	27.52
35	550	2.06	132.29	111.68	551	2.17	270.52	216.5
50	563	2.7	—	—	564	—	—	—
ZO6-B ($n_0 = 2.0 \times 10^{18}$ cm^{-3}) ($\gamma_{p\perp} = 406$ cm^{-1}; $\gamma_{p\parallel} = 350$ cm^{-1}; $\gamma_{f\perp,\parallel} = 21$ cm^{-1})								
30	555	1.86	—	197.25	557	1.94	—	410.19
35	576	2.26	—	—	574	2.38	741.25	720.1
50	585	—	—	—	582	2.85	872.47	853.34

the increasing of the magnitude of the external magnetic field. The "start" frequency and the behavior of this "pure" phonon dispersion branch depend on carrier concentration and the magnitude of the magnetic field. The application of the magnetic field results leads also to the transformation of the low-frequency dispersion branch, whereas the high-frequency one remains unchangeable within the accuracy of the experiment. Besides, the application of magnetic field results also in the variation of damping coefficient values.

7.4 PLASMON–PHONON EXCITATIONS IN ZNO SINGLE CRYSTALS WITH HIGH CARRIER CONCENTRATION STUDIED WITH IR REFLECTION SPECTROSCOPY

In the previous sections, the effect of anisotropy of different types of oscillations on reflection coefficient $R(\nu) = I(\nu)/I_0(\nu)$ for uniaxial optically anisotropic materials using surface polariton spectroscopy was discussed. One more application of infrared spectroscopy is the investigation of the effect of different excitations on the reflection coefficient $R(\nu)$. Earlier, the authors of this chapter showed that the $R(\nu)$ depends significantly on the crystal lattice anisotropy, on the anisotropy of the effective masses of free carriers, on phonon (γ_f) and plasmon (γ_p) damping coefficients. More results were obtained for hexagonal 6H-SiC crystals (see for instance, Melnichuk, 1999; Melnichuk et al., 1998; Melnichuk & Pasechnik, 1992, 1998; Venger et al., 2001). Later, these effects were studied for ZnO crystals. However, the first investigation of the interaction of electromagnetic waves simultaneously with a strong contribution of phonon subsystem and a weak contribution of plasmon subsystems in optically anisotropic uniaxial crystals was reported by Venger, Melnichuk, Melnichuk, and Pasichnyk (2000) along with the effect of anisotropy on $R(\nu)$ behavior. In this section, these effects will be described in detail.

7.4.1 Effect of Plasmon–Phonon Coupling on the IR Reflection Coefficient

We will consider the peculiarities of the reflection of electromagnetic waves of the infrared range from the surface of uniaxial polar semiconductor such as ZnO. Let the wave fall from the vacuum onto the smooth surface of the "semi-infinite" crystal. Let us denote the refractive index n via $n_{\perp}(\nu)$ and $n_{||}(\nu)$, respectively, for the case when the electric field vector $\boldsymbol{E}$ in the incident wave is perpendicular (case 1) and parallel (case 2) to the plane containing the wave vector $\boldsymbol{K}$ and the C-axis of ZnO crystal. In general case, the coefficient of normal reflection $R(\nu)$ can be considered as

$$R(\nu) = \frac{\left(n_{\perp,||}(\nu) - 1\right)^2 + k^2{}_{\perp,||}(\nu)}{\left(n_{\perp,||}(\nu) + 1\right)^2 + k^2{}_{\perp,||}(\nu)} \tag{7.17}$$

$$\begin{cases} n_{\perp}^2 = \varepsilon_{\perp}(\nu) \\ n_{||}^2 = \dfrac{\varepsilon_{\perp}(\nu) \times \varepsilon_{||}(\nu)}{\varepsilon_{\perp}(\nu) \times \sin^2\theta + \varepsilon_{||}(\nu) \times \cos^2\theta} \end{cases} \tag{7.18}$$

where n is refractive index, k is absorption coefficient, θ is the angle between C-axis and wave-vector $\boldsymbol{K}$, $\varepsilon_{\perp}(\nu)$ and $\varepsilon_{||}(\nu)$ are main components of the tensor of dielectric permittivity that are parallel and perpendicular to the C-axis in the range of plasmon–phonon interaction. Taking into account the damping of plasmon and phonon subsystems, the $\varepsilon_{\perp}(\nu)$ and $\varepsilon_{||}(\nu)$ can be determined as

$$\varepsilon_{\perp,||}(\nu) = \varepsilon_{\infty\perp,||} + \frac{\varepsilon_{\infty\perp,||}\left(\nu_{L\perp,||}^2 - \nu_{T\perp,||}^2\right)}{\nu_{T\perp,||}^2 - \nu^2 + i\nu\gamma_{f\perp,||}} - \frac{\varepsilon_{\infty\perp,||}\nu_{p\perp,||}^2}{\nu(\nu + i\nu\gamma_{p\perp,||})} \tag{7.19}$$

where $\varepsilon_{\infty\perp,||}(\nu)$ are the components of the tensor of dielectric permittivity across and along the C-axis at $\nu \to \infty$; $\nu_{T\perp,||}$; $\nu_{p||} = 1430\ \text{cm}^{-1}$ and $\nu_{L\perp,||}$are the frequencies of transverse (T) and longitudinal (L) optical phonons polarized along and across the C-axis of ZnO crystal. If $n^2{}_{\perp,||}(\nu) >> k^2{}_{\perp,||}(\nu)$, then

$$R(\nu) = \frac{\left(n_{\perp,||}(\nu) - 1\right)^2}{\left(n_{\perp,||}(\nu) + 1\right)^2} \tag{7.20}$$

For most semiconductor crystals, the condition $n^2{}_{\perp,||}(\nu) >> k^2{}_{\perp,||}(\nu)$ is fulfilled in wide spectral range where $h\nu > E_g$ (E_g is semiconductor bandgap). The reflection spectrum can be calculated if in Eq. (7.20) instead of $n_{\perp,||}(\nu)$the corresponding values of Eq.(7.18) will be used for $\boldsymbol{E}\perp\boldsymbol{C}$ or $\boldsymbol{E}||\boldsymbol{C}$.

Let us consider the case when $E\perp C$ and $\theta = 0$ and $\pi/2$ (case 1). Substituting (7.18) and (7.19) into (7.20), for the case $\gamma_f \equiv 0$ and $\gamma_p \equiv 0$, for anisotropic ZnO we will obtain the result which coincides with that of the isotropic case. There frequency diapason where refractive index of ZnO ($n_{\perp,||}(\nu)$) takes imaginary values. In such ranges, the crystal is opaque and, consequently, R(ν) = 1. For highly doped ZnO, two frequency bands for which $R(\nu) = 1$ appears in the specular reflection spectra at such orientation. These results will remain also for the case $\boldsymbol{E}||\boldsymbol{C}$ and $\theta = 0$ and $\pi/2$. However, for $\theta = \pi/2$, it is necessary to replace the index $\perp$ by the index $||$ and take into account that for *ZnO* $\nu_{p||} = 1.1 \cdot \nu_{p\perp}$ (Melnichuk et al., 1994).

Let us consider the behavior of $R(\nu)$ for the case $\boldsymbol{E}||\boldsymbol{C}$, $0 < \theta < \pi/2$ (case 2). With this orientation, the extraordinary wave exists in ZnO. The reflection coefficient $R(\nu)$ is determined, similar to the case 1, by Eq. (7.20) taking into account (7.18). However, the refractive index depends on the coupling of the electromagnetic wave with the phonon and plasmon subsystems of this optically anisotropic crystal and on the angle θ.

Figure 7.13 shows theoretical reflection spectra $R(\nu)$ performed for ZnO with high electron concentration (sample ZC1M, $n_0 = 4.2 \times 10^{18}$ cm^{-3}, Table 7.1) for the case $\boldsymbol{E}||\boldsymbol{C}$ and for $\theta = 30°$ and 60° using Eqs. (7.19) and (7.20).

It is seen that the reflection spectra have four ranges of total reflection (opacity) and four transparency ranges. When $0 < \theta < \pi/2$, the number of such ranges increases twice in comparison with the case $\theta = 0$ and $\pi/2$, when the crystal behaves as an isotropic medium. This is caused by the coupling of electromagnetic, plasma, and optical oscillations leading to the appearance of four SPPP dispersion branches in the optically anisotropic crystal at $\theta \neq 0$ and $\theta \neq \pi/2$.

The low-frequency boundaries of the transparency ranges coincide with the cutoff frequencies (Table 7.7). The high-frequency boundaries of the first three transparency regions ($n_{||} \to \infty$, $R(\nu) = 1$) coincide with the resonance frequencies ν_R of the longitudinal transverse wave ($\nu = \nu_{R1,2,3}$) being determined by the solution of the equation

$$\varepsilon_{\perp}(\nu) \times \sin^2\theta + \varepsilon_{||}(\nu) \times \cos^2\theta = 0 \tag{7.21}$$

that is the resonance frequencies are the function of both ν_p and θ, whereas the cutoff frequencies are the functions of ν_p only.

The ν_{R1} and ν_{R3} correspond to the low-frequency and high-frequency coupled longitudinal transverse plasmon–phonon excitations and depend on the electron concentration in the conduction band and the angle θ. The ν_{R2} changes from the frequency $\nu_{T\perp}$ at $\theta \to 0$ up to $\nu_{T||}$ at $\theta \to \pi/2$.

The IR reflection spectrum for the sample ZC1M has four onsets at the frequencies of longitudinal plasmon–phonon excitations and three resonances at the frequencies of longitudinal-transverse plasmon–phonon excitations (Table 7.8). With the θ increase, the ν_{R1} increases from 310.9 cm^{-1} (θ = 30°) up to 315.1 cm^{-1} (θ = 60°).

The zero values of $R(\nu)$ were determined solving Eq. (7.17). For $\theta = 30°$, $R_{min}(\nu) = 0$ when $\nu_{01} = 307.1$ cm^{-1}, $\nu_{02} = 322.6$ cm^{-1}, $\nu_{03} = 798.8$ cm^{-1}, $\nu_{04} = 899.1$ cm^{-1} (Figure 7.13a). With the angle increase up to $\theta = 60°$, the $\nu_{01} = 308.7$ cm^{-1}, $\nu_{02} = 320.1$ cm^{-1}, $\nu_{03} = 787.5$ cm^{-1}, and

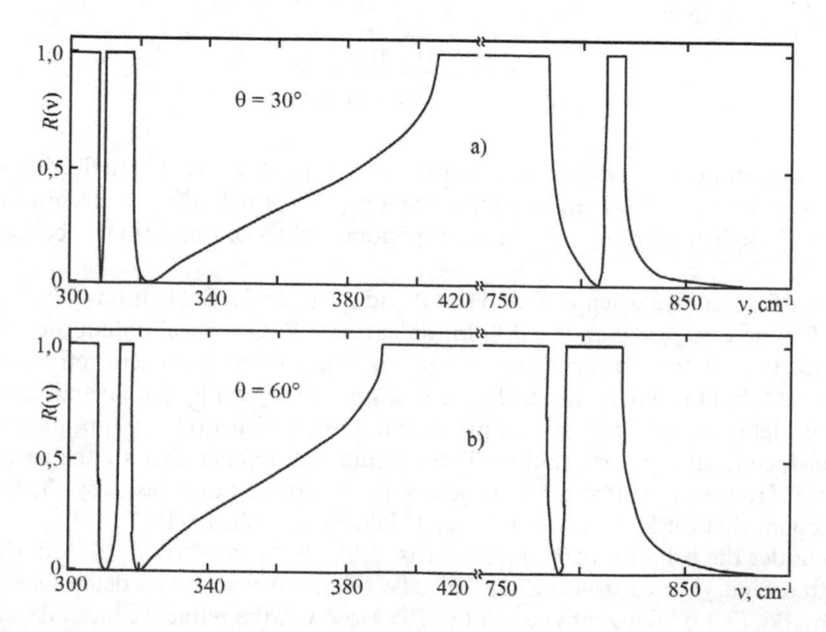

FIGURE 7.13 IR reflection spectra $R(\nu)$ of ZnO crystal with high electron concentration (sample ZC1M, $n_0 = 4.2 \times 10^{18}$ cm^{-3}) at $\boldsymbol{E}||\boldsymbol{C}$ and $\theta = 30°$ (**a**) and 60° (**b**)

TABLE 7.7
The cutoff frequencies $R(\nu)$ versus ν_p in different ZnO crystals

			Cutoff frequencies			
Sample	n_0, cm^{-3}	$\nu_{p\perp}$, cm^{-1}	ν_{L+}, cm^{-1}	Ω_{L+}, cm^{-1}	ν_{L-}, cm^{-1}	Ω_{L-}, cm^{-1}
ZOA-1	4.0×10^{15}	1	566.4	589.9	0.74	0.70
ZOA-2	1.9×10^{16}	50	567.9	591.0	36.8	34.9
ZO2-3	9.3×10^{16}	90	571.2	593.5	65.9	62.5
ZO1-3	6.6×10^{17}	240	602.3	616.4	166.5	160.4
ZO6-B	2.0×10^{18}	420	684.5	677.7	256.5	255.3
ZC1M	4.2×10^{18}	605	817.3	782.7	309.4	318.4
ZOB-3	1.1×10^{19}	1000	1186.4	1098.7	352.4	374.9
ZOB-4	2.3×10^{19}	1500	1705.3	1562.5	367.7	395.5
ZO-9	5.0×10^{19}	2115	2365.1	2158.3	373.8	403.7

TABLE 7.8
Dependence of resonance frequencies on the ν_p and θ in ZnO crystals

Sample	ZO6-B			ZC1M		
θ	ν_{R1}, cm^{-1}	ν_{R2}, cm^{-1}	ν_{R3}, cm^{-1}	ν_{R1}, cm^{-1}	ν_{R2}, cm^{-1}	ν_{R3}, cm^{-1}
10°	265.5	411.2	684.3	309.6	411.3	816.2
30°	256.3	404.5	682.8	310.9	405.5	808.7
60°	255.7	388.6	679.4	315.1	389.9	791.5
85°	255.3	380.3	677.8	318.3	380.3	783.1

ν_{03} = 912.7 cm^{-1} are observed (Figure 7.13b). The four regions of total reflection in the IR reflection spectra are located respectively in the ranges 0-ν_L^-, ν_{R1}-Ω_L^-, ν_{R2}-Ω_L^+, ν_{R3}-ν_L^+. The width of the transparency regions depends on the direction of propagation of the electromagnetic wave, that is, on the angle θ, which is related to the angular dependence of the resonance frequencies at constant cutoff frequencies. At the same time, the increase of carrier concentration also results in the shift of cutoff frequencies Ω_L^- (Ω_L^+). It turned out that for $\theta = 0°$, the Ω_L^- (Ω_L^+) increases from 0.69 cm^{-1} (589.98 cm^{-1}) to 403.73 cm^{-1} (2158.31 cm^{-1}) with $\nu_{p\perp}$ change from 1 cm^{-1} to 2115 cm^{-1}. When $\theta = \pi/2$, the ν_L^- (ν_L^+) vary from 0.74 cm^{-1} (566.47 cm^{-1}) to 373.79 cm^{-1} (2365.12 cm^{-1}), respectively. The frequencies at which $R_{min}(\nu) = 0$ also increase. Thus, the increase of the concentration of free charge carriers in ZnO single crystals results in the broadening of the total reflection ranges and in the narrowing of the transparency ones in IR reflection spectra. Along with this, low-frequency boundaries of the transparency regions coincide with the cutoff frequencies $\Omega_L^\pm$ and $\nu_L^\pm$. This behavior of resonant frequencies of ZnO crystal on carrier concentration and on the angle θ was found more specific than that reported for 6H-SiC (Venger et al., 2000), for which $\nu_{p\parallel} = \nu_{p\perp}/2.7$. In particular, at θ = 0 and π/2, the electron concentration increase results in the shift of the cutoff frequencies Ω_L^- and ν_L^- toward the values $\nu_{T\perp}$ and $\nu_{T\parallel}$, respectively.

The analysis of the $R(\nu)$ dependencies showed that although the total reflection interval ν_{R1}–Ω_L^- is small enough in lightly doped ZnO crystals ($\nu_{p\perp}$ = 240 cm^{-1}), it also demonstrates some decrease with the θ rise. At the same time, the transparency range increases similar to that observed in highly doped ZnO samples. Along with this, the cutoff frequency for this total reflection range is constant (Ω_L^- = 160.4 cm^{-1}), whereas the ν_{R1} changes from

158.7 cm^{-1} (θ = 10°) up to 160.4 cm^{-1} (θ = 85°), that is, when the angle θ increases, $\nu_{R1} \rightarrow \Omega_L^-$ and the transparency and opacity intervals approach the values of ≈2 cm^{-1} and 0 cm^{-1}, respectively.

Possible variants of the allocation of cutoff frequencies for the case $0 < \theta < \pi/2$ for longitudinal-transverse wave for ZnO crystals showed that this relative placement depends strongly on the free carrier concentration, that is:

(i) $\Omega_L^- < \nu_L^- < \Omega_L^+ < \nu_L^+ (372 < \nu_{p\perp} < 443 \text{ cm}^{-1})$;
(ii) $\nu_L^- < \Omega_L^- < \Omega_L^+ < \nu_L^+ (443 \text{ cm}^{-1} \leq \nu_{p\perp} \leq 2150 \text{ cm}^{-1})$;
(iii) $\Omega_L^- < \nu_L^- < \nu_L^+ < \Omega_L^+ (\nu_{p\perp} \leq 372 \text{ cm}^{-1})$;
(iv) $\nu_L^- < \Omega_L^- < \nu_L^+ < \Omega_L^+ (\nu_{p\perp} < 2500 \text{ cm}^{-1})$;

It should be noted in advance that the interrelation of the ν_L and ν_T values affects also the relative placement of cutoff frequencies (Tarkhanyan & Uzunoglu, 2006). For instance, in the materials that demonstrate the equal values of ν_L and ν_T or, even, $\nu_T > \nu_L$, option (iv) for allocation of cutoff frequencies can be observed. However, in ZnO materials, option (iv) cannot be realized, in spite of theoretical probability, even at very high carrier concentration since the ν_T value is significantly less than ν_L (Table 7.2).

Figure 7.14 represents the ranges of total reflection (opacity) (ranges 1–4) and the ranges of transparency for ZnO crystals with carrier concentration $n_0 = 9.3 \times 10^{16}\text{cm}^{-3}$ [$\nu_{p\perp}$ = 420 cm^{-1} (Figure 7.14a)] and $4.2 \times 10^{18} cm^{-3}$ [$\nu_{p\perp}$ = 605 cm^{-1} (Figure7.14b)], respectively.

The width of the ranges of total reflection (dashed segments) and transparency depends on the relative allocation of the cutoff frequencies. For instance, for $\nu_{p\perp} < 372\,\text{cm}^{-1}$ (option (iii) for cutoff frequencies allocation), the width of total-reflection range 1 changes from 0 up to

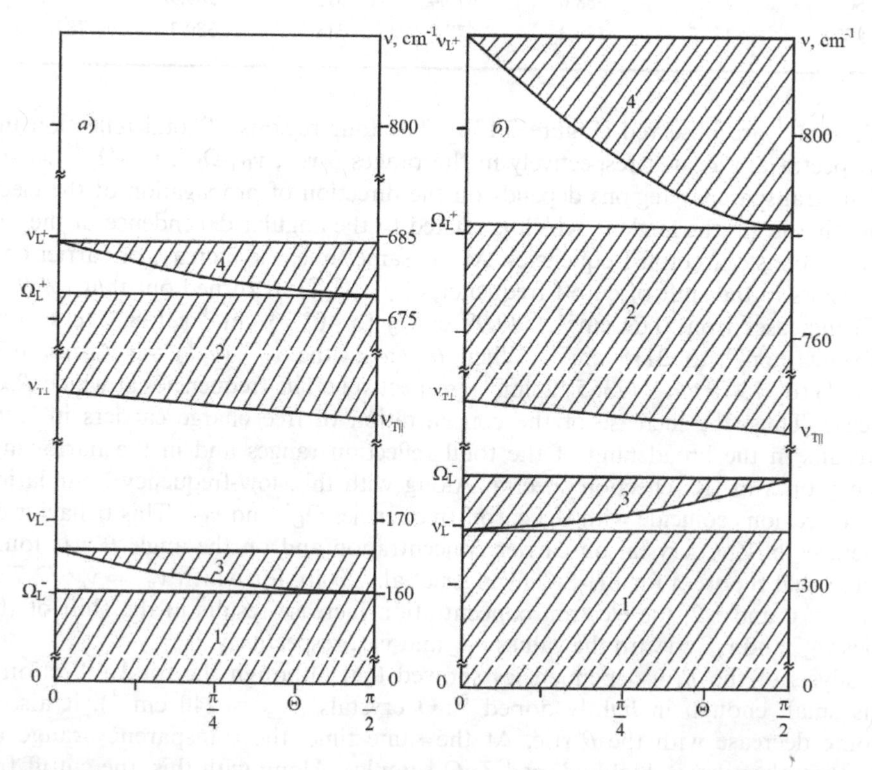

FIGURE 7.14 The ranges of total reflection (dashed segments) and transparency in ZnO with $n_0 = 9.3 \times 10^{16}$ cm^{-3} **(a)** and 4.2×10^{18} cm^{-3} **(b)** at $\boldsymbol{E \perp C}$

232 cm^{-1}, whereas the width of the range 2 increases from $\nu_{T\perp} = 412$ cm^{-1}($\theta = 0$) ($\nu_{T\|} = 380$ cm^{-1}($\theta = \pi/2$)) up to 657 cm^{-1}. With θ and $\nu_{p\perp}$ rise from 0 up to π/2 and from 1 up to 372 cm^{-1}, respectively, the transition from option (iii) to (i) occurs, causing the transformation of IR reflectance spectra. The width of the range 4 decreases from 24 cm^{-1} down to 0 cm^{-1}.

Figure 7.14a shows that for $\theta = 0$ ($\nu_{T\|} = 380$ cm^{-1} for $\theta = \pi/2$), range 1 changes from 0 to 264 cm^{-1}, range 2 – from $\nu_{T\perp} = 412$ cm^{-1} up to 688 cm^{-1}. Ranges 3 and 4 also become broader with the rise of θ. The largest width of about 5 cm^{-1} $\nu_{p\|} = 380$ cm^{-1}) and 10 cm^{-1} ($\nu_{p\perp} = 442$ cm^{-1}) can be detected. However, with the $\nu_{p\perp}$ increase, the width of range 3 approaches zero value, whereas the width of range 4 expands.

The transition from option (i) to option (ii) is shown in Figure 7.14b. When $\nu_{p\perp}$ increases (443 cm^{-1} $\le \nu_{p\perp} \le$ 620 cm^{-1}), range 3 also increases from 0 ($\nu_{p\perp} = 443$ cm^{-1}) up to 30 cm^{-1} ($\nu_{p\perp} = 2115$ cm^{-1}). Besides, range 3 becomes narrower with Θ increase and for Θ = π/2 the transparency range appears to be the largest. This transparency range expands with the carrier concentration rise (Figure 7.14b) up to 30 cm^{-1} ($\nu_{p\perp} = 605$ cm^{-1}) and 207 cm^{-1} ($\nu_{p\perp} = 2115$ cm^{-1}). This finding shows that the total reflection range 3 changes when options (i) and (ii) are realized for cutoff frequencies.

Figure 7.15 demonstrates the dependences $R(\nu)$ obtained using Eq. (7.20), in which the values of Eq. (1.17) were used instead of $n^2{}_{\perp,\|}(\nu)$. All curse were obtained for $\mathbf{E}\|\mathbf{C}$ and

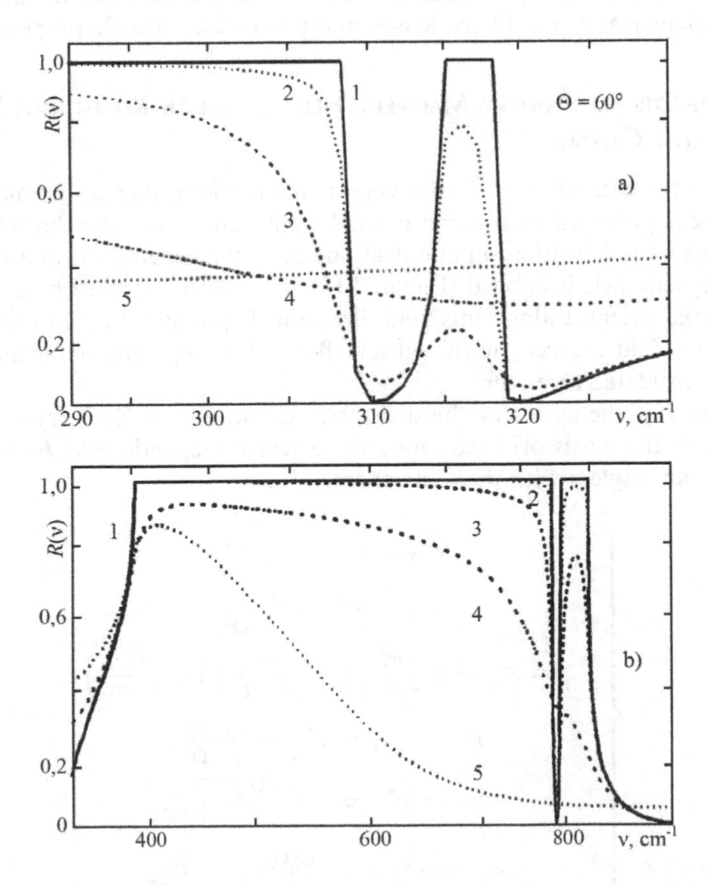

FIGURE 7.15 The $R(\nu)$ for highly doped ZnO in the range 290–330 cm^{-1} (**a**) and 330–900 cm^{-1} (**b**); $\theta = 60°$. The calculation was performed for: 1 – $\gamma_{f\perp,\|}$ and $\gamma_{p\perp,\|} \equiv 0$; 2 – and $\gamma_{p\perp,\|} = 1$ cm^{-1}; 3 – $\gamma_{f\perp,\|} = 0$ and $\gamma_{p\perp,\|} = 10$ cm^{-1}; 4 – $\gamma_{f\perp,\|} = 14$ cm^{-1}, $\gamma_{p\|} = 92$ cm^{-1}and $\gamma_{p\perp} = 100$ cm^{-1}; 5 – $\gamma_{f\perp,\|} = 14$ cm^{-1}, $\gamma_{p\|} = 920$ cm^{-1} and $\gamma_{p\perp} = 1020$ cm^{-1}

$\theta = 60°$ for the sample ZC1M ($\nu_{p\perp} = 605$ cm^{-1}) taking into account different values of phonon and plasmon damping coefficients. Curve 1 corresponds to the $R(\nu)$ dependence when these damping coefficients are equal to zero. In total, this spectrum has four reflection and transparency ranges with clearly seen minima $R(\nu_{min}) = 0$. The dashed lines 2 and 3 are the $R(\nu)$ spectra obtained for the case $\gamma_{f\perp,||} = 0$ and $\gamma_{p\perp,||} \neq 0$, that is, $\gamma_{p\perp,||} = 1$ cm^{-1} (curve 2) and $\gamma_{p\perp,||} = 10\,\text{cm}^{-1}$ (curve 3).

The existence of the plasmon and phonon damping results in the significant decrease of the $R(\nu)$ values, especially in the low-frequency range as shown by curves 4 and 5 (Figure 7.15). Besides, the total reflection ranges become to be narrower in 290–310 and 380–790 cm^{-1} intervals and the red shift of their high-frequency edge occurs. These results are in a good agreement with those described by Ukhanov (1977). The appearance of total reflection ranges (at 314–318 cm^{-1} and 800–820 cm^{-1}) is the signature of the increase of the damping coefficients of plasmon and phonon subsystems. Moreover, for highly doped ZnO the difference between transparency ranges and total reflection ones is negligible because $R(\nu) < 1$ that was reported by Venger et al. (1995).

Thus, the analysis of Figures 7.14 and 7.15 allows to conclude that for anisotropic ZnO crystals, the number of ranges of total reflection and transparency increases twice in comparison with isotropic case described above for $0 < \theta < \pi/2$. The dependence of the width of each range can be attractive for the application of such anisotropic materials for the production of different optical elements such as filters, lenses, and prisms with specific properties.

7.4.3 Effect of Strong Uniform Magnetic Field on the IR Reflections Spectra of ZnO Single Crystals

Here, we consider the case when ***E***⊥***B*** (the Voigt configuration), that is, the magnetic birefringence when a linear polarized light being normally directed toward the direction of the magnetic field changes its polarization on elliptical one after the penetration in an isotropic layer, on which the magnetic field is applied (Figure 7.15). This effect is caused by an optical anisotropy of the material oriented along this field direction. It should be noted that only the case, when the magnetic field induced by the current flowing through the semiconductor is insignificant, will be considered hereafter.

The components of the tensor of the dielectric permittivity of ZnO crystal in the system of coordinates with the z-axis oriented along the external magnetic field ***B*** (which lies in the xz-plane and has an angle φ with the C-axis) are:

$$\begin{cases} \varepsilon_{xx} = \varepsilon_{xx}^{\infty} - \frac{\mu_{xx}\omega_0^2}{\omega^2-\Omega^2}, \\ \varepsilon_{yy} = \varepsilon_{\perp}^{\infty} - \frac{\mu_{\perp}\omega_0^2}{\omega^2-\Omega^2}, \\ \varepsilon_{zz} = \varepsilon_{zz}^{\infty} - \mu_{zz}\frac{\omega_0^2}{\omega^2}\left[1 + \frac{\Omega^2}{\omega^2-\Omega^2}\left(1 - \frac{\mu_{\perp}\mu_{||}}{\mu_{xx}\mu_{zz}}\right)\right], \\ \varepsilon_{xz} = \varepsilon_{zx} = \varepsilon_{xz}^{\infty} - \frac{\mu_{xz}\omega_0^2}{\omega^2-\Omega^2}, \\ \varepsilon_{xy} = \varepsilon_{yx}^{*} = -i\frac{\omega_0^2\Omega\sqrt{\mu_{\perp}\mu_{xx}}}{\omega\left(\omega^2-\Omega^2\right)}, \\ \varepsilon_{yz} = \varepsilon_{zy}^{*} = i\frac{\omega_0^2\Omega\mu_{xz}}{\omega\left(\omega^2-\Omega^2\right)}\sqrt{\frac{\mu_{\perp}}{\mu_{xx}}}. \end{cases} \tag{7.22}$$

Here, $\Omega_e = \frac{eB}{mc}\sqrt{\mu_{\perp}\mu_{xx}}$ is the cyclotron frequency, $\mu_{\perp}$, $\mu_{||}$ are the components of the dimensionless tensor of inverse effective mass μ_{ij}.

$$\begin{cases} \mu_{xx} = \mu_{\perp}\cos^2\theta + \mu_{\parallel}\sin^2\theta, \\ \mu_{yy} = \mu_{\perp}, \\ \mu_{zz} = \mu_{\perp}\sin^2\theta + \mu_{\parallel}\cos^2\theta, \\ \mu_{xz} = \mu_{zx} = \left(\mu_{\parallel} - \mu_{\perp}\right)\sin\theta\cos\theta, \\ \mu_{xy} = \mu_{yx} = \mu_{yz} = \mu_{zy} = 0, \end{cases} \tag{7.23}$$

where θ is the angle between the C-axis and z-axis. Similar expressions also exist for the components of the high-frequency permittivity tensor of the crystal lattice $\varepsilon_{ij}^{\infty}$.

By solving the dispersion equation

$$\left|\varepsilon_{ij} + n^2\left(s_i s_j - \delta_{ij}\right)\right| = 0, \quad \vec{s} = \frac{\vec{k}}{k}, \quad n = \frac{ck}{\omega} \tag{7.24}$$

one can see that an extraordinary (longitudinal transverse) wave with $\boldsymbol{E} \perp \boldsymbol{B}$, $\boldsymbol{H} \parallel \boldsymbol{B}$ can propagate across the static magnetic field, and at $\theta = 0$, the refractive index of such a wave is considered as

$$n^2 = \varepsilon_{\perp}^{\infty} \frac{\left(\omega^2 - \omega_+^2\right)\left(\omega^2 - \omega_-^2\right)}{\omega^2\left(\omega^2 - \omega_p^2\right)} \tag{7.25}$$

where $\omega_{\pm}^2 = \omega_{0\perp}^2 + \frac{\Omega^2}{2}\left(1 \pm \sqrt{1 + 4\frac{\omega_{0\perp}^2}{\Omega^2}}\right)$, and $\omega_p^2 = \Omega^2 + \omega_{0\perp}^2$.

Since in the electromagnetic wave $\boldsymbol{E} \perp \boldsymbol{B}$, then

$$R = \left|\frac{1 - n}{1 + n}\right|^2 \tag{7.26}$$

Taking into account that the frequency dependence of the dielectric permittivity in the range of plasmon–phonon interaction is determined as

$$\varepsilon_j(\nu) = \varepsilon_{1j}(\nu) + i\varepsilon_{2j}(\nu) = \varepsilon_{\infty j}\left(1 + \frac{\nu_{Lj}^2 - \nu_{Tj}^2}{\nu_{Tj}^2 - \nu^2 - i\nu\gamma_{fj}} - \frac{\nu_{pj}^2}{\nu\left(\nu + i\gamma_{pj}\right)}\right), \tag{7.27}$$

the variation of the refractive index versus magnetic field can be obtained for different ZnO crystals (Figure 7.16).

One can see from Figure 7.16a for ZnO crystal with $n_0 = 9.3 \times 10^{16}$ cm^{-3}, the increase of magnetic field value results in the increase of the number of minima and maxima in the $R(\nu)$ dependence as well as their shift toward higher frequencies. Besides, the minimum resonance frequency (ν_{min}) also decreases, whereas its magnitude $R_{min} \neq 0$. At the same time, the maximum resonance frequency (ν_{max}) increases, while $R_{max} \neq 1$. It should be noted that the increase of magnetic field does not affect the IR reflection spectra in the range 355–1200 cm^{-1}. Contrary to this, in the samples with higher carrier concentration ($n_0 = 2.0 \times 10^{18}$ cm^{-3}), it was detected in two regions (370–640 cm^{-1} and 840–1200 cm^{-1}) where magnetic field does not affect the $R(\nu)$ values (Figure 7.16b). Most prominent variation of $R(\nu)$ was found for the 250–380 cm^{-1} frequency range (see the inset in Figure 7.16b). Besides, the increase of magnetic field values results also in the appearance of the extrema near 300 cm^{-1}, whereas their shift is negligible.

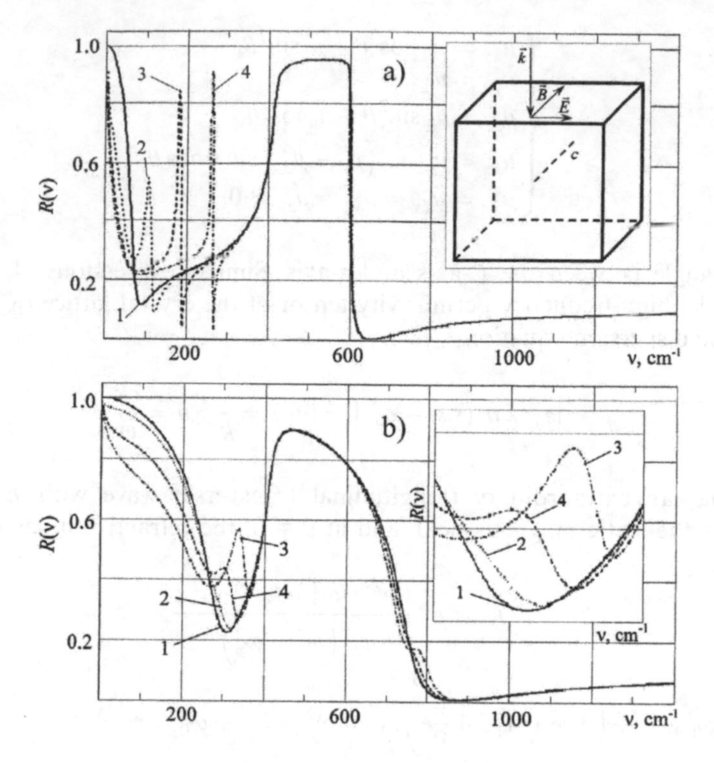

FIGURE 7.16 Dependence of $R(\nu)$ on magnetic field obtained for ZnO crystals with $n_0 = 9.3 \times 10^{16}$ cm^{-3} (**a**) and 2.0×10^{18} cm^{-3} (**b**). Curves 1–4 correspond to the magnetic field value: 1 Oe (1), 30 kOe (2), 65 kOe (3), and 100 kOe (4). Inset in (**a**) represents the direction of magnetic field vector and optical axis of ZnO crystal (*C*-axis). Inset in (**b**) shows in detail the variation of $R(\nu)$ in the range 250–380 cm^{-1}

Figure 7.17 demonstrates the effect of optical phonon damping coefficient on the $R(\nu)$ curves for the same ZnO crystals. It is seen that for the case of low carrier concentration ($n_0 = 9.3 \times 10^{16}$ cm^{-3}, $\nu_p = 100$ cm^{-1}, and $\gamma_p = 100$ cm^{-1}), the γ_f increase from 11 up to 30 cm^{-1} results in the appearance of additional extrema at about 180 cm^{-1} and 190 cm^{-1} as well as in the decrease of the $R(\nu)$ values in the 390–600 cm^{-1} frequency range (Figure 7.17a). At the same time, for ZnO sample with higher electron concentration ($n_0 = 2.0 \times 10^{18}$ cm^{-3}, $\nu_p = 500$ cm^{-1}and $\gamma_p = 500$ cm^{-1}), the γ_f rise causes both the increase of $R_{min}(\nu)$ from 0.295 up to 0.36 in the 250–380 cm^{-1} frequency range and the decrease of $R_{max}(\nu)$ from 0.925 down to 0.875 in the 400–570 cm^{-1} frequency range (Figure 7.17b).

Figure 7.18 demonstrates the $R(\nu)$ dependence on the magnitude of uniform magnetic field at the frequencies corresponding to the ν_{min} and ν_{max} values obtained for ZnO single crystal ($n_0 = 9.3 \times 10^{16}$ cm^{-3}). The comparison of the obtained results showed that the most significant effect of magnetic field on specular IR reflection spectra occurs for ZnO crystals with low carrier concentration.

Thus, the results presented above showed that magnetic field affects significantly the IR reflection spectra of ZnO crystals. The most significant transformation such as anomaly of the reflection coefficient was found for the crystals with low carrier concentration. When $\boldsymbol{E} \perp \boldsymbol{B}$, additional extrema appear in the spectra that can be attractive for the development of reference materials for the detection of magnetic field value. The absolute values of reflection coefficient were observed to be depended on magnetic field amplitude. This fact can be used to obtain information about physical and chemical properties of the materials

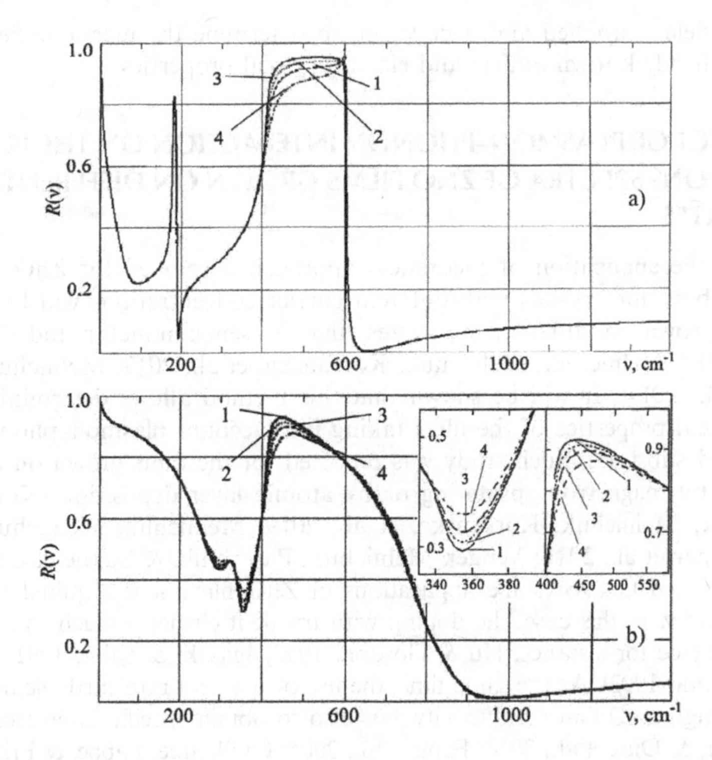

FIGURE 7.17 Effect of the phonon damping coefficient (γ_f) on the $R(\nu)$ for ZnO crystal with $n_0 = 9.3 \times 10^{16}$ cm^{-3}, $\nu_p = 100$ cm^{-1}, and $\gamma_p = 100$ cm^{-1} (**a**) and 2.0×10^{18} cm^{-3}, $\nu_p = 500$ cm^{-1}, and $\gamma_p = 500$ cm^{-1} (**b**). Curves 1–4 correspond to the γ_f value: 11 cm^{-1} (1), 15 cm^{-1} (2), 20 cm^{-1}, (3) and 30 cm^{-1} (4). $H = 65$ kOe

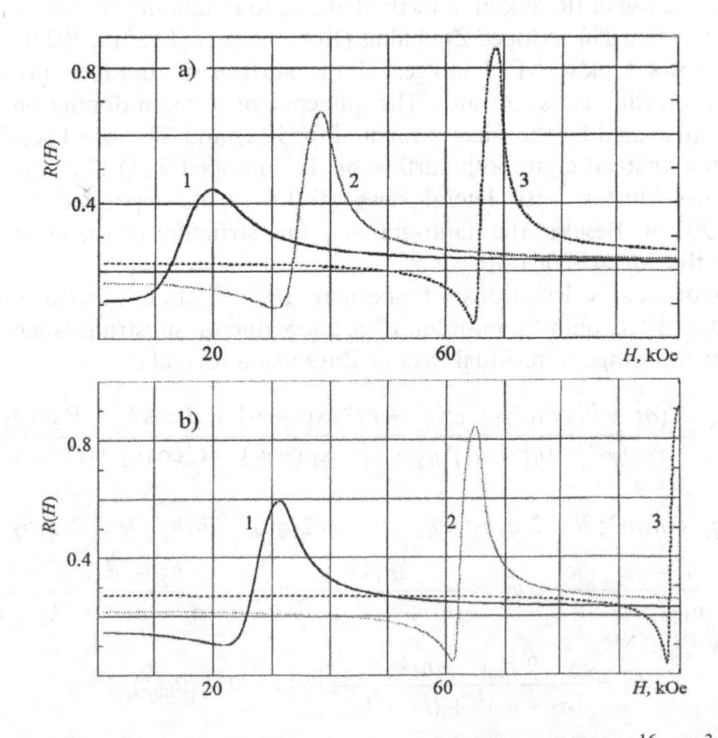

FIGURE 7.18 Effect of magnetic field on the $R(\nu)$ of ZnO crystal ($n_0 = 9.3 \times 10^{16}$ cm^{-3}) taken at different values of ν_{min} and ν_{max}: (**a**) ν_{min} = 88 (1), 119 (2), and 186 (3) cm^{-1}; (**b**) ν_{max} = 104 (1), 177 (2), and 259 (3) cm^{-1}. Corresponding $R(H)$ spectra are presented in Figure 7.16a

when magnetic field is applied and, vice versa, to determine the magnetic field value using materials with already known optical and electrophysical properties.

7.5 THE EFFECT OF PLASMON–PHONON INTERACTION ON THE IR REFLECTIONS SPECTRA OF ZNO FILMS GROWN ON DIFFERENT SUBSTRATES

In this section, the application of theoretical approach developed for ZnO bulk materials and described above for crystals with different carrier concentrations will be demonstrated for ZnO films grown on different substrates, that is, semiconductor and dielectric (Korsunska et al., 2019; Melnichuk, Melnichuk, Korsunska, et al., 2019; Melnichuk, Melnichuk, Tsykaniuk, et al., 2019). It will be shown that this method allows determining the optical and electrophysical properties of the films taking into account plasmon–phonon interaction between film and substrate. Such study was reported for the films grown on Si, Al_2O_3, and SiO_2 substrates by magnetron sputtering or by atomic layer deposition (Korsunska et al., 2019; Melnichuk, Melnichuk, Korsunska, et al., 2019; Melnichuk, Melnichuk, Tsykaniuk, et al., 2019; Venger et al., 2016; Venger, Melnichuk, Pasechnik, & Sukhenko, 1997).

It should be noted that for some applications of ZnO films, it is required to achieve high electron conductivity. In this case, the doping with trivalent elements (such as Al, Ga, and In) was usually used (see for instance, Hu & Gordon, 1992; Igasaki & Saito, 1991; Minami, Sato, Imamoto, & Takata, 1992). At the same time, the use of trivalent rare-earth elements allows not only to achieve high ZnO film conductivity, but also to obtain specific luminescence properties (Elfakir, Douayar, & Diaz et al., 2014; Fang et al., 2005; Guillaume, Labbe, & Frilay et al., 2018; Hastir, Kohli, & Singh, 2017; Korsunska et al., 2019, 2020; Ziani, Davesnne, & Labbé et al., 2014). However, in most cases such films are highly textured (Figure 7.19) that causes some difficulties in the observation of the correlation between doping level and film conductivity. Recently, it was shown that the use of IR reflection method allows to obtain this information, for instance, for Tb-doped and/or (Eu,Tb)-codoped ZnO films (Korsunska et al., 2019, 2020).

Figure 7.19 shows typical AFM images of the surface of undoped (**a**) and Tb-doped films (**b–d**) grown on different substrates. The influence of terbium doping on surface topography is clearly illustrated by the images for undoped (**a**) and Tb-doped (**c**) films grown on Si (001) substrates. Instead of smooth surface of the undoped ZnO film, the film grown on Si substrate shows clusters with lateral sizes of 0.5–1.6 μm separated by trenches with a depth of 50–190 nm. Besides, the nano-granular fine structure of the clusters can also be distinguished on the surfaces of both films.

Generally, theoretical calculations of specular IR reflection spectra inherent to the absorbing film deposited onto "semi-infinite" semiconductor substrate (such as silicon) are performed within the range of residual rays of ZnO using formula

$$R(\nu) = \frac{(q_1^2 + h_1^2)\exp(\gamma_2) + (q_2^2 + h_2^2)\exp(-\gamma_2) + A\cos\delta_2 + B\sin\delta_2}{\exp(\gamma_2) + (q_1^2 + h_1^2)(q_2^2 + h_2^2)\exp(-\gamma_2) + C\cos\delta_2 + D\sin\delta_2} \tag{7.28}$$

where $A = 2(q_1q_2 + h_1h_2)$; $B = 2(q_1h_2 - q_2h_1)$; $C = 2(q_1q_2 - h_1h_2)$; $D = 2(q_1h_2 + q_2h_1)$;

$$q_1 = \frac{n_1^2 - n_2^2 - k_2^2}{(n_1 + n_2)^2 + k_2^2};\ h_1 = \frac{2n_1k_2}{(n_1 + n_2)^2 + k_2^2};\ q_2 = \frac{n_2^2 - n_3^2 + k_2^2 - k_3^2}{(n_2 + n_3)^2 + (k_2 + k_3)^2};$$

$$h_2 = \frac{2(n_2k_3 - n_3k_2)}{(n_2 + n_3)^2 + (k_2 + k_3)^2};\ \gamma_2 = (4\pi k_2 d_{\text{film}})/\lambda;$$

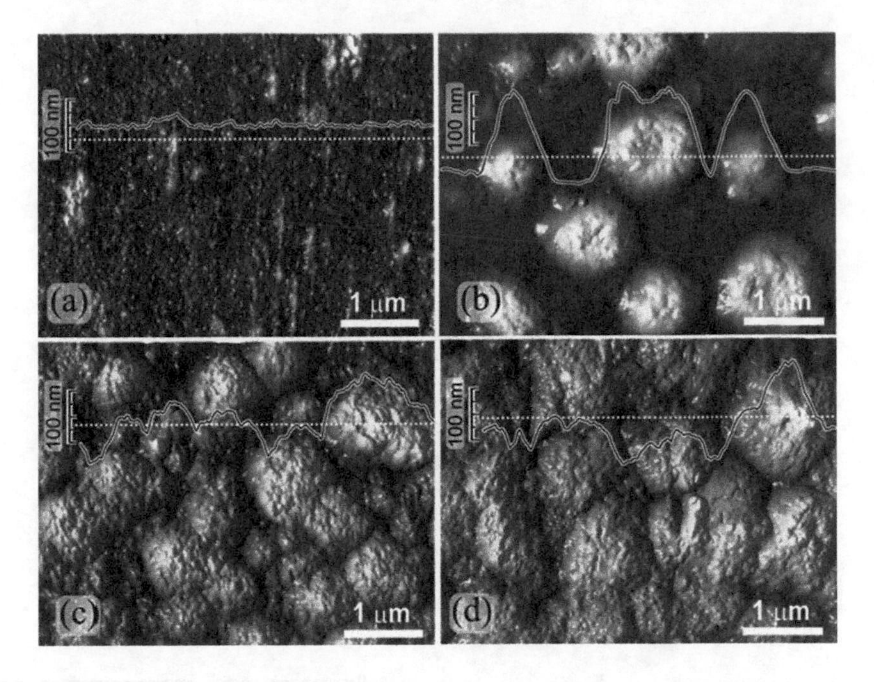

FIGURE 7.19 AFM images of the surface of different samples: ZnO/Si (**a**), ZnO-Tb/Al_2O_3 (**b**), ZnO-Tb/Si (**c**), and ZnO-Tb/SiO_2 (**d**)

$\delta_2 = (4\pi n_2 d_{\text{film}})/\lambda$ and $n_1(k_1)$, $n_2(k_2)$, and $n_3(k_3)$ are the refractive indexes (extinction coefficient) of air, ZnO, and substrate, respectively; d_{film} is the thickness of ZnO film.If the ZnO film is grown on Si substrate, then for the calculation of the n_2 and n_3, the dielectric permittivity is considered as a function with additive contribution of active optical phonons (ν_T) and plasmons (ν_p) in the film and substrate according to Eq. (7.27). For further calculation, the parameters for ZnO were taken from Table 7.2, whereas for Si substrate $\varepsilon_0 = \varepsilon_\infty = 11.4$ and $\nu_L = \nu_T = 520.9$ cm^{-1} are considered.

For the simulation of IR reflection spectra for the ZnO film situated on sapphire (Al_2O_3) substrate, the dielectric permittivity of Al_2O_3 versus the frequency can be taken accounted for four-oscillator model

$$\varepsilon(\nu) = \varepsilon_\infty + \sum_{j=1}^{4} \frac{\Delta\varepsilon_j \nu_{Tj}^2}{\nu_{Tj}^2 - \nu^2 + i\nu\gamma_{fj}}, \tag{7.29}$$

where ε_∞ – high-frequency dielectric constant of sapphire for $\boldsymbol{E} \perp \boldsymbol{C}$; $\Delta\varepsilon_j$ – the strength of the *j*th oscillator; ν_{Tj} – the frequency of transversal optical vibration of *i*th oscillator; γ_{fj} – damping constant of the *j*th oscillator. In the case of SiO_2 substrate, the model with four oscillators for noncrystalline quarts was also used. The parameters for sapphire and quarts substrates are collected in Table 7.9.

The simulation of the $R(\nu)$ spectra can be performed versus carrier concentration, film thickness, plasmon, and phonon damping coefficients. As an example, Figure 7.20 shows simulated $R(\nu)$ spectra versus carrier concentration. The simulation was performed with Eqs. (7.28) and (7.29) for the ZnO/Al_2O_3 structure for the film thickness of 500 nm (Figure 7.20a) and 675 nm (Figure 7.20b) that correspond to experimental values. Phonon damping

TABLE 7.9
The parameters of Al_2O_3 and SiO_2 substrates used for the R(ν) simulation

Substrate	Parameter *the j*th oscillator		
	ν_{Tj}, cм$^{-1}$	$\Delta\varepsilon_j$	γ_{fj}/ν_{Tj}
Al_2O_3 single crystal, $E\perp C$			
$j = 1$	384	0.2	0.015
$j = 2$	442	2.8	0.01
$j = 3$	571	3.1	0.2
$j = 4$	634	0.2	0.02
SiO_2 (JGS1 type)			
$j = 1$	457	0.95	0.015
$j = 2$	810	0.05	0.1
$j = 3$	1072	0.6	0.06
$j = 4$	1160	0.15	0.04

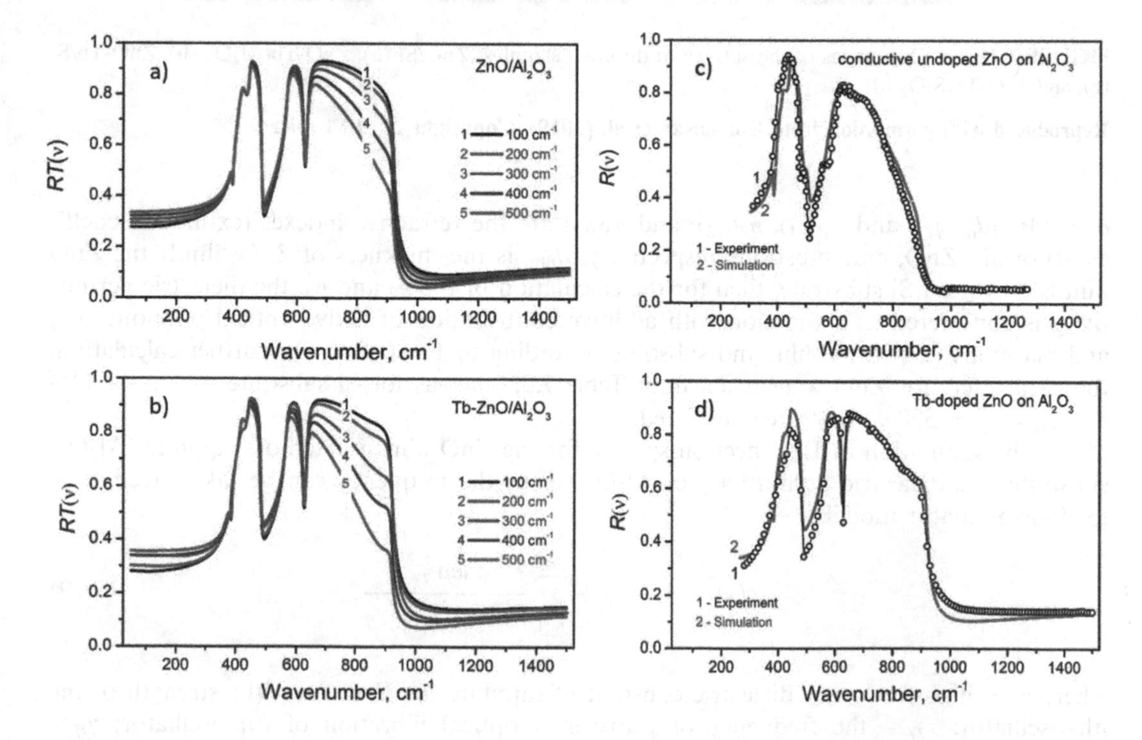

FIGURE 7.20 Simulated (lines) and experimental (symbols) R(ν) spectra for conductive undoped (**a, c**) and Tb-doped ZnO (**b, d**) films. Parameters of simulation are as follows: the thickness and phonon damping coefficient are d = 500 nm, $\gamma_{f\perp}$ = 20 cm^{-1} (**a, c**) and d = 675 nm, $\gamma_{f\perp}$ = 20 cm^{-1} (**b, d**), respectively. (**a, b**) $\nu_{p\perp}$ = $\gamma_{p\perp}$ = 100–500 cm^{-1} (curves 1–5); (**c**) $\nu_{p\perp}$ = 400 cm^{-1}, $\gamma_{p\perp}$ = 870 cm^{-1}; (**d**) $\nu_{p\perp}$ = 350 cm^{-1} and $\gamma_{p\perp}$ = 250 cm^{-1}

coefficient was fixed $\gamma_{f\perp} = 20$ cm^{-1} for both films. The frequency and damping coefficient of plasmons were varied from $\nu_{p\perp} = 100$ cm^{-1} and $\gamma_{p\perp} = 100$ cm^{-1} (curve 1) up to $\nu_{p\perp} = 500$ cm^{-1} and $\gamma_{p\perp} = 500$ cm^{-1} (curve 5) with a scanning step of 100 cm^{-1}.

It is seen that the increase of carrier concentration (variation of the ν_p) in conductive undoped film results in the significant deformation of the $R(\nu)$ dependence in the range of 100–1200 cm^{-1} (Figure 7.21a) followed by the decrease of the $R(\nu)$ value, for instance, down to 0.8 at $\nu = 680$ cm^{-1} and to 0.4 at $\nu = 510$ cm^{-1}. This fact can be applied for the estimation of the concentration and mobility of free carriers in ZnO films (Melnichuk, Melnichuk, Tsykaniuk, et al., 2019). At the same time, the variation of phonon damping coefficient γ_f in ZnO film affects the reflection spectra mainly in the vicinity of TO phonon frequency ($\nu_T = 412$ cm^{-1}), whereas in other spectral ranges its influence is negligible.

For Tb-doped ZnO film (Tb content is about 3 at.%), the increase of the $\nu_{p\perp}$ and $\gamma_{p\perp}$ values causes the increase of the slope of $R(\nu)$ curves in the 600–900 cm^{-1} spectral range that corresponds to an increase of electron concentration. The appearance of a shoulder at about 890–900 cm^{-1} is due to higher film thickness (Melnichuk, Melnichuk, Tsykaniuk, et al., 2019).

The experimental and simulated spectra are compared in Figure 7.20c, d. For conductive undoped ZnO film, the best fitting was obtained for $\gamma_{f\perp} = 25$ cm^{-1}, $\nu_{p\perp} = 400$ cm^{-1}, and $\gamma_{p\perp} = 870$ cm^{-1}(Figure 7.20c). The electron concentration, mobility, and conductivity were estimated to be $n_0 = 1.83 \times 10^{18}$ cm^{-3}, $\mu = 258$ cm^2 V^{-1} × s^{-1} and $\sigma = 76\,\Omega^{-1} \times$ cm^{-1}, respectively. For the Tb-doped ZnO film (Figure 7.20d), these parameters were as follows: $\gamma_{f\perp} = 20$ cm^{-1}, $\nu_{p\perp 1} = 350$ cm^{-1}, and $\gamma_{p\perp 1} = 250$ cm^{-1}, $n_0 = 1.03 \times 10^{18}$ cm^{-3}, $\mu = 89.7$ cm^2 V^{-1} s^{-1} and $\sigma = 202\,\Omega^{-1} \times$ cm^{-1}. It is seen that the electron concentration is nearly the same in both films. This allows concluding that Tb incorporation does not change essentially the electron concentration in ZnO in spite of the fact that Tb^{3+} is the donor. This effect was explained by the formation of some defects acting as charge compensator for Tb^{3+} ions (Melnichuk, Melnichuk, Korsunska, et al., 2019). Another reason can be the dependence of the conductivity on Tb concentration as it was reported by Elfakir et al. (2014). It was shown that the increase of Tb content up to 1 at.% results in the higher film conductivity, while for higher Tb content (up to 5 at.%), the decrease of conductivity was observed. It should be noted that both experimental and theoretical curves are in good agreement. This confirms the reliability of the parameters of bulk ZnO and the possibility of their use for the study of textured ZnO films.

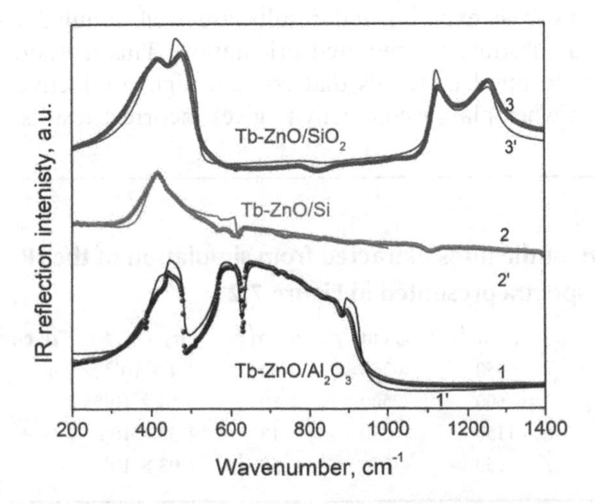

FIGURE 7.21 The IR reflection spectra of Tb-doped ZnO films grown on Al_2O_3 (1), Si (2), and SiO_2 (3) substrates and their corresponding simulation (curves 1′–3′). The spectra are shifted in vertical direction for clarity

It is interesting that direct current conductivity was also studied by Melnichuk, Melnichuk, Korsunska, et al. (2019). However, significantly different values were obtained. The planar electrical conductivity was about 10^{-7}–$10^{-8}\ \Omega^{-1} \times cm^{-1}$, which is much lower than the values obtained from the measurements of IR reflectance spectra. This difference was explained by the presence of high-resistive spaces between ZnO columns that was confirmed by structural characterization of the films (Korsunska et al., 2019; Melnichuk, Melnichuk, Korsunska, et al., 2019). Thus, IR reflectance spectra are useful for the investigation of conductivity of nonhomogeneous systems. Note also that in the case of Si substrate (in contrast to insulated one), the sheet resistance can be influenced by the substrate conductivity, while IR reflectance allows extracting the conductivity values of ZnO film and substrate, separately.

The experimental IR reflection spectra of Tb-doped ZnO films grown on different substrates are shown in Figure 7.21. To simulate these spectra in the range of residual rays, the approach described above was also applied.

The best fitting of experimental data for Tb-doped film grown on different substrates was obtained using the parameters shown in Table 7.10. The calculation of carrier concentrations and their mobility was performed with the uncertainty less than 1%.

It should be noted that all Tb-doped samples were prepared in the same deposition run. The analysis of the IR reflection spectra shows that the concentration of electrons in Tb-ZnO films on different substrates lies in the range of $n_0 = 10^{18}$–10^{19} cm^{-3}. It was found that Tb^{3+} luminescence does not correlate with the Tb^{3+} content. It was concluded that the Tb incorporation into ZnO host requires Tb ions' association with other charge-compensating lattice defects or impurities. Besides, such doping provokes some film disordering and formation of intrinsic defects in ZnO lattice. The higher is the Tb concentration, the higher is the number of compensating defects, thus resulting in unchanged conductivity.

The results described above show the effect of substrate material, including Si, SiO_2, and Al_2O_3, on structural, optical, and electrical properties of Tb-doped ZnO films. The doping with Tb ions results in the formation of highly textured films with the micro-clusters of 0.5–1.6 μm diameter consisted of ZnO columns oriented perpendicularly to substrate surface and separated by deep tranches. The width of these tranches depends on the type of substrate, being the largest in the Tb-ZnO/Al_2O_3 one. Besides, the Tb-ZnO/Al_2O_3 film is found to consist of two layers with different morphologies, only the upper part being of column structure.The investigation described above shows the utility of this nondestructive approach for the estimation of the conductivity of ZnO columns as well as the concentration and mobility of free carriers. The experimental results are well simulated using self-consistent parameters of ZnO and substrate for required orientation. This method allows analyzing the electrical parameters in textured materials that contain high-conductive crystallites separated by high-resistive regions when planar conductivity gives incorrect results.

TABLE 7.10

Parameters of the films extracted from simulation of the IR reflection spectra presented in Figure 7.21

Sample	ν_p, cm^{-1}	γ_p, cm^{-1}	γ_f, cm^{-1}	n_0, cm^{-3}	μ, $cm^2\ V^{-1} \times s^{-1}$
ZnO-Tb/Al_2O_3	350	250	30	1.4×10^{18}	89
ZnO-Tb/SiO_2	600	2500	80	4.14×10^{18}	89
ZnO-Tb/Si	1150	880	18	1.52×10^{19}	254
ZnO/Si	950	650	50	1.03×10^{19}	344

7.6 CONCLUSIONS

Infrared spectroscopy is a powerful tool for the investigation of optical properties of different materials and structures. It allows to study the elementary excitations (plasmons, phonons, and polaritons) and their interactions because their coupling affects significantly the shape of specular and attenuated total infrared reflection spectra in the range of residual rays when the plasmon frequency is approaching the frequency of longitudinal optical phonon. Besides, this technique allows to study the effect of magnetic field on elementary excitations and their coupling. The theoretical approaches developed by the authors permit to determine electro-physical parameters of materials. The modeling of experimental spectra gives a hand for estimating the concentration and mobility of free carriers, as well as material conductivity. The developed method is fast, contactless, and sensitive, and it can be applied not only for the investigation of different semiconductor single crystals, but also for textured films grown on both dielectric and semiconductor substrates.

REFERENCES

Agranovich, V. M., & Dubovskii, O. A. (1965). Surface excitons in uniaxial crystals. *Fizika Tverddogo Tela (Leningrad)*, *7*, 2885–2889.

Agranovich, V. M., & Mills, D. L. (Eds.). (1985). *Surface polaritons: Electromagnetic waves on surfaces and media interfaces.* Moscow: Nauka.

Brion, J. J., Wallis, R. F., Hartstein, A., & Burstein, E. (1973). Interaction of surface magnetoplasmons and surface optical phonons in polar semiconductors. *Surface Science, 34*, 73–80.

Bryksin, V. V., Mirlin, D. N., & Firsov, Y. A. (1974). Surface optical phonons in ionic crystals. *Soviet Physics Uspekhi*, *17*, 305–325.

Dmitruk, N. L., Litovchenko, V. G., & Stryshevskii, V. L. (1989). *Surface polaritons in semiconductors and dielectrics.* Kyiv: Naukova Dumka.

Elfakir, A., Douayar, A., Diaz, R., Chaki, I., Prieto, P., Loghmarti, M., … Abd-Lefdil, M. (2014). Elaboration and characterization of sprayed Tb-doped ZnO thin films. *Sens Transducers*, *27*, 161–164.

Ellmer, K., Klein, A., & Rech, B. (2008). *Transparent conductive zinc oxide: Basics and applications in thin film solar cells* (446 p). Springer, Hahn-Meitner-Institut Berlin GmbH Abt., Berlin, Germany.

Fang, Z. B., Tan, Y. S., Gong, H. X., Zhen, C. M., He, Z. W., & Wang, Y. Y. (2005). Transparent conductive Tb-doped ZnO films prepared by rf reactive magnetron sputtering. *Materials Letters*, *59*, 2611–2614.

Fortunato, E., Barquinha, P., Pimentel, A., Gonçalves, A., Marques, A., Pereira, L., & Martins, R. (2005). Recent advances in ZnO transparent thin film transistors. *Thin Solid Films*, *487*, 205–211.

Gieraltowska, S., Wachinski, L., Witkowski, B. S., Godlewski, M., & Guziewicz, E. (2012). Atomic layer deposition grown composite dielectric oxides and ZnO for transparent electronic applications. *Thin Solid Films, 520*, 4694–4697.

Guillaume, C., Labbe, C., Frilay, C., Doualan, J.L., Lemarié, F., Khomenkova, L., Borkovska, L., & Portier X. (2018). Thermal treatments and photoluminescence properties of ZnO and ZnO:Yb films grown by magnetron sputtering. *Physica Status Solidi A*, *216*, 1800203.

Gurevich, L.E., & Ipatova I.P. (1959). The faraday effect in semiconductors due to free carriers in a strong magnetic field. *Journal of Experimental and Theoretical Physics (U.S.S.R.)*, *37*, 1324–1329 (in Russian). (English translation: Soviet Physics JETP, 1960, 10, 943-946.).

Gurevich, L. E., & Tarkhanyan, R. G. (1975). Surface plasmon-polaritons in uniaxial semiconductors. *Fizika Tverdogo Tela (Leningrad) 17: 1944–1949 (English Translation: Sov. Phys. Solid State, 17*, 1273–1278.

Haider, T. (2017). A review of magneto-optical effects and its application. *International Journal of Electromagnetics Applications*, *7*, 17–24. doi:10.5923/j.ijea.20170701.03

Hastir, A., Kohli, N., & Singh, R. C. (2017). Comparative study on gas sensing properties of rare earth (Tb, Dy and Er) doped ZnO sensor. *Journal of Physics and Chemistry of Solids*, *105*, 23–34.

Hopfield, J. J. (1958). Theory of the contribution of excitons to the complex dielectric constant of crystals. *Physical Review.*, *112*, 1555–1567.

Hu, J., & Gordon, R. G. (1992). Atmospheric pressure chemical vapor deposition of gallium doped zinc oxide thin films from diethyl zinc, water, and triethyl gallium. *Journal of Applied Physics, 72*, 5381–5392.

Huang, K. (1951). Lattice vibrations and optical waves in ionic crystals. *Nature, 167*, 779–780.

Igasaki, Y., & Saito, H. (1991). The effects of deposition rate on the structural and electrical properties of ZnO:Al films deposited on oriented sapphire substrates. *Journal of Applied Physics, 70*, 3613–3619.

Kapustina, A. B., Petrov, B. V., Rodina, A. V., & Seisyan, R. P. (2000). Magnetic absorption of hexagonal crystals CdSe in strong and weak fields: Quasi-cubic approximation. *Physics of the Solid State, 42*, 1242–1252.

Korsunska, N., Borkovska, L., Khomenkova, L., Gudymenko, O., Kladko, V., Kolomys, O., … Melnichuk, L. (2020). Transformations in the photoluminescent, electrical and structural properties of Tb^{3+} and Eu^{3+} co-doped ZnO films under high-temperature annealing. *Journal of Luminescence, 217*, 116739.

Korsunska, N., Borkovska, L., Polischuk, Y., Kolomys, O., Lytvyn, P., Markevich, I., … Portier, X. (2019). Photoluminescence, conductivity and structural study of terbium doped ZnO films grown on different substrates. *Materials Science in Semiconductor Processing, 94*, 51–56.

Krichevtsov, B. B., & Weber, H.-J. (2004). Reciprocal and nonreciprocal magnetic linear birefringence in the γ-Dy_2S_3 sesquisulfide. *Physics of the Solid State, 46*, 502–509.

Landau, L., & Lifhsitz, E. (1984). *Course of theoretical physics: Electrodynamics of continuous media* (Vol. 8). Oxford: Pergamon Press.

Kuzmina, I.P., & Nikitenko, V.A. (1984). *Zinc oxide:Processing and optical properties.* Moscow: Nauka. in Russian.

Liang, Q., Venkatramani, A. V., Cantu, S. H., Nicholson, T. L., Gullans, M. J., Gorshkov, A. V., … Vuletić, V. (2018). Observation of three-photon bound states in a quantum nonlinear medium. *Science, 359*, 783–786.

Martin, B. G., Maradudin, A. A., & Wallis, R. F. (1978). Theory of damped surface magnetoplasmons in n-type InSb. *Surface Science, 77*, 416–426.

Melnichuk, A.V., Melnichuk, L. Y., & Pasechnik, Y.A. (1994). Anisotropy of the electro physical properties of zinc oxide single crystals. *Physics of the Solid State, 36*, 1430–1435.

Melnichuk, A. V. (1998). Optical and electrophysical properties of thin doped ZnO/SiC-6H films from the IR reflection spectra. *Ukrainian Journal of Physics, 43*, 1310–1315.

Melnichuk, A. V. (1999). Effect of plasmon-phonon excitations on the coefficient of reflection from the surface of hexagonal silicon carbide. *Journal of Experimental and Theoretical Physics, 89*, 344–348.

Melnichuk, A. V., Melnichuk, L. Y., & Pasechnik, Y. A. (1996). The effect of anisotropy on the dispersion curves of surface plasmon-phonon-polaritons in zinc oxide. *Phys Solid State, 38*, 362–363.

Melnichuk, A. V., Melnichuk, L. Y., & Pasechnik, Y. A. (1998). Surface plasmon-phonon polaritons of hexagonal zinc oxide. *Technical Physics, 43*, 52–55.

Melnichuk, A. V., & Pasechnik, Y. A. (1992). Anisotropy of effective mass of electrons in silicon carbide. *Physics of the Solid State, 32*, 423–428. in Russian.

Melnichuk, A. V., & Pasechnik, Y. A. (1996). Damping of surface plasmon-phonon polaritons in zinc oxide. *Physics of the Solid State, 38*, 1289–1290.

Melnichuk, A. V., & Pasechnik, Y. A. (1998). Influence of anisotropy on the dispersion of surface plasmon-phonon polaritons in silicon carbide. *Physics of the Solid State, 40*, 582–585.

Melnichuk, O., Melnichuk, L., Korsunska, N., Khomenkova, L., & Venger, Y. (2019). Optical and electrical properties of Tb–ZnO/SiO_2 structure in the infrared spectral interval. *Ukrainian Journal of Physics, 64*, 434–441.

Melnichuk, O., Melnichuk, L., Tsykaniuk, B., Tsybrii, Z., Lytvyn, P., Guillaume, C., … Korsunska, N. (2019). Investigation of undoped and Tb-doped ZnO films on Al_2O_3 substrate by infrared reflection method. *Thin Solid Films, 673*, 136–140.

Minami, T., Sato, H., Imamoto, H., & Takata, S. (1992). Substrate Temperature Dependence of Transparent Conducting Al-Doped ZnO Thin Films Prepared by Magnetron Sputtering. *Japanese Journal of Applied Physics, 31*, L257–L261.

Otto, A. Z. (1968). Excitation of nonradiative surface plasma waves in silver by the method of frustrated total reflection. *Physik, 216*, 398–410.

Palik, E. D., Kaplan, R., Gammon, R. W., Kaplan, H., Wallis, R. F., & Quinn, J. J. (1976). Coupled surface magnetoplasmon-optic-phonon polariton modes on InSb. *Physical Review B, 13*, 2497–2506.

Reshina, I. I., Mirlin, D. N., & Banshchikov, A. G. (1976). Determination of unharmonious parameters and optical constants of single crystals using surface polariton spectra. *Physics of the Solid State, 15*, 506–510. (in Russian).

Shramkova, O. V. (2004). Attenuation of electromagnetic waves in a semiconductor superlattice in a magnetic field. *Technical Physics, 49*, 232–237.

Sizov, F. F., & Ukhanov, Y. I. (1979). *Magnetooptical faraday and voigt effects as applied to semiconductors.* Kiev: Naukova Dumka. (in Russian).

Sugano, S., & Kojima, N. (2000). *Magneto-optics, Springer series in solid-state sciences.* Berlin: Springer.

Tarkhanyan, R. H., & Uzunoglu, N. K. (2006). *Radiowaves and polaritons in anisotropic media.* John Wiley & Sons Inc.

Tolpygo, K. B. (1950). Physical properties of a rock salt lattice made up of deformable ions. *Zhurnal Eksperimentalnoi i Teoreticheskoi Fiziki, 20*, 497–509. (in Russian). English translation: *Ukrainian Journal of Physics.* 2008, 53, Special Issue, 93–102. Retrieved from http://archive.ujp.bitp.kiev.ua/files/journals/53/si/53SI21p.pdf

Ukhanov, Y. I. (1977). *Optical properties of semiconductors.* Nauka. (in Russian).

Venger, E. F., Ievtushenko, A. I., Melnichuk, L. Y., & Melnichuk, O. V. (2008). Investigation of ZnO single crystals subjected to high uniform magnetic field in IR spectral range. *Semiconductor Physics, Quantum Electronics and Optoelectronics, 11*, 6–10.

Venger, E. F., Melnichuk, A.V., Melnichuk, L. Ju., & Pasechnik, Ju. A. (1995). Anisotropy of the ZnO single crystals reflectivity in the region of region of resiqual rays. *Physica Satus Solidi (B), 188*, 118–127.

Venger, E. F., Ievtushenko, A. I., Melnichuk, L. Y., & Melnichuk, O. V. (2010). Effect of strong magnetic field on surface polaritons in ZnO. *Semiconductor Physics, Quantum Electronics and Optoelectronics, 13*, 16–22.

Venger, E. F., Melnichuk, A. V., Pasechnik, J. A., & Sukhenko, E. I. (1997). IR spectroscopy of the zinc-oxide-on-sapphire structure. *Ukrainian Journal of Physics, 42*, 1357–1360. (in Ukrainian).

Venger, E. F., Melnichuk, O. V., & Pasychnik, Y. A. (2001). *Spectroscopy of residual rays.* Kyiv: Naukova Dumka.

Venger, E. V., Melnichuk, L. Y., Melnichuk, O. V., & Pasichnyk, Y. A. (2000). Influence of the plasmon-phonon coupling on the reflectance coefficient in one-axis polar ZnO semiconductor. *Ukrainian Journal of Physics, 45*, 976–984. (in Ukrainian).

Venger, Y. F., Melnichuk, L. Y., Melnichuk, O. V., & Semikina, T. V. (2016). Investigation of zinc oxide thin films grown using ALD by the methods of IR spectroscopy. *Ukrainian Journal of Physics, 61*, 1053–1060.

Vinogradov, E. A., & Dorofeev, I. A. (2009). Thermally stimulated electromagnetic fields of solids. *Physics Uspekhi, 52*, 425–460.

Ziani, A., Davesnne, C., Labbé, C., Cardin, J., Marie, P., Frilay, C., Boudin, S., & Portier, X. (2014). Annealing effects on the photoluminescence of terbium doped zinc oxide films. *Thin Solid Films, 553*, 52–57.

Zvezdin, A. K., & Kotov, V. A. (1997). *Modern magneto-optics and magneto-optical materials: Studies in condensed matter.* New York: Taylor & Francis Group, CRC Press.

8 Zinc Oxide Grown by Atomic Layer Deposition

A Versatile Material for Microelectronics

Elżbieta Guziewicz

8.1 INTRODUCTION

Zinc oxide, which has been studied since the 1930s, still remains a mysterious material. Investigations performed in the last few years clearly show that structural and native point defects, which play a much more important role in ZnO than in other compound semiconductors, contribute to this fact to a large extent (Bauer et al., 2009; Du & Biswas, 2011; Frodason, Johansen, Bjorheim, Svenson, & Alkauskas, 2018; Janotti & van de Walle, 2009; Lyons et al., 2017; Oba, Choi, Togo, & Tanaka, 2011; Shimizu et al., 2018). A specific mechanism of damage buildup in ion bombarded single crystals, which is significantly different than in other semiconducting materials, is a fingerprint of this fact (Turos et al., 2017).

A wide and direct energy gap, 3.47 eV at room temperature, along with a high exciton binding energy (60 meV) predestinates zinc oxide for optoelectronic devices. However, the old-age problem with acceptor doping means that such applications are still out of reach. On the other hand, an easily achievable high level of unintentional *n*-type doping, interfering with *p*-type conversion, greatly facilitates obtaining ZnO films with high electrical conductivity, which can be further increased by Al or Ga doping (Luka et al., 2012). Because of this, zinc oxide successfully competes with other wide bandgap materials such as indium tin oxide in the field of transparent conductive oxides (Zhang, Dandeneau, Zhou, & Cao, 2009). Probably the most important area of applications of transparent conductive oxide materials is transparent conductive electrodes for solar cells, the market, which is expected to exceed 1 TW global energy production by 2022. The important advantage of zinc oxide in this field is its feasibility to grow thin films and a wide range of nanostructures by inexpensive chemical methods including atomic layer deposition (ALD), which allows covering of a large area of substrates necessary for photovoltaics. However, a potential application spectrum of zinc oxide in transparent electronics is much wider and also includes active parts of devices and heterojunctions, provided that electron concentration is stabilized at a moderate level of 10^{16}–10^{17} cm^{-3}. If the latter requirement is fulfilled, zinc oxide can be successfully used in transparent organic electronics, especially since it can be easily obtained at low temperatures compatible with thermally sensitive organic substrates.

Other advantages of zinc oxide are high radiant resistance, high heat capacity, low thermal expansion coefficient, and high thermal conductivity, which are important for good heat dissipation in electronic devices. The wurtzite crystal structure and related lack of inversion symmetry are responsible for the strong piezoelectricity of ZnO and opens the possibility to construct piezoelectric devices, while the strongly reactive surface of this material enables its use in chemical or radiation sensors.

Last but not least is the environmental safety of zinc oxide, which can also be electrically doped for higher *n*-type or for *p*-type using biologically safe elements (alumina and nitrogen, respectively).

Zinc oxide films can be grown by many different techniques, such as molecular beam epitaxy (Anderson et al., 2009; Heo, Norton, & Pearton, 2005; Look et al., 2002), radio-frequency magnetron sputtering (Fei et al., 2017; Fortunato et al., 2008; Grochowski et al., 2012; Hong et al., 2009), chemical vapor deposition (CVD) (Barreca et al., 2018; Pakkala & Putkonen, 2010; Roro et al., 2008), and pulsed laser deposition (Liu et al., 2012; Schubert et al., 2015; Suchea, Christoulakis, Katharakis, Vidakis, & Koudoumas, 2009). In this chapter, we show possibilities that are created for zinc oxide grown by ALD. This growth technique is especially dedicated to applications that require low deposition temperature, precise control of thickness at the nanometer scale, possibility of deposition on large substrates (even about 1 m^2), and uniform coverage substrates with highly developed morphology. The polycrystalline ZnO films grown at low temperature are of considerable importance for hybrid organic/inorganic electronics and 3D memory devices, which posted severe restrictions on deposition temperature that it should not exceed 150°C (Huby, Tallarida et al., 2008; Katsia et al., 2009; Kim, Kang, Kim, Byun, & Yoon, 2019; Luka et al., 2014; Na & Yoon, 2019; Stakhira et al., 2010). Probably one of the most interesting advantages of thin ZnO-ALD films is the ability to adopt electrical parameters to a specific application without any external doping. Several papers reported that conductivity of polycrystalline ZnO-ALD films can be regulated up to four orders of magnitude, with related electron concentration ranging from $\sim 10^{16}$ to $\sim 10^{20}$ cm^{-3}, only by changing growth temperature within the 100–200°C range (Bang et al., 2009; Gong et al., 2010; Krajewski, Guziewicz et al., 2009; Krajewski, Luka et al., 2009; Kwon et al., 2009; Makino, Miyake, Yamada, Yamamoto, & Yamamoto, 2009; Min, An, Kim, Song, & Hwang, 2010; Przezdziecka et al., 2009). Electrical properties of ZnO films can be additionally modified by post-growth annealing; however, in the annealed ZnO films also the previously applied growth temperature clearly influences the conductivity of the films. This remark, obviously, refers to the ZnO layers obtained by various techniques (Bouderbala, Hamzaoui, Adnane, Sahraoui, & Zerdali, 2009; Hamad, Braunstein, Patil, & Dhere, 2005; Krajewski, Guziewicz et al., 2009). This chapter is focused on structural, electrical, and optical properties of ZnO-ALD films. Most attention will be paid to the electrical properties and possibility of their modeling via ALD process parameters, external doping, and post-growth annealing.

8.2 ALD OF ZINC OXIDE

ALD was invented in Finland by Suntola in the mid-1970s (Suntola & Antson, 1977), primarily for the growth of polycrystalline and amorphous zinc sulfide and dielectric oxide films dedicated to large area electroluminescence displays (Suntola & Antson, 1977). One of them was used for several years at Helsinki airport (Goodman & Pessa, 1986).

By 1980, ALD was regarded as a niche deposition method and applied for the growth of only few materials, mainly some II–VI compounds such as CdTe, CdMnTe, ZnS, or ZnMnS (Goodman & Pessa, 1986), but over the years it greatly extended the scope of its application. In the 1980s and 1990s, ALD was used for growing low-dimensional structures of semiconducting III–V as well as II–VI compounds dedicated to integrated circuits and optoelectronic devices (Leskela & Ritala, 2003; Niinisto & Leskela, 1993; Puurunen, 2005; Leskela & Ritala, 2002). The growing interest in ALD has started from the beginning of the present century when the first high-k dielectric oxides were obtained using ALD (Buchanan, 1999; Lee, Shin, Chae, et al., 2002; Puurunen, 2005), which is related to a miniaturization trend in the microelectronic industry. The attempt to proceed technology from 65 nm to 45 nm generation processors had encountered a major obstacle in the form of a leakage current that appeared in metal–oxide–semiconductor field-effect transistors (MOSFETs), when SiO_2 gate dielectric had to be as thin as

1.2 nm. Because of that, microelectronic companies investigated several growth methods in order to achieve an appropriate technology of a material with a high value of dielectric constant that could act as a new generation gate oxide instead of SiO_2. The problem was that dielectric constants of the investigated 2–3 nm thick films showed considerably lower ε values as compared with bulk materials because of their heterogeneity and the presence of pinholes. The problem had been resolved in 2007 by Intel, which announced that their 45 nm node processor (Penryn, Model QX9650) had been constructed based on a high-k HfO_2 gate dielectric grown by ALD. The application of ALD for high-k dielectrics in novel integrated circuits caused a booming interest in this method that can be observed up to present. This is because ALD guarantees flat, conformal, and uniform films with low stress, uniform stoichiometry, low defect density, and reproducible thickness at the nanometer scale.

8.2.1 ZnO Films Grown by ALD

ALD can be regarded as a kind of CVD method. The difference between them is that, unlike CVD, ALD is a sequential growth process, in which there is a certain time interval between the introductions of the subsequent reagents.

In the ALD process, reactants (called here precursors) are alternatively introduced into a growth chamber and precursors' doses are interrupted by purging the reaction chamber with an inert gas. In Figure 8.1, we schematically show four ALD cycles and pressure of each chemical component during the process. Because of time intervals between doses, precursors do not have any possibility to react in the volume of the chamber and can meet only at the surface. Therefore, the ALD growth process originates from self-limiting chemical reactions that take place at the surface of the growing film (George, 2010; Goodman & Pessa, 1986; Suntola, 1994, 1992; Suntola & Antson, 1977). Moreover, very reactive precursors can be used in ALD and thus low deposition temperatures are possible. For this reason, ALD is a deposition method that is dedicated to the low-temperature growth.

In fact, the growth temperature used for deposition of a peculiar material depends on the kind of chemical reactants applied in the ALD process. Generally, organic precursors have higher vapor pressure, so they reach the partial pressure necessary for thin

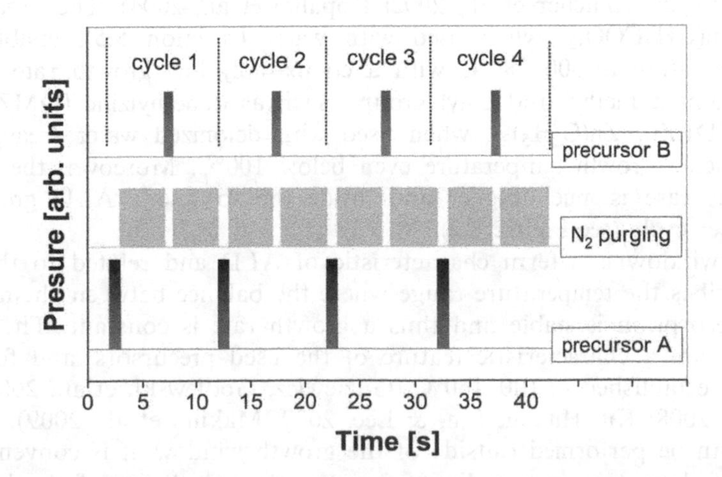

FIGURE 8.1 Schematic view of the sequential procedure of the ALD process. Four ALD cycles are shown. For better understanding, both precursors' doses and purging times are shown the same, which is not usually the case in real processes (Guziewicz et al., 2019, reproduced with permission).

film deposition even when heated at lower temperatures. The selection of appropriate temperature plays a major role in the ALD technique, as it assures a delicate balance between the surface chemical reaction and physical absorption/desorption phenomena. For this reason, in ALD processes, the growth temperature (i.e., temperature of the substrate) should be at least slightly higher than the precursor temperature, and the difference usually exceeds 20°C.

Zinc oxide in the ALD method can be obtained in several types of chemical reactions: synthesis from elemental precursors, single chemical exchange, and double chemical exchange. Examples of chemical reactions that can be used in ALD are:

$$\text{Synthesis: } 2Zn+\frac{1}{2}O_2 \rightarrow ZnO \tag{8.1}$$

$$\text{Single chemical exchange: } Zn+H_2O \rightarrow ZnO+H_2 \tag{8.2}$$

$$ZnCl_2+\frac{1}{2}O_2 \rightarrow ZnO + Cl_2 \tag{8.3}$$

$$\text{Double chemical exchange: } ZnCl_2+H_2O \rightarrow ZnO+H_2O \tag{8.4}$$

$$Zn(CH_3COO)_2+H_2O \rightarrow ZnO + 2CH_3COOH \tag{8.5}$$

$$Zn(CH_3)_2+H_2O \rightarrow ZnO + 2CH_3 \tag{8.6}$$

$$Zn(C_2H_5)_2+H_2O \rightarrow ZnO + 2C_2H_6 \tag{8.7}$$

Because of high efficiency, reaction 8.7 is the most frequently used in ALD processes. Inorganic precursors (e.g., Zn and O or $ZnCl_2$ and water, reactions 8.1–8.4) require growth temperature in the 400–500°C range and result in a rather low growth rate of about 0.5 Å per cycle (Butcher et al., 2002; Kopalko et al., 2004). The organic precursor zinc acetate, $Zn(CH_3COO)_2$, when used with water (reaction 8.5), enables decreasing deposition temperature to 300–380°C with a comparably low growth rate. Only organic precursors containing methyl and ethyl groups, such as dimethylzinc (DMZn, $Zn(CH_3)_2$) or diethylzinc (DEZn, $Zn(C_2H_5)_2$), when used with deionized water (reactions 8.6 and 8.7), allow reducing growth temperature even below 100°C. Moreover, the ALD process efficiency in this case is much higher and inside the so-called "ALD growth window" exceeds 1.8 Å per cycle (see Figure 8.2).

The growth window is a term characteristic of ALD and related to the self-limited growth. It describes the temperature range where the balance between chemical reactivity and physical desorption is stable and thus a growth rate is constant. The limits of the growth window are a characteristic feature of the used precursors, and for DEZn and water they were established as 100–160°C (Guziewicz, Godlewski, et al., 2009, Guziewicz, Kowalik, et al., 2008; Ku, Huang, Lin, & Lee, 2009; Makino et al., 2009). Although the ALD process can be performed outside of the growth window, it is convenient to maintain it inside of these temperature limits, because then thickness of the layer does not depend on temperature fluctuations, so deposited film is perfectly uniform even at the atomic scale.

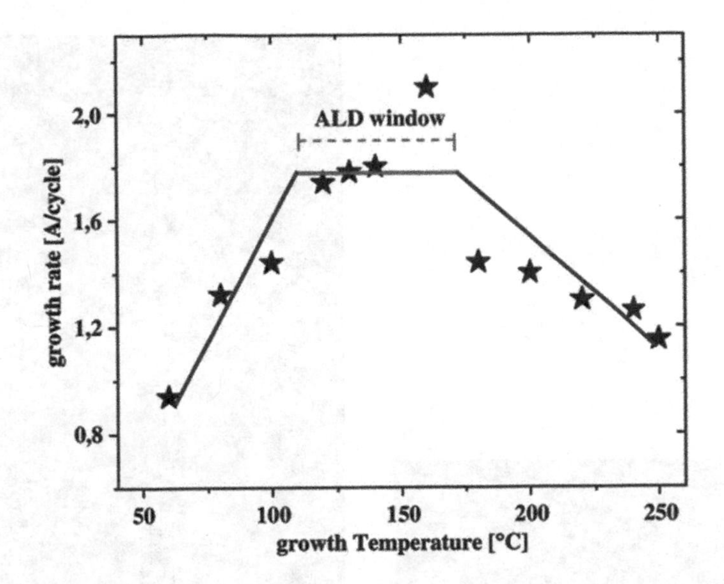

FIGURE 8.2 The growth rate versus growth temperature for ZnO-ALD films deposited on a silicon substrate in 1000 cycles with DEZn and H_2O precursors. (Guziewicz et al., 2019, reproduced with permission).

The self-limited surface ALD reaction results in a uniform coverage of substrates with a highly developed morphology. Even in the case when the so-called aspect ratio (depth of a hole to its diameter) is as high as 100, the substrate can be uniformly covered during the ALD process, provided the parameters of the process are properly chosen (George, 2010). The example of such coverage can be seen in Figure 8.3, where ZnTe-ZnO core-shell nanowires (NWs) are presented. The structure was obtained by a combination of two growth techniques: molecular beam epitaxy for the growth of 1–1.5 μm long ZnTe NWs with a diameter of about 30 nm and ALD for shell deposition. As can be seen, all the NWs are uniformly covered with the 20 nm thick ZnO shell.

Because of the possibility of uniform coating of nanometer-sized structures, ALD fulfills challenging requirements of present semiconductor processing. Therefore, the *International Technology Roadmap for Semiconductors* (ITRS) included ALD for copper diffusion barriers as well as for high-k dielectric gate oxide used in the MOSFET structures (ITRS, 2007).

8.2.2 Structure and Morphology of ZnO Films Grown by ALD

For a low growth temperature (T_g) range (100–200°C), commonly used in ALD processes, a polycrystalline growth is observed regardless the type of the substrate used (Figure 8.4). The main factor determining a preferential orientation is the growth temperature (Pearton, Norton, Ip, Heo, & Steiner, 2003; Tanskanen, Bakke, Pakkanen, & Bent, 2011; Wojcik, Godlewski, Guziewicz, Minikayev, & Paszkowicz, 2008), although some dependences on the purging time and film thickness have also been found (Kowalik et al., 2009). Typically, for growth temperature of 200°C and higher, the *c* axis tends to be perpendicular to the substrate, while at lower deposition temperature, the orientation with the *c* axis parallel to the substrate is predominant. This tendency can be observed not only for ZnO-ALD films, but also for films grown with other methods such as sputtering (Kajikawa, 2006) or CVD (Funakubo, Mizutani, Yonetsu, Saiki, & Shinozaki, 1999). However, influence of deposition parameters other than temperature is neither negligible nor straightforward (see Figure 8.5), and such a factor as construction of the reactor also plays a role.

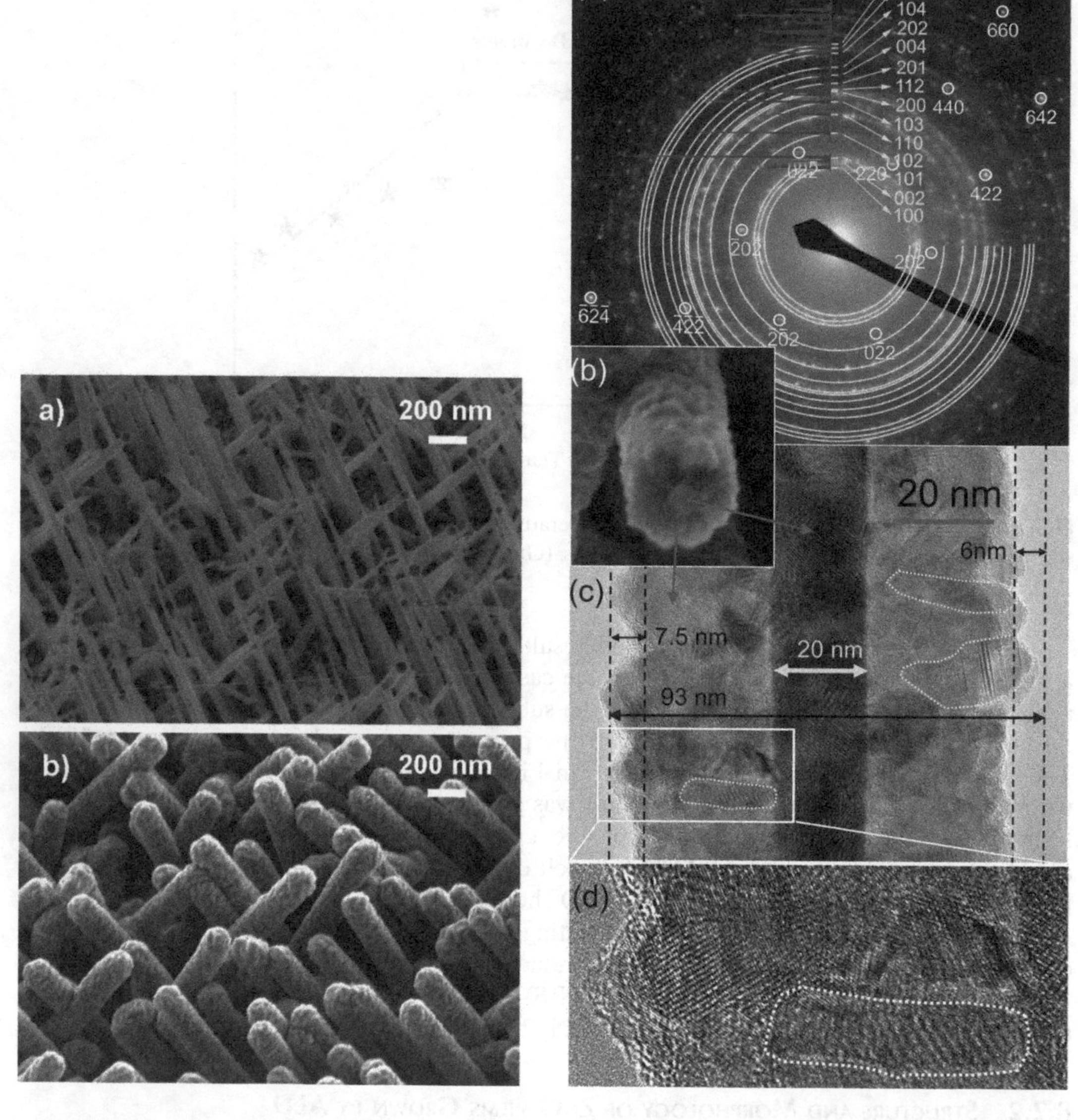

FIGURE 8.3 (Left) SEM images of (a) ZnTe NWs grown on (110)-oriented GaAs; (b) ZnO-coated ZnTe NWs. (Right) electron microscopy image of the core-shell ZnTe-ZnO NW heterostructure: (a) the powder selected area diffraction analysis of individual nanostructures, (b) SEM image of a broken nanostructure; arrows indicate the core and the shell, (c) high-resolution transmission electron microscopy (HR TEM) image of some part of the nanostructure, (d) magnification part of (c) showing details of the ZnO shell (Janik et al., 2010, reproduced with permission)

Moreover, thickness of the film influences the X-ray diffraction (XRD) pattern as well. Therefore, it is impossible to clearly derive the set of specific orientations, together with the preferred one, only from deposition parameters themselves (Cheng, Yuan, & Chen, 2016; Gong et al., 2010; Kowalik et al., 2009; Pearton et al., 2003; Wojcik et al., 2008). The same concerns size of the grains.

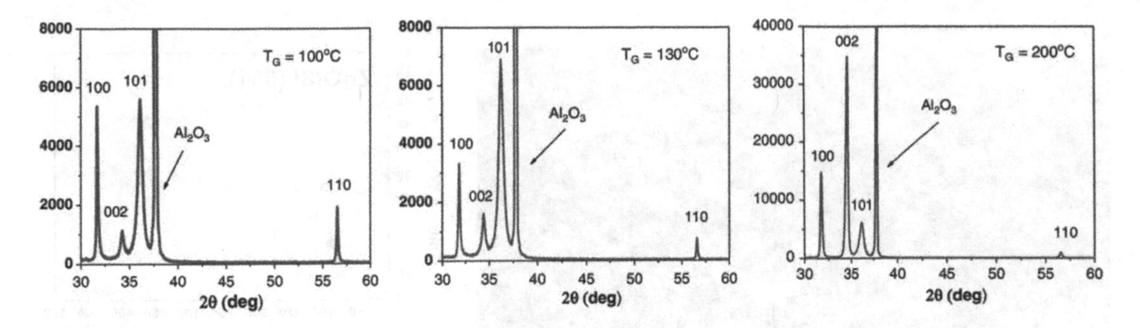

FIGURE 8.4 Diffractograms of ~900 nm thick ZnO-ALD films deposited on sapphire at temperatures 100, 130, and 200°C (Guziewicz et al., 2019, reproduced with permission).

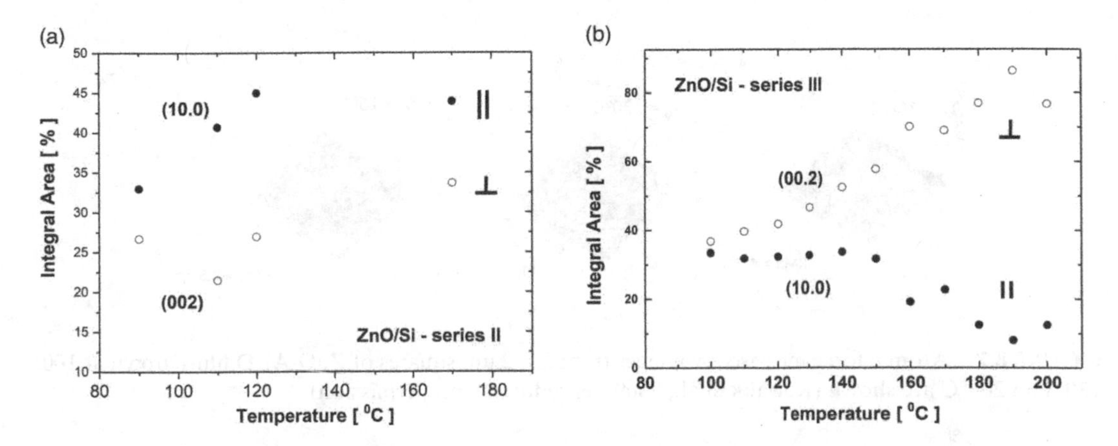

FIGURE 8.5 Integral area of the XRD 10.0 and 00.2 diffraction maxima obtained for two different series of ZnO samples as a function of growth temperature (Kowalik et al., 2009, reproduced with permission).

Polycrystalline films deposited within the ALD window show granular surface microstructure and columnar growth. The size of the columns varies between 20 and 60 nm, as derived from XRD and confirmed by TEM, depending on both the layer thickness and growth temperature. Thicker layers and higher deposition temperature result in larger crystallites observed in XRD and wider column visible in TEM images [see Figure 8.6, (Kowalik et al., 2009)].

Crystallographic structure of the layers is closely related with their surface morphology. The atomic force microscopy measurements in the tapping mode show that the root mean square (RMS) value of the surface roughness correlates with the preferred orientation (Kowalik et al., 2009; Wojcik et al., 2008) and takes the lowest values when crystallites are oriented along the [001] direction, that is, for deposition temperature of 200°C. Typically, the RMS value of about 90 nm thick ZnO-ALD films varies between 0.5 and 4, as shown in Figure 8.7.

However, the RMS value strongly depends on the thickness of the films (see Figure 8.8b). Therefore, smooth surfaces of ALD films can be achieved for film thickness of up to 100 nm, while for thicker films it is hardly possible. Fortunately, industrially used semiconducting films are usually even thinner, while dielectric films are typically 2–3 nm thick.

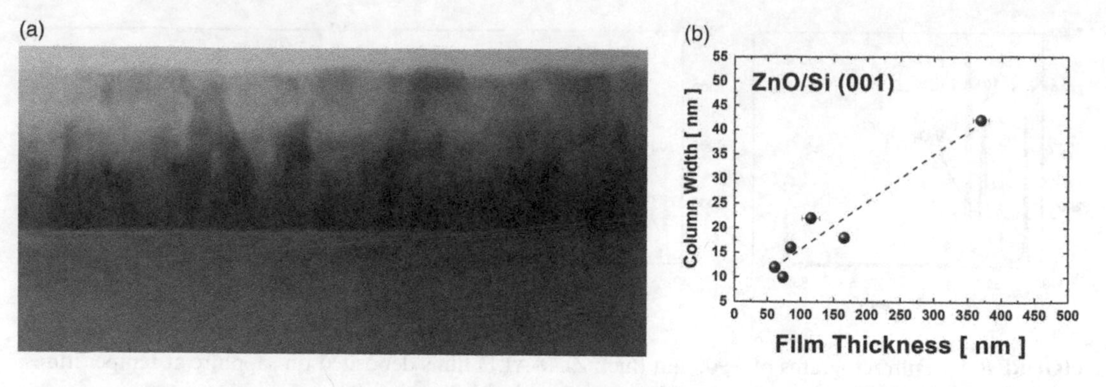

FIGURE 8.6 (Left) TEM image of ZnO samples grown at 170°C (83 nm) (right) and column width as the film thickness (Kowalik et al., 2009, reproduced with permission).

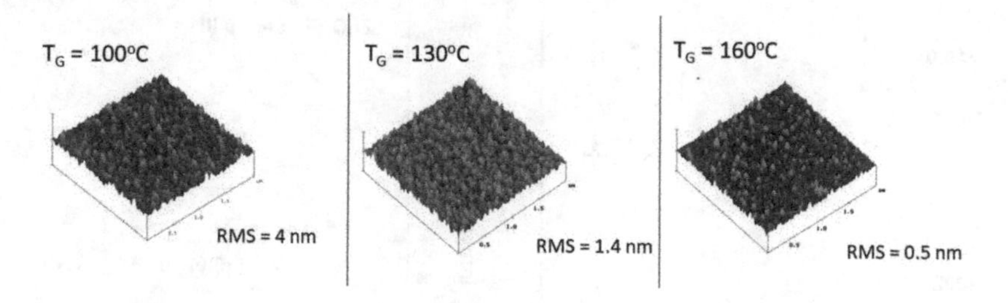

FIGURE 8.7 Atomic force microscopy images from 2 × 2 μm squares of ZnO-ALD films grown at 100, 130, and 200°C are shown (Kowalik et al., 2009, reproduced with permission).

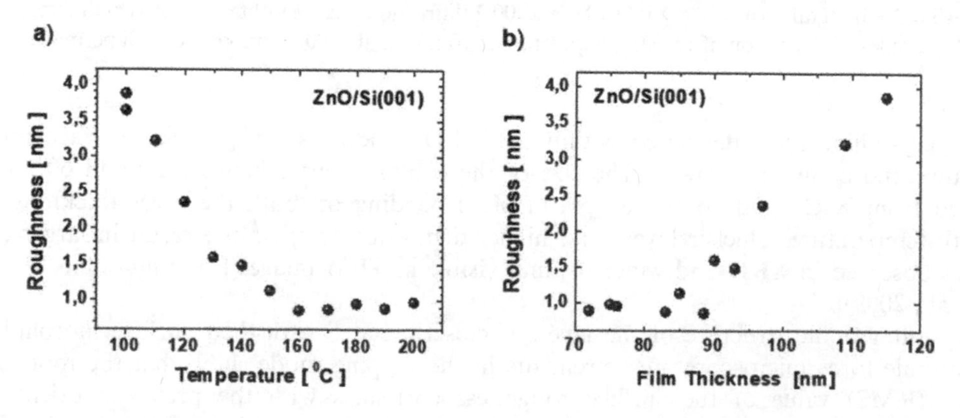

FIGURE 8.8 The RMS values determined from 10 × 10 μm regions as a function of growth temperature (a) and thickness (b) of ZnO films (Kowalik et al., 2009, with a permission).

It should be mentioned that for T_g reaching 300°C, the epitaxial growth of ZnO film can be achieved, provided the deposition is performed on a lattice-matched substrate such as GaN or SiC. In this case, effectiveness of the ALD process is about twice lower than that within the growth window, but quality of the film is surprisingly good, taking into account the simplicity of the growth process. The Rutherford backscattering channeling (RBS/c)

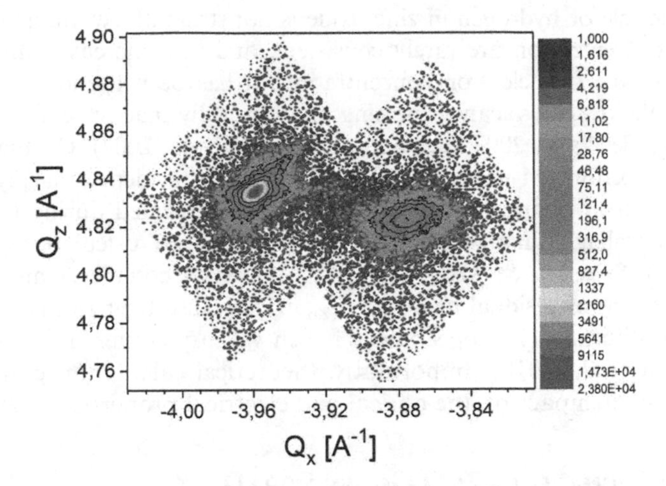

FIGURE 8.9 Reciprocal space map collected for (−1–1.4) reflection with anti-scattered slit 1/8 before a detector (Wachnicki, Lukasiewicz, et al., 2010, reproduced with permission).

spectra of crystalline ZnO films deposited on GaN/sapphire substrate reveal the $\chi_{\min}$ parameter at the level of 3.0%, which is comparable with the values reported for single ZnO crystals grown by the hydrothermal method (Ratajczak et al., 2013). High-resolution XRD confirmed that ZnO/GaN films are monocrystalline with a full width at half maximum rocking curve of the 00.2 reflection equaled to 0.07° (Wachnicki, Krajewski, et al., 2010, Wachnicki, Lukasiewicz, et al., 2010), see Figure 8.9.

Epitaxial ZnO-ALD films might be promising material for optoelectronics; however, because of their relatively high conductivity and elevated growth temperature, they are not appropriate for such applications as hybrid organic/inorganic junctions or 3D electronics, where back-end-of-line technology is required. For such purposes, polycrystalline ZnO films deposited within the ALD window reveal much more appealing characteristics.

8.3 ELECTRICAL PROPERTIES OF UNDOPED ZnO FILMS

The origin of electron concentration in nominally undoped ZnO is still under debate. Although the wide energy gap (3.37 eV at RT) suggests its very low intrinsic value (at the level of $n \sim 10^6$ cm^{-3}), the experimentally obtained electron concentration exceeds the intrinsic level by more than ten orders of magnitude. The measured n value depends not only on the growth method, but on applied parameters as well and it varies between 10^{16} and 10^{20} cm^{-3} (Beh et al., 2017; Krajewski, Luka, Wachnicki et al., 2011, 2014 and references therein). The lower n, at the level of 10^{14} cm^{-3}, is only reported for single zinc oxide crystals grown by the hydrothermal method where group I elements are incorporated as unintentional dopants, likely contributing to a self-compensation effect (Jiang et al., 2012; Maeda, Sato, Niikura, & Fukuda, 2005).

The source of high level of electron doping in nominally undoped ZnO is widely investigated for last few decades. Initially, it has been assigned to native point defects, but density functional theory (DFT) calculations show that they are either deep, as oxygen vacancy, or have high formation energy, as zinc interstitial or Zn_O (Janotti & van de Walle, 2009 and references herein). For this fact, the native point defects alone cannot provide a large number of carriers justifying high conductivity at room temperature. Hydrogen, which is unintentional contaminant in most of the growth processes, has been commonly considered as a dominant donor in zinc oxide (Kolb & Laudise, 1965 and references therein). However, investigations reported in the last few years show that the role of hydrogen in zinc oxide is not straightforward, as the hydrogen content and electron concentration are rarely consistent, and in some cases amount of hydrogen exceeds considerably the free electron concentration. It has been found that the H_2 molecule can be trapped in the oxygen vacancy forming an electrically inactive complex (Du & Biswas, 2011; Janotti & van de Walle, 2007; Shi, Sabotakin, & Stavola, 2004). On the other hand, theoretical calculations performed in the last few years give an evidence that hydrogen can form a variety of complexes involving zinc vacancy, that can be created during the growth process and post-grown annealing (Shimizu et al., 2018). Hybrid DFT calculations show (Frodason, Johansen, Bjorheim, Svenson, & Alkauskas, 2018) that such complexes are highly stable and can involve a few atoms of residual hydrogen $V_{Zn}{\cdot}nH$, because their formation energy is lower than individual constituents. The appearance of such vacancy clusters has been experimentally evidenced (Shimizu et al., 2018). Importantly, theoretical calculation points out that such complexes can have an impact on the optical and electrical properties of zinc oxide.

8.3.1 Electrical Properties of ZnO Grown by ALD

Taking into account the recent theoretical and experimental studies, as well as the results of the previous theoretical density functional calculation on formation energies of native point defects under oxygen- and zinc-rich conditions (Janotti and van de Walle, 2007; Janotti and van de Walle, 2009), it can be expected that growth conditions might have key implications for ZnO electrical conductivity. In this context, electrical conductivity and electron concentration of undoped ZnO performed by ALD, which may differ a few orders of magnitude depending on growth conditions, are of great interest.

In case of polycrystalline ZnO layers grown by ALD within the ALD window, the measured electron concentration n varies between $\sim 10^{16}$ and $\sim 10^{19}$ cm^{-3} when T_g changes from 100 to 200°C. It means that n strongly depends on T_g, even though for a deposition process performed in the ALD window the chemical reactions at the surface are chemically well balanced, and the same growth rate ensures conformal coverage of a substrate.

Free electron concentration depends also on other ALD parameters such as pulsing and purging times, but these dependences are much weaker. Moreover, electron concentration versus growth temperature dependence is stronger for thinner films (see Figure 8.10).

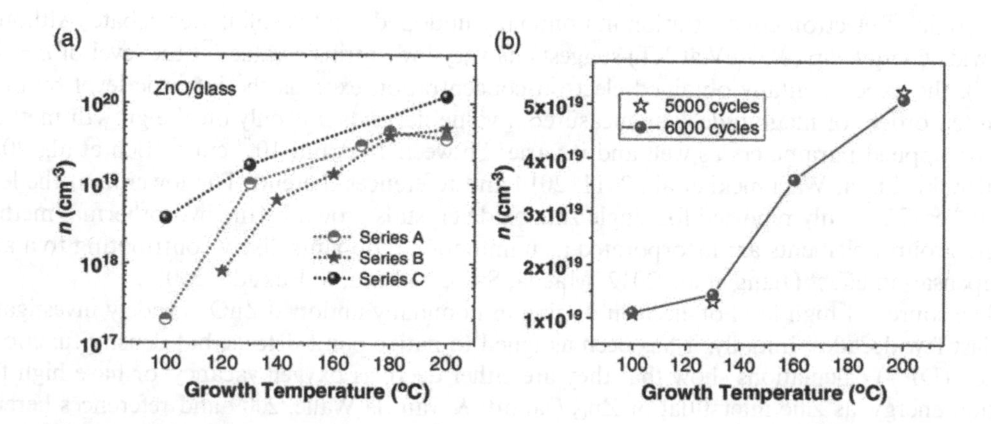

FIGURE 8.10 (a) Electron concentration versus growth temperature for ZnO-ALD films grown with 1000 cycles; series A, B, and C were obtained with different pulsing and purging times; (b) Electron concentration versus growth temperature for ZnO-ALD films obtained with 5000 and 6000 cycles (Guziewicz et al., 2019, reproduced with permission).

In case of ZnO-ALD with thickness less than 400 nm, the carrier concentration can differ up to three orders of magnitude as a result of T_g increase (Figure 8.10a), while for layers ~1 μm thick this dependence is less significant and stay at the same order of magnitude (Figure 8.10b) (Krajewski, Luka, et al., 2010, Krajewski, Luka, Wachnicki et al., 2011; Przezdziecka et al., 2009).

Concerning, in turn, the free electron mobility, μ value increases significantly when T_g changes from 100 to 200°C (Krajewski, Luka, Wachnicki et al., 2011b; Przezdziecka et al., under consideration). The mobility value of polycrystalline films also depends on the thickness of ZnO layer. For ~100 nm thick layers deposited at 100°C, mobility is at the level of 1–5 $cm^2 V^{-1} s^{-1}$, while for ~400 nm thick or thicker films, mobility exceeds 30 $cm^2 V^{-1} s^{-1}$.

8.3.2 Impurities

In the ALD processes, hydrogen and carbon impurities can be expected, because the used precursors contain hydrocarbon and hydroxyl chemical groups. The Secondary Ion Mass Spectroscopy (SIMS) and X-ray Photoemission Spectroscopy (XPS) measurements show that carbon contamination varies between 1 and 2%, and it is lower in ZnO samples deposited at higher temperature (Guziewicz et al., 2012). Therefore, it is unlikely that carbon is a reason for observed conductivity deference between films deposited at different temperatures. Hydrogen concentration, as measured by SIMS, reveals an unexpected dependence on growth temperature. Hydrogen is commonly regarded as electron dopant in ZnO. However, in ZnO-ALD films hydrogen concentration is higher in films deposited at 100°C, where electron concentration takes the values much lower than in films grown at 200°C (Guziewicz et al., 2012; Krajewski, Dybko, et al., 2014, Krajewski, Guziewicz, et al., 2009). Such a dependence between hydrogen content and electrical conductivity has been reported by many groups for polycrystalline ZnO-ALD films deposited at low temperature regime (Beh et al., 2017; Huang, Ye, Sun, Kiang, & de Groot, 2017; Thomas & Cui, 2012).

The high hydrogen content accompanying low carrier concentration can be understood in terms of complexes involving hydrogen and native point defects. It must be taken into account that the high hydrogen content is accompanied by oxygen-rich conditions and the associated low formation energy of zinc vacancy. For this reason, it can be expected that the complexes involving hydrogen and zinc vacancies are easily created. As theoretical calculations show, such complexes, $V_{Zn}{\cdot}nH$, have an influence on the ZnO conductivity (Shimizu et al., 2018). This problem will be more thoroughly discussed in the next section.

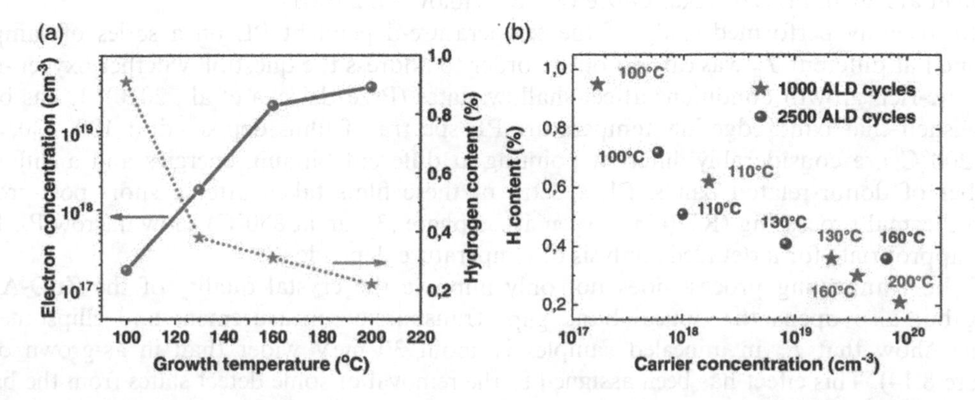

FIGURE 8.11 Electron concentration and hydrogen content for a series of ZnO films grown at temperature range between 100 and 200°C (Guziewicz et al., 2019, reproduced with permission).

8.4 STOICHIOMETRY

The results of stoichiometry measurements of ZnO-ALD films shed some light onto the origin of the large conductivity difference. These kind of investigations are much more difficult than measurements of impurity concentration, because of the possibility of systematic errors; therefore, they ought to be performed on a series of ZnO samples prepared under exactly the same experimental conditions. The stoichiometry studies of ZnO-ALD films were performed both by RBS and XPS and led to the same conclusion. They show that the O:Zn ratio decreases with deposition temperature, that is, the oxygen content in ZnO films obtained at 100°C is higher than in films obtained at higher temperature (Guziewicz et al., 2012; Ratajczak et al., 2013).

The above finding points at different growth conditions as the origin of observed conductivity difference. When growth temperature approaches 100°C, the conditions are more oxygen-rich while more zinc-rich conditions are achieved for 200°C. As shown by DFT calculations (Janotti and van de Walle, 2007; Janotti and van de Walle, 2009), the formation energy of native point defects is quite different under oxygen-rich and zinc-rich conditions. In the former case, the formation energy of such defects as zinc vacancy, V_{Zn}, or oxygen interstitial, O_i, is lower, while under zinc-rich conditions, zinc interstitial, Zn_i, or oxygen vacancy, V_O, can be easily created.

8.5 OPTICAL PROPERTIES

Optical studies provide valuable information on deep and shallow defect levels. The knowledge of shallow defect states that can be thermalized at room temperature, thus providing carriers participating in electrical conductivity, is particularly valuable.

It is well established that growth temperature, which determines oxygen- and zinc-rich conditions, considerably influence the energy of defect-related photoluminescence (PL) bands of ZnO-ALD films situated between 1.9 and 2.4 eV. The resistive ZnO samples obtained at temperature 100–125°C show a broad PL band at ~1.9 eV, which is ascribed to the acceptor-like zinc vacancy defect. In turn, the conductive ZnO samples obtained at 175–225°C show a defect PL at ~2.4 eV, which is assigned to the donor-like oxygen vacancy (Beh, Hiller, & Salava, et al., 2018; Beh, Hiller, & Zachariars, 2018; Krajewski et al., 2017). The post-growth annealing experiments, showing a variation of the defect PL band depending on the annealing atmosphere, support the above assignments (Beh, Hiller, Salava et al., 2018). However, deep defect-related PL bands were reported for ZnO-ALD layers with thickness 30–200 nm, while for samples thicker than 500–700 nm, only a band-edge PL is visible (see Figure 8.13) (Przezdziecka et al., 2020; Przezdziecka, Guziewicz, & Witkowski, 2018).

The recently performed study of the temperature-dependent PL on a series of samples obtained at different T_g was carried out in order to address the question whether oxygen-rich and zinc-rich growth conditions affect shallow states (Przezdziecka et al., 2020). It has been established that band-edge low temperature PL spectra of films deposited at 100, 130, 160 and 200°C are considerably different pointing at different binding energies and a different number of donor-related bands. PL spectra of these films taken after a short post-grown rapid thermal processing (RTP) in oxygen atmosphere (3 min at 800°C) show narrow PL features appropriate for a detailed analysis of temperature dependence.

A short annealing process does not only improve the crystal quality of the ZnO-ALD films, but also opens the optical band gap. Transmission measurement and ellipsometric studies show that E_g in annealed samples is about 30 meV wider than in as-grown ones (Figure 8.14). This effect has been assigned to the removal of some defect states from the band gap. However, it should be noticed that even after RTP annealing, the optical band gap depends on the previously applied growth temperature, and for samples deposited at 200°C the band

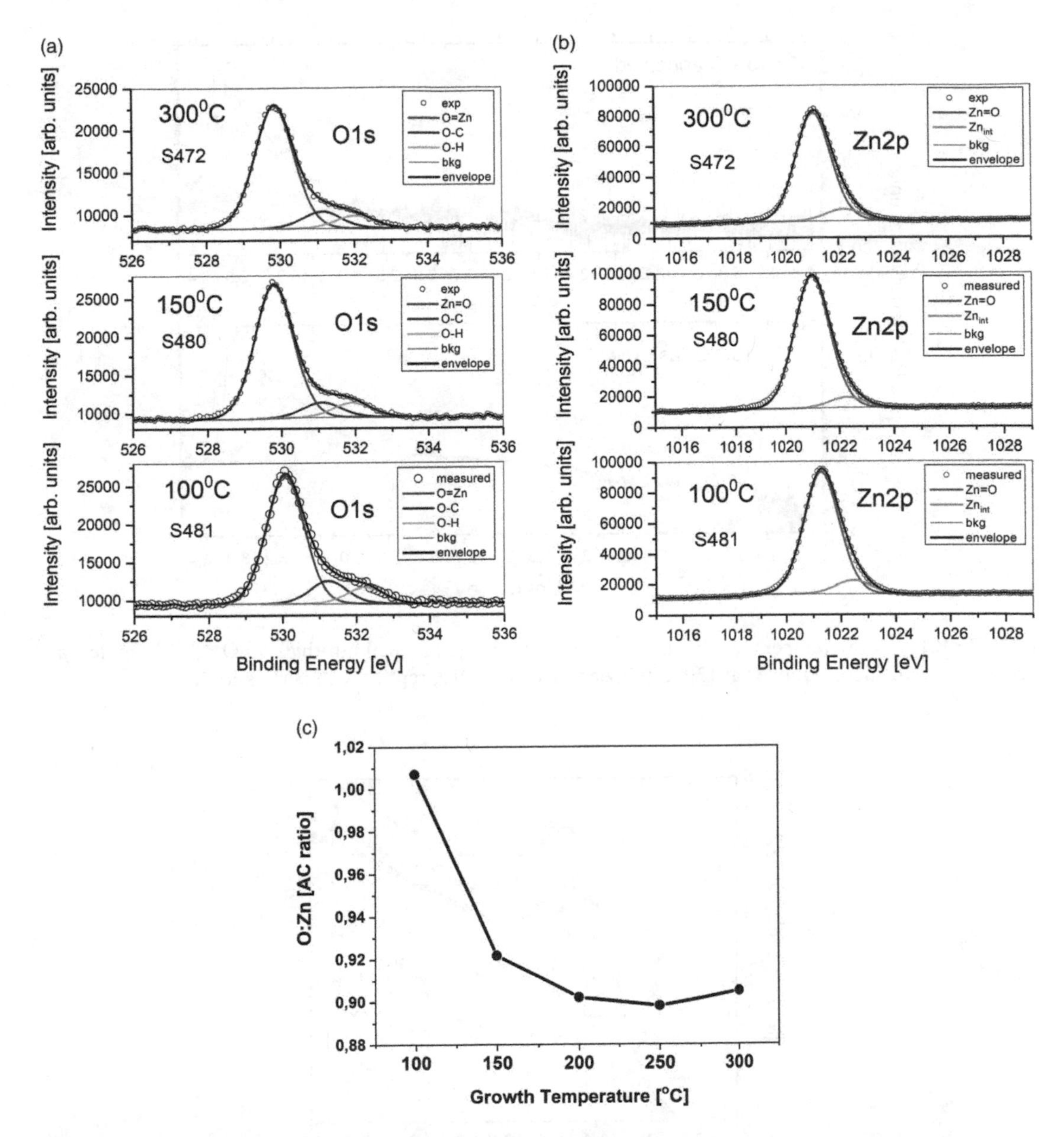

FIGURE 8.12 (Left) The O1s and Zn2p core level spectra measured for a series of polycrystalline ZnO films grown at 100°C (bottom), 150°C (middle), and 300°C (top); (right) the O:Zn atomic concentration ratio in ZnO-ALD films grown at temperatures 100–300°C as derived from high-resolution XPS measurements. The O:Zn values are normalized to the O:Zn value (0.716) found on the commercial ZnO crystal (MaTecK™) (Guziewicz et al., 2012, reprint with permission).

gap is about 20 meV wider than in that grown at 100°C (see Figure 8.8). Both for the series of as-grown and annealed samples E_g scales with $n^{2/3}$, where n is electron concentration, so the last phenomenon has been assigned to the Burstein–Moss effect (Guziewicz et al., 2019).

The band-edge PL spectra of the annealed polycrystalline ZnO films depend on their growth temperature (see Figure 8.14). For samples deposited at 100 and 130°C, two donors can be distinguished, while for samples deposited at 160 and 200°C, only one is clearly

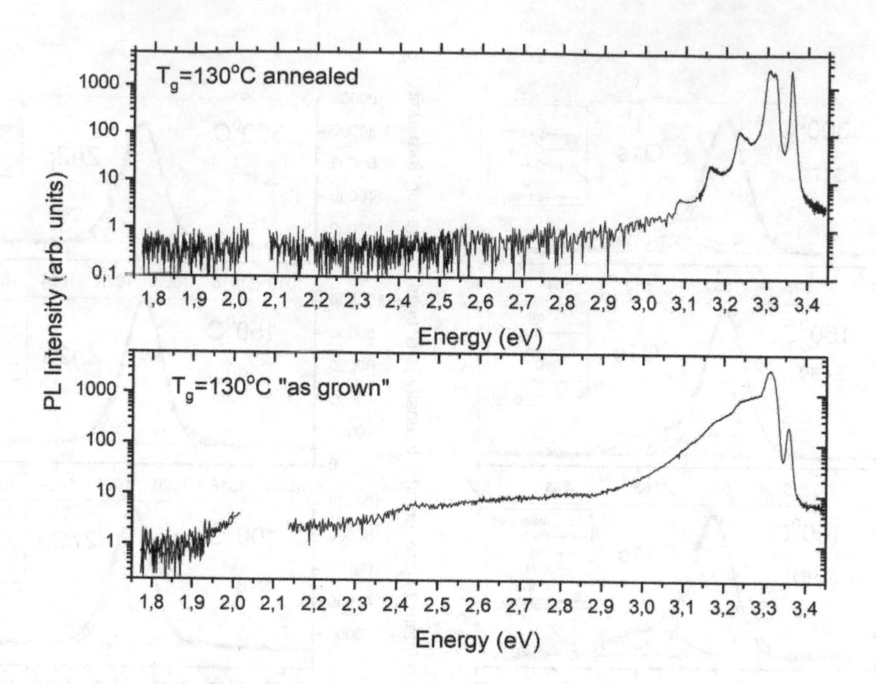

FIGURE 8.13 Low-temperature PL in a wide spectra range of ~800 nm thick ZnO-ALD samples, as grown and annealed, obtained at 130°C (Guziewicz et al., 2019, reprint with permission).

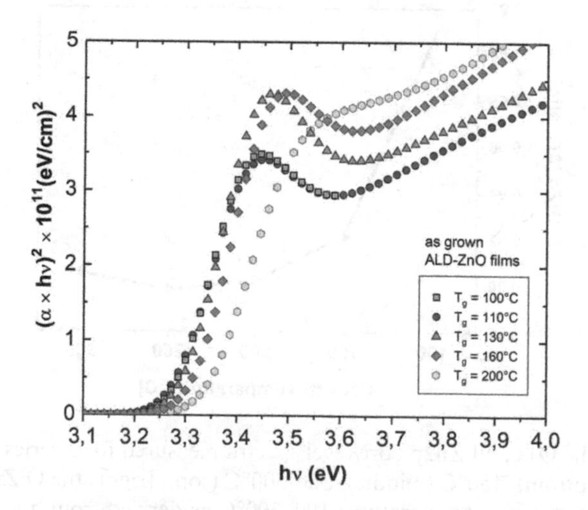

FIGURE 8.14 Tauc plot for the ALD-ZnO films deposited at different (indicated) temperatures. The absorption coefficient α is taken from ellipsometry measurements (Guziewicz et al., 2019, with permission).

visible. Moreover, the intensity ratio of acceptor-related PL band, situated between 3.30 and 3.32 eV, and donor-related PL band, situated very close to the band edge, changes a lot for the films deposited at different T_g (Figure 8.15) (Przezdziecka et al., 2020). This suggests a higher number of acceptor-related states in ZnO films deposited at low temperature i.e., when growth conditions are oxygen-rich. The predominance of acceptor-related PL over the donor-related PL is the most pronounced for the ZnO film grown at 100°C.

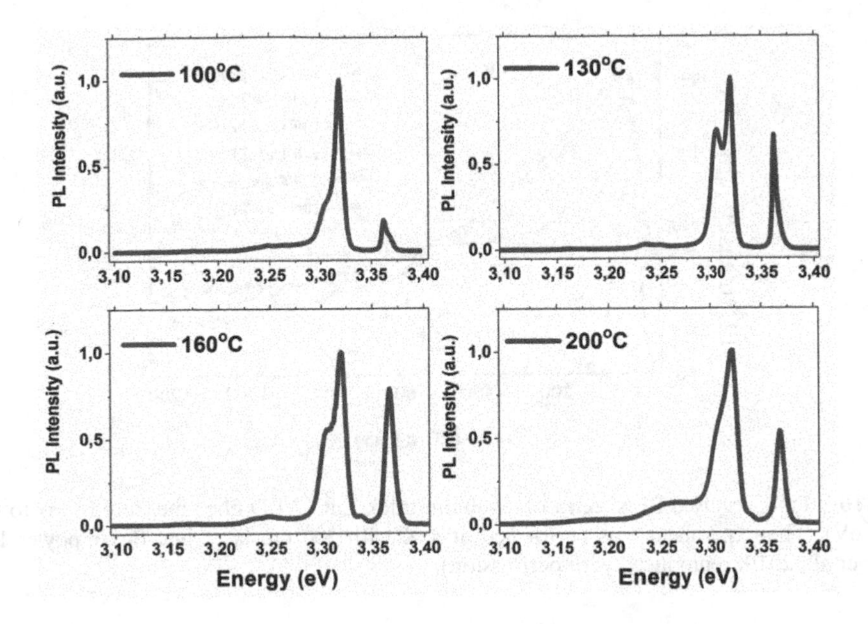

FIGURE 8.15 The band-edge PL spectra of ~800 nm thick ZnO-ALD films deposited at 100, 130, 160 and 200°C and post-grown annealed in oxygen for 3 min at 800°C (Guziewicz et al., 2019).

It has been found that donor binding energies, calculated based on donor localization energies and the Haynes' rule, vary from 49 to 58 eV and are different for samples deposited at different T_g. Acceptor binding energy is similar for deposition temperature up to 160°C and then equals 127 meV, while it is slightly lower (123–125 meV) when growth temperature rises to 200°C. However, probably the most pronounced difference was found in the relative intensities of acceptor- to donor-related states, which is much higher for low deposition temperature. This suggests that the number of shallow acceptors versus the number of shallow donors increases when growth conditions are moved toward oxygen-rich condition.

From the results presented above, it is evident that the polycrystalline ZnO films deposited at 100°C, which in ALD means oxygen-rich conditions, are more resistive and contain significantly more acceptor states versus donor states as compared to films grown at temperature 160°C and higher, where conditions are moved towards zinc-rich. The question arises whether the quality of these films, which are grown at temperature much below that required for epitaxial growth, is suitable for their use in electronics. In the last part of the chapter, we will attempt to address this question showing the optical decay dynamics measurements, ability to achieve the *p*-type conduction, and an electrical junction based on ZnO obtained at low temperature limits.

Interesting insight into the quality of ZnO-ALD films grown between 100 and 200°C provides time-resolved photoluminescence measurements (Guziewicz et al., 2019). Time-resolved PL measurements were performed to compare the optical decay dynamics and to reveal the lifetimes of ZnO-ALD films obtained at 100, 130 and 200°C. ZnO samples were grown with 2500 ALD cycles, which resulted with a thickness of about 400 nm. The time-resolved PL spectra were measured with a streak camera at temperature of 4 K using the 300 nm laser line with 100 μW average power.

The PL decays of the 3.36 eV excitonic peak of as-grown and RTP-annealed ZnO-ALD films are shown in Figure 8.16. The luminescence decay times are short and for the as-grown samples they range between 70 and 168 ps (dot lines in Figure 8.16). More than twice

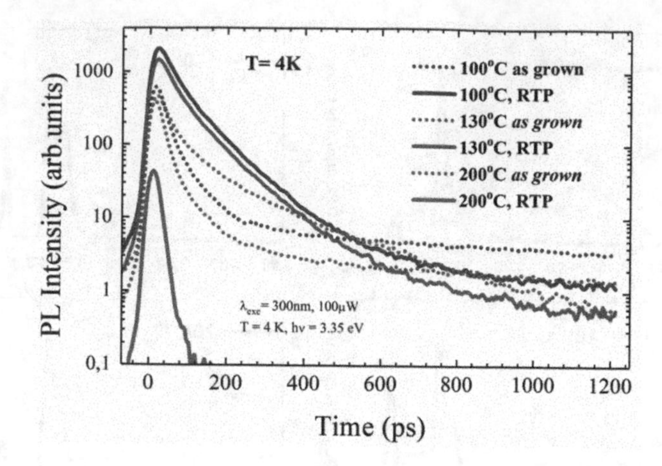

FIGURE 8.16 Time-resolved PL spectra of ~400 nm thick ZnO-ALD films measured near to the band edge (3.35 eV). The experiment was performed at 4 K with 300 nm laser line (laser power 100 μW), (Guziewicz et al., 2019, reproduced with permission).

longer decay times are observed for samples deposited at 100 and 130°C (150 and 168 ps, respectively) than for the sample deposited at 200°C (70 ps). After RTP annealing, all the PL spectra are much sharper and the PL decays are closer to be exponential (solid lines in Figure 8.10). Moreover, in case of samples deposited at 100 and 130°C, the intensity of the PL band at 3.36 eV increases more than three times after the RTP annealing process. The PL decay times after annealing are shorter, but the samples grown at lower temperature reveal much longer decay times (100 and 134 ps) than the sample grown at 200°C (only 24 ps), similarly as it was for the as-grown samples. It should be mentioned that the decay times of the samples deposited at 100 and 130°C are comparable to the decay time of the 3.36 eV PL of the reference commercial single crystalline ZnO (from MaTeck), which was determined as 160 ps.

From the results presented above, it is evident that the samples grown at the temperature range 100–130°C reveal better optical characteristics than the polycrystalline ZnO films obtained at much higher temperature (200°C) and, at some points, comparable with crystalline zinc oxide. This phenomenon might be understood in terms of growth conditions, which are changed from oxygen-rich toward zinc-rich when deposition temperature is increased.

8.6 DOPED ZnO-ALD FILMS

The doping procedure in ALD technology is relatively simple. This is accomplished through alternative introduction of several precursors during the growth process.

8.6.1 Donor Doping

As can be seen in Figure 8.10, ZnO films show higher carrier concentration for growth temperature of 200°C. Electron mobility of these films is at the level of 28–30 $cm^2\ V^{-1}\ s^{-1}$, thus resistivity is ~10^{-3} Ω cm. Such values belong to the lowest reported in the literature for undoped ZnO films grown by various methods, that is, ZnO films obtained by radio-frequency magnetron sputtering (Schropp, 1989) or pulsed laser deposition (Lorenz et al., 2003). For some applications, such as inorganic solar cells, these resistivity values can still be too high for efficient device operation (Ellmer, Klein, & Rech, 2008). In those cases, additional doping of ZnO films, for example, by aluminum or another group III element, is necessary. In this way the ZnO films resistivity as low as 2–3 × 10^{-4} Ω cm can be obtained. These values are comparable with those reported for Indium–Tin–Oxide films (1 × 10^{-4} Ω cm).

8.6.2 Acceptor Doping

Obtaining *p*-type ZnO is a very challenging task, because this semiconductor compound is naturally *n*-type and self-compensation phenomena are likely to occur, nullifying the doping effect. After almost two decades of intensive search for the best acceptor dopant, the initial idea of nitrogen as the best *p*-type dopant in ZnO has returned.

In the ALD technology, nitrogen doping can be easily realized by replacing deionized water with an ammonia water precursor in a part of the ALD cycles.

At temperature of the ALD process, which remains in majority of cases not lower than 100°C, ammonia water decomposes into ammonia and water: $NH_4OH \rightarrow NH_3 + H_2O$. Then, the chemical reactions at the surface can be expressed as follows:

$$NH_3 \text{ phase: Surface} - Zn - C_2H_5 + NH_3 \rightarrow \text{surface} - Zn - NH_2 + C_2H_6 \uparrow \quad (8.8)$$

$$H_2O \text{ phase: Surface} - Zn - NH_2 + C_2H_5 - Zn - C_2H_5 \rightarrow \text{surface} - Zn - NH - C_2H_5 + C_2H_6 \uparrow \quad (8.9)$$

The SIMS measurements show that the amount of nitrogen that can be introduced in this way into the ZnO lattice depends on the number of ammonia water versus deionized water cycles and can vary up to ~10^{21} at. cm^{-3} in case when all cycles are performed with ammonia water with NH_3 concentration of 25% (Snigurenko, Kopalko, Krajewski, Jakiela, & Guziewicz, 2015). As a result of nitrogen doping, the ZnO films exhibit the concentration of free electrons much lower than that of the undoped ZnO films, which means that the ZnO layers with higher nitrogen content are more compensated. It has been reported that carrier concentration can be lowered from 2 × 10^{19} to 5.4 × 10^{15} cm^{-3} for ~2 μm thick ZnO:N films doped with every second ammonia water versus ionized water ALD cycle (Snigurenko et al., 2015).

According to Eq. 8.9, nitrogen is introduced as the –NH group onto the oxygen site of the ZnO lattice, so it can be regarded as a kind of co-doping process. This fact explains a surprisingly high nitrogen concentration that can be realized with ammonia water precursor and DEZn. According to theoretical calculations, nitrogen and hydrogen co-doping is expected to be effective, because (NH) as a whole has six valence electrons, identical to oxygen, so the formation of hole-killer defects will be suppressed (Zhang, Wei, & Zunger, 2001).

When the nitrogen doping described above is applied to the ZnO-ALD films deposited under oxygen-rich conditions (T_g = 100°C), conductivity conversion is achieved after the RTP annealing process. The highest hole mobility, 17.3 cm^2 V^{-1} s^{-1}, has been observed when nitrogen concentration was at the level 10^{19} cm^{-3}. The related hole concentration was measured as 4.5 × 10^{16} cm^{-3} (Guziewicz et al., 2017).

Low temperature PL (see Figure 8.17) and cathodoluminescence (CL) measurements confirmed activation of *p*-type conductivity after the RTP annealing process (Guziewicz et al., 2017).

Conductivity conversion is also independently seen in low temperature (10 K) microphotoluminescence spectra measured from a cross-section of the ZnO film (see Figure 8.18), in which a He-Cd laser beam emitting at 325 nm was focused to a spot with a diameter of about 1 μm. The signal was collected with the same objective and focused onto an entrance slit of a 500 mm focal length monochromator coupled with a charge-coupled device detector.

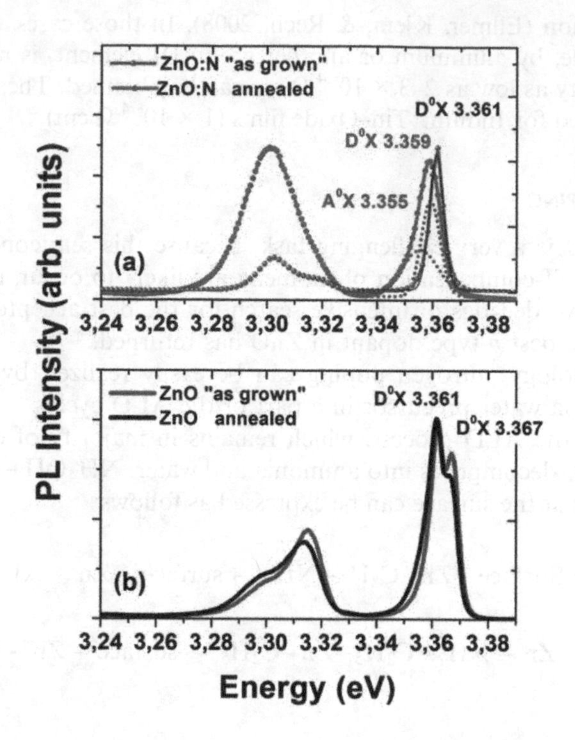

FIGURE 8.17 Low temperature PL spectra of the "as-grown" and annealed (a) ZnO:N and (b) ZnO films. The intensities are normalized to the most intensive line (Guziewicz et al., 2017 reproduced with permission).

A few important conclusions can be derived from the presented CL images. The first one concerns a spatial distribution of acceptor- and donor-related regions in undoped and N-doped samples. It is clearly seen that acceptor and donor emissions are derived from different spatial regions of the samples. It is also evident that acceptor-related luminescence does not derive from grain boundaries. The latter conclusion is also supported by the XRD study (Guziewicz et al., 2017), which reveals a larger grain size in the nitrogen-doped sample. This means that an area of grain boundaries in the ZnO:N film is lower than in the undoped ZnO film, so grain boundaries cannot decide on the observed enhancement of acceptor luminescence. Moreover, in the annealed ZnO film both acceptor- and donor-related regions occupy similar spatial areas that are randomly distributed over the sample cross-section, while in the ZnO:N film the regions with acceptor emission prevail, are arranged along the columns of growth, and average acceptor-related PL and CL are considerably more intensive, pointing at much higher concentration of acceptor states.

Such a spatial distribution of acceptor regions would allow, at least partially, to explain the problem of ambiguous results of Hall effect measurements, which often provide results depending on the sample geometry. In fact, in Hall measurements, the related electrical characteristics are taken across the samples, where acceptor and donor regions are randomly distributed, and hence one might obtain ambiguous results. Similar situation was already reported in scanning capacitance spectroscopy measurements of ZnO samples co-doped with nitrogen and arsenic (Dadgar et al., 2005; Krtschil et al., 2005). On the other hand, contrary to these results, good electrical characteristics can be obtained for ZnO-based homojunction obtained with ALD, as will be shown in the next paragraph. In case of such a homojunction, the carriers' transport is realized across the ZnO films, that is, along the columns.

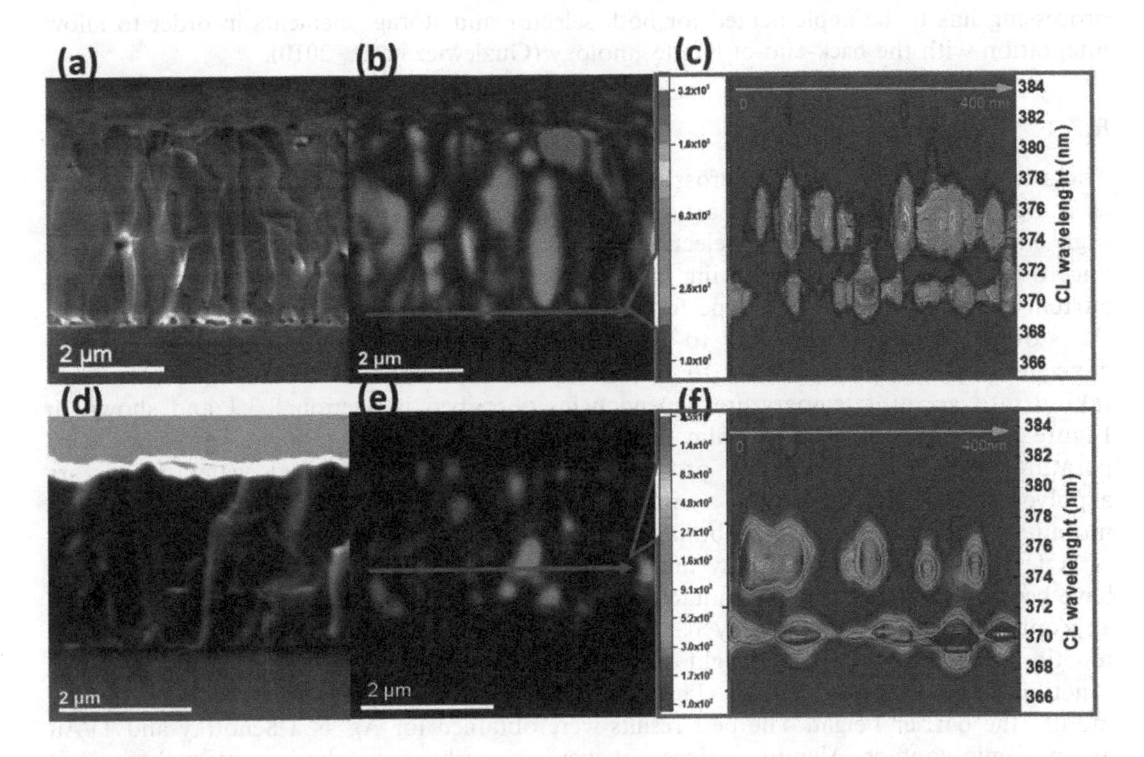

FIGURE 8.18 Cross-section view of annealed ZnO:N (top panel) and ZnO (bottom panel) samples: (a) and (d) SEM images; (b) and (e) low temperature CL maps – red color represents CL at 375 nm (3.305 eV) and the green one at 370 nm (3.36 eV); (c) and (f) two-dimensional maps of CL (5 kV) come from the points along the blue lines marked in (b) and (e), the color scale shows intensity of the PL signal (Guziewicz et al., 2017, reproduced with permission).

8.7 DEVICES BASED ON ZnO-ALD Films

The correlation between free electron concentration and growth temperature observed for undoped zinc oxide films grown by ALD with DEZn and water precursors is of great importance from the application point of view. This relation means that one can control electron concentration of ZnO films without any external doping but only by choosing appropriate parameters of the ALD process. In fact, different types of applications require various electrical parameters of the ZnO films. Theoretical calculations show (Pra et al., 2008) that electron concentration of zinc oxide film dedicated to Schottky junction should not exceed 2×10^{17} cm^{-3} while electron mobility should not be lower than 10 cm^2 V^{-1} s^{-1}. When ZnO is dedicated to a *p–n* junction, electron concentration at the level of 10^{18} cm^{-3} is needed, and for the ZnO films used as a transparent electrode for solar cells a very high *n*, even at the level of 10^{21} cm^{-3}, is required.

Below, we present examples which show that zinc oxide obtained at a temperature of 100°C by ALD process can really work as an active element of rectification junctions, so that this material can be successfully applied in microelectronics.

Rectification junctions based on ZnO deposited at low temperature are elements that can be used as selectors in highly integrated nonvolatile memories based on 3D crossbar architecture (Huby, Tallarida, Kutrzeba et al., 2008; Katsia et al., 2009; Lee, Park, et al., 2007, Lee, Seo, et al., 2007; Oh et al., 2011; Park, Cho, Kim, & Kim, 2011). Low temperature processing has to be implemented for both selector and storage elements in order to allow integration with the back-end-of-line technology (Guziewicz et al., 2010).

8.7.1 Schottky Diode

The selector element in the 3D crossbar memory needs to fulfil specific requirements such as a high rectification ratio and a high forward current. The rectification ratio should be high enough to ensure proper selection of the appropriate memory cell without disturbance due to leakage current flowing through the remaining memory cells. A high forward current density is necessary to perform various operations (reading and writing) within the storage element. According to theoretical calculations (Pra et al., 2008), electron concentration in ZnO dedicated to a Schottky junction should not exceed $10^{17}cm^{-3}$, so taking into account temperature dependencies described in section 8.3.1 and shown in Figure 8.10 (right), ZnO-ALD film dedicated to a Schottky junction was grown at temperature 100°C. In the ALD process with DEZn and deionized water precursors a polycrystalline ZnO film with an intrinsic free carrier concentration $n = 10^{17}cm^{-3}$ and mobility of 17 cm^2 V^{-1} $s^{-1}$are obtained.

A few possibilities of Schottky and ohmic contacts have been tested. Pt, Au, and Ag have been checked as Schottky contacts and Al and Ti/Au as ohmic contacts. The choice of appropriate metal for a Schottky barrier is an important issue in a ZnO-based Schottky junction, because the barrier height at the ZnO/metal interface does not follow the work function values (Ip et al., 2006). This is because of the ZnO surface states that strongly modify the barrier height. The best results were obtained for Ag as a Schottky and Ti/Au as an ohmic contact. Aluminum does not work properly as an ohmic contact, because it leads to a high leakage current as it is shown in Figure 8.19 (left). Al is a known donor dopant in zinc oxide (Fan & Freer, 1995), so probably Al diffuses from the contact into a ZnO layer leading to a higher electron concentration and thus to a high leakage current due to increased tunneling through a thinner barrier. The change of an ohmic contact from Al to Ti/Au leads to the decrease of leakage current by almost two orders of magnitude and subsequent increase of the junction rectification ratio from 3×10^3 to 10^5 (at 2V) (Figure 8.19, left). The oxidation of silver at the Ag/ZnO interface causes raised series resistance. In order to minimize this effect, a rectification Ag contact was deposited at the

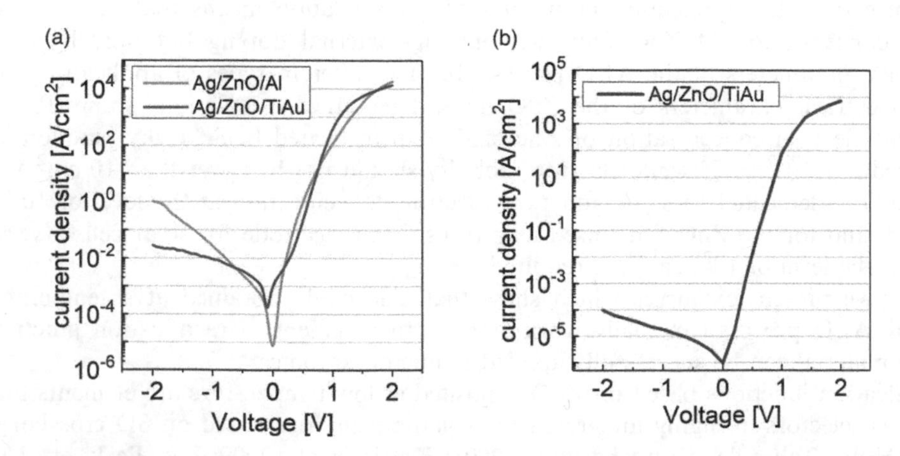

FIGURE 8.19 The *I–V* characteristics of the Ag/ZnO-ALD Schottky junction obtained at 100°C: contact optimization (left) and optimization of ALD growth parameters (right).

bottom of a structure and an ohmic contact at the top. Both *I–V* characteristics presented in Figure 8.19 are obtained for such a structure. Details of the contact optimization have been described elsewhere (Allen & Dubin, 2008; Fan & Freer, 1995; Tallarida et al., 2009).

Further improvement of the leakage current has been achieved by optimization of the ZnO-ALD growth parameters. As it is presented in Figure 8.19 (right), for the same growth temperature various electron concentrations can be obtained depending on the used ALD parameters such as pulses and purging times. The optimization procedure is based on an unique approach that relies on the correlation of optical and electrical properties. For the ZnO deposition process, such growth parameters that lead to ZnO film with very low defect-related photoluminescence (Tallarida et al., 2009).

Optimization of ZnO parameters results in a further decrease of leakage current. In this way, a rectification ratio at the level of 10^8 at 2V has been achieved as is presented in Figure 8.19 (right). Such a very high I_{ON}/I_{OFF} value fulfils requirements for a switching element dedicated to a crossbar memory. It assures the proper functioning of the crossbar array at very large integration scale. Moreover, the diode exhibit a high current density of 10^4 A cm^{-2}. These Schottky junction properties are among the best results published so far for the diodes obtained at low temperature regime (Allen & Dubin, 2008). The 10 kb crossbar memory array with a NiO-based MIM (metal–insulator–metal) memory element and a Ag/ZnO-ALD/TiAu switching element was successfully constructed and showed appropriate switching properties (Fan & Freer, 1995). It should be noted that the high rectification ratio might be further increased by deposition of a 2–3 nm thick HfO_2 layer, which positively influences the I_{ON}/I_{OFF} ratio as was reported (Krajewski, Luka, Gieraltowska et al., 2011).

8.7.2 Homojunction

Based on the *p*-type ZnO:N, we constructed a ZnO-based homojunction, which has been fully prepared by the ALD technique (Snigurenko et al., 2015, Figure 8.20). The full ALD-ZnO rectifying structure was built on a highly resistive silicon substrate, covered with an approximately 40 nm thick *n*-ZnO buffer layer followed by the 8 nm "bottom" dielectric (Al_2O_3) film. These steps, made prior to the *p*-type ZnO deposition, were aimed at the elimination of the possible substrate-related influence on the junction *I–V* characteristics. Before growing of the second part of the homojunction (*n*-ZnO obtained at 130°C), the *p*-type ZnO:N layer was capped with 4 nm of Al_2O_3 ("upper dielectric") to prevent nitrogen out-diffusion. Ohmic contacts to the structure were formed by the Ti/Au bilayer. The I_{ON}/I_{OFF} ratio of the ZnO-ALD homojunction obtained in this way was 4×10^4 at ±2 V. It was more than twice higher than for the ZnO-ALD homojunction without any Al_2O_3 protective layer (Snigurenko et al., 2015).

8.7.3 Other Electronic Devices

II–VI semiconductors are nowadays studied as prospective candidates for thin-film transistors dedicated to applications that require low processing temperature, such as transparent electronics or active matrix displays where the silicon-based technology becomes less popular. Polycrystalline ZnO presents a few advantages over amorphous silicon such as much higher electron mobility and transparency. ALD is beneficial in this case, as not only the ZnO films, but also the best quality dielectric layers, especially high-k oxides, can be obtained by this method. In that way, few parts of field effect transistor, channel, and gate dielectric, can be deposited using the same technology.

High-quality films of high-dielectric oxides such as HfO_2 and Al_2O_3 can be grown by ALD at temperature that does not exceed 100°C (Gierałtowska et al., 2011). The ALD method provides unique possibility to control thickness of the films at the nanometer scale.

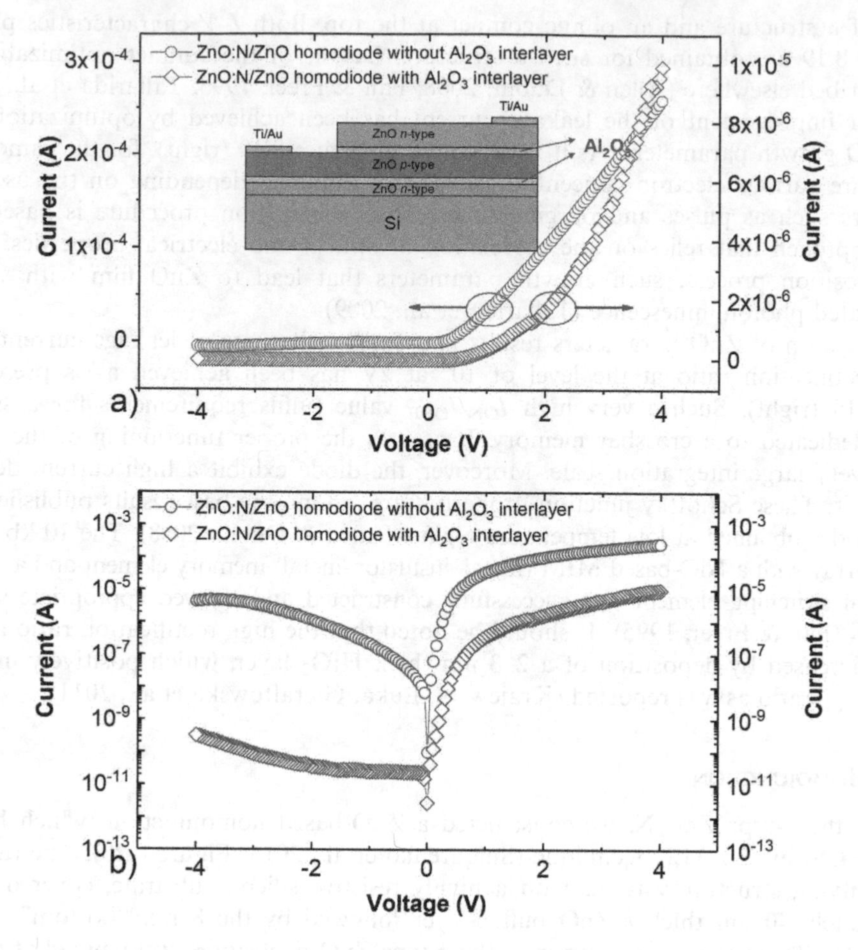

FIGURE 8.20 The I–V characteristics of ZnO-based homojunctions: without Al_2O_3 ultrathin films ($I_{ON}/I_{OFF} = 10^2$) and with the ultrathin Al_2O_3 film ($I_{ON}/I_{OFF} = 4 \times 10^4$), (Snigurenko et al., 2015, with permission).

This is especially important for gate dielectrics used in MOSFET dedicated to highly integrated circuits where dielectric thickness is 3–5 nm. The MOSFET transistor with ZnO-ALD as a channel material and Al_2O_3 as a gate dielectric has been reported (Huby, Ferrari, Guziewicz, Godlewski, & Osinniy, 2008). The ZnO layer deposited by ALD, which acts as a channel there, was obtained at 100°C and has carrier concentration below 10^{18} cm^{-3}. The device features a high I_{ON}/I_{OFF} ratio of 10^7.

Transparent thin-film transistor where gate, gate dielectric, and channel were obtained by ALD at low temperature regime was also reported (Gieraltowska, Wachnicki, Witkowski, Godlewski, & Guziewicz, 2012). Transistor structure was deposited on a glass substrate and two ZnO films with various conductivities were used as a gate and as a channel, whereas a dielectric composite layer $Al_2O_2/HfO_2/Al_2O_3$ was used as a gate dielectric.

8.8 SUMMARY

In this review, the properties of undoped ZnO-ALD films grown under low temperature regime (between 100 and 200°C) have been presented. The ZnO films obtained with DEZn and water precursors are atomically flat and polycrystalline with the grain size varying between 20 and 60 nm. It has been shown that the considerable conductivity difference between films grown at 100 and 200°C is reflected in different binding energy of donors, as well as different ratio between donor- and acceptor-related shallow defect states was observed in low temperature photoluminescence. This phenomenon can be ascribed to changing growth conditions from oxygen-rich to zinc-rich and creation of complexes involving native point defects and hydrogen, which is found to be abundant in the ALD processes.The ZnO films grown at 100°C, that is, under oxygen-rich conditions, exhibit abundant acceptor states. Moreover, their PL decay time is at the level of 150 ps, so it is more than twice longer than the decay time of ZnO films deposited under zinc-rich conditions (i.e., grown at 200°C).

The electrical conductivity of nitrogen-doped ZnO films grown under oxygen-rich conditions can be converted to *p*-type, provided an appropriate post-growth annealing is applied. Such *p*-type layers are suitable as a partner of a ZnO-ALD homojunction, revealing promising electrical characteristics.

In conclusion, the polycrystalline ZnO films grown under oxygen-rich conditions reveal promising electrical characteristic and, doped with nitrogen, might create a new way for achieving *p*-type conductivity of ZnO.

ACKNOWLEDGMENT

This work was supported by the Polish National Science Centre project DEC-2018/07/B/ST3/03576.

REFERENCES

Allen, M. W., & Durbin, S. M. (2008). Influence of oxygen vacancies on Schottky contacts to ZnO. *Applied Physics Letters*, *92*, 122110.

Anderson, T., Ren, F., Pearton, S., Kang, B. S., Wang, H.-T., Chang, C.-Y., & Lin, J. (2009). Advances in Hydrogen, Carbon Dioxide, and Hydrocarbon Gas Sensor Technology Using GaN and ZnO-Based Devices. *Sensors*, *9*(6), 4669–4694.

Bang, S., Lee, S., Park, J., Park, S., Jeong, W., & Jeon, H. (2009). Investigation of the effects of interface carrier concentration on ZnO thin film transistors fabricated by Atomic Layer Deposition. *Journal of Physics D: Applied Physics*, *42*, 235102.

Barreca, D., Carraro, G., Maccato, C., Altantzis, T., Kaunisto, K., & Gasparotto, A. (2018). Controlled growth of supported ZnO inverted nanopyramids with downward pointing tips. *Crystal Growth & Design*, *18*(4), 2579–2587.

Bauer, J., Hewson, A.C., & Dupuis, N. (2009). Dynamical mean-field theory and numerical renormalization group study of superconductivity in the attractive Hubbard model. *Physical Review B*, *79*, 214518.

Beh, H., Hiller, D., Bruns, M., Welle, A., Becker, H.-W., Berghoff, B., … Zacharias, M. (2017). Quasi-metallic behavior of ZnO grown by Atomic Layer Deposition: The role of hydrogen. *Journal of Applied Physics*, *122*, 025306.

Beh, H., Hiller, D., Salava, J., Trojánek, F., Zacharias, M., Malý, P., & Valenta, J. (2018). Photoluminescence dynamics and quantum yield of intrinsically conductive ZnO from Atomic Layer Deposition. *Journal of Luminescence*, *201*, 85–89.

Beh, H., Hiller, D., & Zacharias, M. (2018). Optimization of ALD-ZnO thin films towards higher conductivity. *Physica Status Solidi (A) Applications and Materials Science*, *215*, 1700880.

Bouderbala, M., Hamzaoui, S., Adnane, M., Sahraoui, T., & Zerdali, M. (2009). Annealing effect on properties of transparent and conducting ZnO thin films. *Thin Solid Films*, *517*(5), 1572–1576.

Buchanan, D. A. (1999). *IBM Journal of Research and Development*, *43*, 245–264.

Butcher, K. S. A., Afifuddin, P., Chen, P. T., Godlewski, M., Szczerbakow, A., Goldys, E. M., … Freitas, J. A., Jr. (2002). *Journal of Crystal Growth*, *246*, 237.

Cheng, Y. C., Yuan, K. Y., & Chen, M. J. (2016). ZnO thin films prepared by atomic layer deposition at various temperatures from 100 to 180 degrees C with three-pulsed precursors in every growth cycle. *Journal of Alloys and Compounds*, *685*, 391–394.

Dadgar, A., Krtschil, A., Bertram, F., Giemsch, S., Hempel, T., Veit, P., … Krost, A. (2005). ZnO MOVPE growth: From local impurity incorporation towards p-type doping. *Superlattices and Microstructures, 38*, 245–255.

Du, M. H., & Biswas, K. (2011). Anionic and hidden hydrogen in ZnO. *Physical Review Letters, 106*, 115502.

Ellmer, K., Klein, A., & Rech, B. (ed). (2008). *Transparent conductive Zinc Oxide: Basics and applications in thin film solar cells.* Berlin and Heidelberg: Springer-Verlag.

Fan, J., & Freer, R. (1995). The roles played by Ag and Al dopants in controlling the electrical properties of ZnO varistors. *Journal of Applied Physics, 77*, 4795.

Fei, C., Hsu, H.-S., Vafanejad, A., Li, Y., Lin, P., Li, D., … Zhou, Q. (2017). Ultrahigh frequency ZnO silicon lens ultrasonic transducer for cell-size microparticle manipulation. *Journal of Alloys and Compounds, 729*, 556–562.

Fortunato, E., Raniero, L., Silva, L., Gonçalves, A., Pimentel, A., Barquinha, P., … Martins, R. (2008). Highly stable transparent and conducting gallium-doped zinc oxide thin films for photovoltaic applications. *Solar Energy Materials and Solar Cells, 92*(12), 1605–1610.

Frodason, Y. K., Johansen, K. M., Bjorheim, T. S., Svenson, B. G., & Alkauskas, A. (2018). Zn vacancy – Donor impurity complexes in ZnO. *Physical Review B, 97*, 104109.

Funakubo, H., Mizutani, N., Yonetsu, M., Saiki, A., & Shinozaki, K. (1999). Orientation control of ZnO thin film prepared by CVD. *Journal of Electroceramics, 4*, 25–32.

George, S. M. (2010). Atomic Layer Deposition: An Overview. *Chemical Reviews, 110*, 111–131.

Gierałtowska, S., Sztenkiel, D., Guziewicz, E., Godlewski, M., Łuka, G., Witkowski, B. S., … Sawicki, M. (2011). Properties and Characterization of ALD Grown Dielectric Oxides for MIS Structures. *Acta Physica Polonica A, 119*, 692–695.

Gieraltowska, S. A., Wachnicki, L., Witkowski, B. S., Godlewski, M., & Guziewicz, E. (2012). Atomic layer deposition grown composite dielectric oxides and ZnO for transparent electronic applications. *Thin Solid Films, 520*, 4694–4697.

Gong, S. C., Bang, S., Jeon, H., Park, H.-H., Chang, Y. C., & Chang, H. J. (2010). Effects of Atomic Layer Deposition temperatures on structural and electrical properties of ZnO films and its thin film transistors. *Metals and Materials International, 16*(6), 953–958.

Goodman, C. H. L. & Pessa, M. V. (1986). *Applied Journal Physics, 60*, R65.

Grochowski, J., Guziewicz, M., Kruszka, R., Borysiewicz, M., Kopalko, K., & Piotrowska, A. (2012). Fabrication and Characterization of p-NiO/n-ZnO Heterojunction Towards Transparent Diode. *35th Int. Spring Seminar on Electronic Technology ISSE*, pp. 488–491.

Guziewicz, E., Godlewski, M., Krajewski, T., Wachnicki, Ł., Szczepanik, A., Kopalko, K., … Ferrari, S. (2009). ZnO grown by atomic layer deposition: A material for transparent electronics and organic heterojunctions. *Journal of Applied Physics, 105*, 122413.

Guziewicz, E., Godlewski, M., Krajewski, T. A., Wachnicki, Ł., Łuka, G., Domagała, J. Z., … Suchocki, A. (2010). Zinc oxide grown by atomic layer deposition - a material for novel 3D electronics. *Physica Status Solidi (B), 247*, 1611.

Guziewicz, E., Godlewski, M., Wachnicki, L., Krajewski, T. A., Luka, G., Gieraltowska, S., … Jablonski, A. (2012). ALD grown zinc oxide with controllable electrical properties. *Semiconductor Science and Technology, 27*(7), 074011.

Guziewicz, E., Kowalik, I. A., Godlewski, M., Kopalko, K., Osinniy, V., Wójcik, A., … Paszkowicz, W. (2008). Extremely low temperature growth of ZnO by atomic layer deposition. *Journal of Applied Physics, 103*, 033515.

Guziewicz, E., Krajewski, T. A., Przezdziecka, E., Korona, K. P., Czechowski, N., Klopotowski, L., & Terziyska, P. (2019). Zinc Oxide Grown by Atomic Layer Deposition: From Heavily n-Type to p-Type Material. *Physica Status Solidi (B)*, 1900472.

Guziewicz, E., Przezdziecka, E., Snigurenko, D., Jarosz, D., Witkowski, B. S., Dluzewski, P., & Paszkowicz, W. (2017). Abundant Acceptor Emission from Nitrogen-Doped ZnO Films Prepared by Atomic Layer Deposition under Oxygen-Rich Conditions. *ACS Applied Materials & Interfaces, 9*, 26143.

Hamad, O., Braunstein, G., Patil, H., & Dhere, N. (2005). Effect of thermal treatment in oxygen, nitrogen, and air atmospheres on the electrical transport properties of zinc oxide thin films. *Thin Solid Films, 489*(1–2), 303–309.

Heo, Y. W., Norton, D. P., & Pearton, S. J. (2005). Origin of green luminescence in ZnO thin film grown by molecular-beam epitaxy. *Journal of Applied Physics, 98*, 073502.

Hong, J., Paik, H., Hwang, H., Lee, S., deMello, A. J., & No, K. (2009). The effect of growth temperature on physical properties of heavily doped ZnO:Al films. *Physica Status Solidi A*, *206*(4), 697–703.

Huang, R., Ye, S., Sun, K., Kiang, K. S., & de Groot, C. H. (2017). Fermi level tuning of ZnO films through supercycled Atomic Layer Deposition. *Nanoscale Research Letters*, *12*, 541.

Huby, N., Ferrari, S., Guziewicz, E., Godlewski, M., & Osinniy, V. (2008). Electrical behavior of zinc oxide layers grown by low temperature atomic layer deposition. *Applied Physics Letters*, *92*, 023502.

Huby, N., Tallarida, G., Kutrzeba, M., Ferrari, S., Guziewicz, E., Wachnicki, L., & Godlewski, M. (2008). New selector based on zinc oxide grown by low temperature Atomic Layer Deposition for vertically stacked non-volatile memory devices. *Microelectronic Engineering*, *85*(12), 2442–2444.

International Technology Roadmap for Semiconductors (ITRS), 2007 Edition. Retrieved from www.itrs.net/.

Ip, K., Thaler, G. T., Yang, H., Han, S. Y., Li, Y., Norton, D. P., … Ren, F. (2006). Contacts to ZnO. *Journal of Crystal Growth*, *287*, 149.

Janik, E., Wachnicka, A., Guziewicz, E., Godlewski, M., Kret, S., Zaleszczyk, W., … Wojtowicz, T. (2010). ZnTe-ZnO core-shell radial heterostructures grown by the combination of molecular beam epitaxy and atomic layer deposition. *Nanotechnology*, *21*, 015302.

Janotti, A., & van de Walle, C. G. (2007). Native point defects in ZnO. *Physical Review B*, *76*, 165–202.

Janotti, A., & van de Walle, C. G. (2009). Fundamentals of zinc oxide as a semiconductor. *Reports on Progress in Physics*, *72*(1 2), 126501.

Jiang, M., Wang, D. D., Zou, B., Chen, Z. Q., Kawasuso, A., & Sekiguchi, T. (2012). Effect of high temperature annealing on defects and optical properties of ZnO single crystals. *Physica Status Solidi A*, *209*(11), 2126–2130.

Kajikawa, Y. (2006). Texture development of non-epitaxial polycrystalline ZnO films. *Journal of Crystal Growth*, *289*, 387–394.

Katsia, E., Huby, N., Tallarida, G., Kutrzeba-Kotowska, B., Perego, M., Ferrari, S., … Luka, G. (2009). Poly(3-hexylthiophene)/ZnO hybrid *pn* junctions for microelectronics applications. *Applied Physics Letters*, *94*, 143501.

Kim, H.-R., Kang, C.-S., Kim, S.-K., Byun, C.-W., & Yoon, S.-M. (2019). Characterization on the operation stability of mechanically flexible memory thin-film transistors using engineered ZnO charge-trap layers. *Journal of Physics D: Applied Physics*, *52*(32), 325106.

Kolb, E. D., & Laudise, R. A. (1965). Properties of lithium-doped hydrothermally grown single crystals of zinc oxide. *Journal of the American Ceramic Society*, *48*, 342.

Kopalko, K., Godlewski, M., Domagala, J. Z., Lusakowska, E., Minikayev, R., Paszkowicz, W., & Szczerbakow, A. (2004). Monocrystalline ZnO films on GaN/Al2O3 by atomic layer epitaxy in gas flow. *Chemistry of Materials*, *16*, 1447.

Kowalik, I. A., Guziewicz, E., Kopalko, K., Yatsunenko, S., Wójcik-Głodowska, A., Godlewski, M., … Paszkowicz, W. (2009). Structural and optical properties of low-temperature ZnO films grown by Atomic Layer Deposition with diethylzinc and water precursors. *Journal of Crystal Growth*, *311*, 1096.

Krajewski, T., Guziewicz, E., Godlewski, M., Wachnicki, L., Kowalik, I. A., Wojcik-Glodowska, A., … Guziewicz, M. (2009). The influence of growth temperature and precursors' doses on electrical parameters of ZnO thin films grown by Atomic Layer Deposition technique. *Microelectronics Journal*, *40*(2), 293–295.

Krajewski, T. A., Dybko, K., Luka, G., Guziewicz, E., Nowakowski, P., Witkowski, B. S., … Godlewski, M. (2014). Dominant shallow donors in zinc oxide layers obtained by low-temperature atomic layer deposition: Electrical and optical investigations. *Acta Materialia*, *65*, 69–75.

Krajewski, T. A., Luka, G., Gieraltowska, S., Zakrzewski, A. J., Smertenko, P. S., Kruszewski, P., … Guziewicz, E. (2011). Hafnium dioxide as a passivating layer and diffusive barrier in ZnO/Ag Schottky junctions obtained by atomic layer deposition. *Applied Physics Letters*, *98*, 263502.

Krajewski, T. A., Luka, G., Wachnicki, L., Guziewicz, E., Godlewski, M., Witkowski, B., … Paszkowicz, W. (2010). Growth conditions and structural properties as limiting factors of electrical parameters of ZnO thin films grown by Atomic Layer Deposition with diethylzinc and water precursors. *Physica Status Solidi C*, *7*(6), 1550–1552.

Krajewski, T. A., Luka, G., Wachnicki, L., Jakiela, R., Witkowski, B., Guziewicz, E., … Tallarida, G. (2009). Optical and electrical characterization of defects in zinc oxide thin films grown by Atomic Layer Deposition. *Optica Applicata*, *39*(4), 865–874.

Krajewski, T. A., Luka, G., Wachnicki, L., Zakrzewski, A. J., Witkowski, B. S., Lukasiewicz, M. I., … Guziewicz, E. (2011). Electrical parameters of ZnO films and ZnO-based junctions obtained by Atomic Layer Deposition. *Semiconductor Science and Technology*, *26*, 085013, 1–6.

Krajewski, T. A., Terzyiska, P., Luka, G., Lusakowska, E., Jakiela, R., Vlakov, E. S., & Guziewicz, E. (2017). Diversity of contributions leading to the nominally n-type behavior of ZnO films obtained by low temperature Atomic Layer Deposition. *Journal of Alloys and Compounds*, *727*, 902–911.

Krtschil, A., Dadgar, A., Oleynik, N., Bläsing, J., Diez, A., & Krost, A. (2005). Local p-type conductivity in zinc oxide dual-doped with nitrogen and arsenic. *Applied Physics Letters*, *87*, 262105.

Ku, C.-S., Huang, J.-M., Lin, C.-M., & Lee, H.-Y. (2009). Fabrication of epitaxial ZnO films by atomic-layer deposition with interrupted flow. *Thin Solid Films*, *518*, 1373.

Kwon, S., Bang, S., Lee, S., Jeon, S., Jeong, W., Kim, H., … Jeon, H. (2009). Characteristics of the ZnO thin film transistor by atomic layer deposition at various temperatures. *Semiconductor Science and Technology*, *24*, 035015.

Lee, M. J., Park, Y., Suh, D. S., Lee, E. H., Seo, S., Kim, D. C., … Park, B. H. (2007). Two series oxide resistors applicable to high speed and high density nonvolatile memory. *Advanced Materials*, *19*, 3919.

Lee, M. J., Seo, S., Kim, D. C., Ahn, S. E., Seo, D. H., Yoo, I. K., … Park, B. H. (2007). A low-temperature-grown oxide diode as a new switch element for high-density, nonvolatile memories. *Advanced Materials*, *19*, 73.

Lee, W. J., Shin, W. C., Chae, B. G., Ryu S.O., You I.K., Cho S.M., Yu B.G., & Shin, B.C. (2002). Electrical properties of dielectric and ferroelectric films prepared by plasma enhanced atomic layer deposition. *Integrated Ferroelectrics*, *46*, 275–284.

Leskela, M., & Ritala, M. (2002). Atomic layer deposition (ALD): from precursors to thin film structures. *Thin Solid Films*, *409*, 138.

Leskela, M., & Ritala, M., (2003). Atomic layer deposition chemistry: recent developments and future challenges. *Angewandte Chemie International Edition*, *42*, 5548–5554.

Liu, W. Z., Xu, H. Y., Zhang, L. X., Zhang, C., Ma, J. G., Wang, J. N., & Liu, Y. C. (2012). Localized surface plasmon-enhanced ultraviolet electroluminescence from n-ZnO/i-ZnO/p-GaN heterojunction light-emitting diodes via optimizing the thickness of MgO spacer layer. *Applied Physics Letters*, *101*, 142101.

Look, D. C., Reynolds, D. C., Litton, C. W., Jones, R. L., Eason, D. B., & Cantwell, G. (2002). Characterization of homoepitaxial p-type ZnO grown by molecular beam epitaxy. *Applied Physics Letters*, *81*(10), 1830–1832.

Lorenz, M., Kaidashev E.M., von Wenckstern H., Riede V., Bundesmann C., Spemann D., Bendorf G., … Grundmann M. (2003). Optical and electrical properties of epitaxial (Mg,Cd) xZn1−xO, ZnO, and ZnO:(Ga,Al) thin films on c-plane sapphire grown by pulsed laser deposition. *Solid State Electronics*, *47*, 2205.

Luka, G., Godlewski, M., Guziewicz, E., Stakhira, P., Cherpak, V., & Volynyuk, D. (2012). ZnO films grown by atomic layer deposition for organic electronics. *Semiconductor Science and Technology*, *27*, 074006.

Luka, G., Witkowski, B. S., Wachnicki, L., Jakiela, R., Virt, I. S., Andrzejczuk, M., … Godlewski, M. (2014). Electrical and mechanical stability of aluminum-doped ZnO films grown on flexible substrates by atomic layer deposition. *Materials Science and Engineering: B*, *186*, 15–20.

Lyons, J. L., Varley, J. B., Steiauf, D., Janotti, A., van de Walle, C. G. (2017). First-principles characterization of native-defect-related optical transitions in ZnO. *Applied Journal Physics*, *122*, 035704.

Maeda, K., Sato, M., Niikura, I., & Fukuda, T. (2005). Growth of 2 inch ZnO bulk single crystal by the hydrothermal method. *Semiconductor Science and Technology*, *20*(4), S49–S54.

Makino, H., Miyake, A., Yamada, T., Yamamoto, N., & Yamamoto, T. (2009). Influence of substrate temperature and Zn-precursors on atomic layer deposition of polycrystalline ZnO films on glass. *Thin Solid Films*, *517*(10), 3138–3142.

Min, Y.-S., An, C. J., Kim, S. K., Song, J., & Hwang, C. S. (2010). Growth and characterization of conducting ZnO thin films by atomic layer deposition. *Bulletin of the Korean Chemical Societ*, *31*(9), 2503–2508.

Na, S.-Y., & Yoon, S.-M. (2019). Impacts of HfO_2/ZnO stack-structured charge-trap layers controlled by atomic layer deposition on nonvolatile memory characteristics of In-Ga-Zn-O channel charge-trap memory thin-film transistors. *IEEE J. Electron Devices Soc.*, *7*(1), 453–461.

Niinisto, L., & Leskela, M. (1993). Atomic layer epitaxy – chemical opportunities and challenges. *Thin Solid Films*, *225*, 130.

Oba, F., Choi, M., Togo, A., & Tanaka, I. (2011). Point defects in ZnO: an approach from first principles. *Science and Technology of Advanced Materials, 12*, 034302.
Oh, B. Y., Kim, Y. H., Lee, H. J., Kim, B. Y., Park, H. G., Han, J. W., ... Seo, D. S. (2011). High-performance ZnO thin-film transistor fabricated by atomic layer deposition. *Science and technology of advanced materials, 26*, 085007.
Pakkala, A., & Putkonen, M.. (2010). *Handbook of deposition technologies for films and coatings – Science, applications and technology*, Chapter 8, P. M. Martin (Ed.). Amsterdam, Boston, Heidelberg, London, New York, Oxford, Paris, San Diego, San Francisco, Singapore, Sydney, Tokyo: Elsevier Inc.
Park, B., Cho, K., Kim, S., & Kim, S. (2011). Nano-Floating Gate Memory Devices Composed of ZnO Thin-Film Transistors on Flexible Plastics. *Nanoscale Research Letters, 6*, 41.
Pearton, S. J., Norton, D. P., Ip, K., Heo, Y. W., & Steiner, T. (2003). Recent progress in processing and properties of ZnO. *Superlattices Microstruct., 34*(1–2), 3–32.
Pra, M., Csaba G., Erlen C., Lugli P. (2008). Simulation of ZnO diodes for application in non-volatile crossbar memories. *Journal of Computational Electronics, 7*, 146–150.
Przezdziecka, E., Guziewicz, E., Jarosz, D., Snigurenko, D., Sulich, A., Sybilski, P., ... Paszkowicz, W. 2020. Influence of oxygen-rich conditions on donor and acceptor states and conductivity mechanism of ZnO films grown by ALD – Experimental studies. *Journal of Applied Physics* 127, 075104 (1-14).
Przezdziecka, E., Guziewicz, E., & Witkowski, B. S. (2018). Photoluminescence investigation of the carrier recombination processes in N-doped and undoped ZnO ALD films grown at low temperature. *Journal Luminescence, 198*, 68–76.
Przezdziecka, E., Wachnicki, L., Paszkowicz, W., Lusakowska, E., Krajewski, T., Luka, G., ... Godlewski, M. (2009). Photoluminescence, electrical and structural properties of ZnO films, grown by ALD at low temperature. *Semiconductor Science and Technology, 24*(10), 105014.
Puurunen, R. L. (2005). Surface chemistry of atomic layer deposition: A case study for the trimethylaluminum/water process. *Journal of Applied Physics, 97*, 121301.
Ratajczak, R., Stonert, A., Guziewicz, E., Gierałtowska, S., Krajewski, T. A., Luka, G., ... Godlewski, M. (2013). RBS/Channeling Analysis of Zinc Oxide Films Grown at Low Temperature by Atomic Layer Deposition. *Acta Physica Polonica A, 123*, 899.
Roro, K. T., Kassier, G. H., Dangbegnon, J. K., Sivaraya, S., Westraadt, J. E., Neethling, J. H., ... Botha, J. R. (2008). Temperature-dependent Hall effect studies of ZnO thin films grown by metalorganic chemical vapour deposition. *Semiconductor Science and Technology, 23*, 055021.
Schropp, R. E. I. (1989). A. Madan properties of conductive zinc oxide films for transparent electrode applications prepared by rf magnetron sputtering. *Journal of Applied Physics, 66*, 2027.
Schubert, S., Schmidt, F., von Wenckstern, H., Grundmann, M., Leo, K., & Müller-Meskamp, L. (2015). Eclipse pulsed laser deposition for damage-free preparation of transparent ZnO electrodes on top of organic solar cells. *Advanced Functional Materials, 25*(27), 4321–4327.
Shi, G. A., Sabotakin, M., & Stavola, M. (2004). "Hidden hydrogen" in as-grown ZnO. *Applied Physics Letters, 85*, 5601.
Shimizu, H., Sato, W., Mihara, M., Fujisawa, T., Fukuda, M., & Matsuta, K. (2018). Temperature-dependent thermal behavior of impurity hydrogen trapped in vacancy-type defects in single crystal ZnO. *Applied Radiation and Isotopes, 140*, 224–227.
Snigurenko, D., Kopalko, K., Krajewski, T. A., Jakiela, R., & Guziewicz, E. (2015). Nitrogen doped p-type ZnO films and p-n homojunction. *Semiconductor Science and Technology, 30*, 015001.
Stakhira, P. I., Pakhomov, G. L., Cherpak, V. V., Volynyuk, D., Luka, G., Godlewski, M., ... Yu, Z. (2010). Hotra, photovoltaic cells based on nickel phtalocyanine and zinc oxide formed by atomic layer deposition. *Central European Journal of Physics, 8*(5), 798–803.
Suchea, M., Christoulakis, S., Katharakis, M., Vidakis, N., & Koudoumas, E. (2009). Influence of thickness and growth temperature on the optical and electrical properties of ZnO thin films. *Thin Solid Films, 517*(15), 4303–4306.
Suntola, T. (1992). Atomic Layer Epitaxy. *Thin Solid Films, 216*, 84–89.
Suntola, T. 1994. *Handbook of crystal growth, vol. 3, part B: Growth mechanisms and dynamics.* Chapter 14. D. T. J. Hurle (Ed.). Amsterdam: Elsevier.
Suntola, T., & Antson, J.. (1977). US Patent #4058430.
Tallarida, G., Huby, N., Kutrzeba-Kotowska, M., Spiga, S., Arcari, M., Csaba, G., ... Bez, R. (2009, May 10–14). Low temperature rectifying junctions for crossbar non-volatile memory devices. *IEEE Int. Memory Workshop*, pp. 6–9. Monterey, CA.

Tanskanen, J. T., Bakke, J. R., Pakkanen, T. A., & Bent, S. F. (2011). Influence of organozinc ligand design on growth and material properties of ZnS and ZnO deposited by atomic layer deposition. *Journal of Vacuum Science and Technology*, *29*(3), 031507.

Thomas, M. A., & Cui, J. B. (2012). Highly tunable electrical properties in undoped ZnO grown by plasma enhanced thermal-atomic layer deposition. *ACS Applied Materials & Interfaces*, *4*(6), 3122–3128.

Turos, A., Jóźwik, P., Wójcik, M., Gaca, J., Ratajczak, R., & Stonert, A. (2017). Mechanism of damage buildup in ion bombarded ZnO. *Acta Materialia*, *134*, 249–256.

Wachnicki, L., Krajewski, T., Luka, G., Witkowski, B., Kowalski, B., Kopalko, K., … Guziewicz, E. (2010). Monocrystalline zinc oxide films grown by atomic layer deposition. *Thin Solid Films*, *518*, 4556–4559.

Wachnicki, L., Lukasiewicz, M., Witkowski, B., Krajewski, T., Luka, G., Kopalko, K., … Guziewicz, E. (2010). Comparison of dimethylzinc and diethylzinc as precursors for monocrystalline zinc oxide grown by atomic layer deposition method. *Physica Status Solidi (B)*, *247*, 1699–1701.

Wojcik, A., Godlewski, M., Guziewicz, E., Minikayev, R., & Paszkowicz, W. (2008). Controlling of preferential growth mode of ZnO thin films grown by atomic layer deposition. *Journal of Crystal Growth*, *310*(2), 284–289.

Zhang, Q., Dandeneau, C. S., Zhou, X., & Cao, G. (2009). ZnO Nanostructures for Dye-Sensitized Solar Cells. *Advanced Materials*, *21*, 4087–4108. Retrieved from www.solarpowereurope.com/.

Zhang, S. B., Wei, S.-H., & Zunger, A. (2001). Intrinsic *n*-type versus *p*-type doping asymmetry and the defect physics of ZnO. *Physical Review B*, *63*, 075205.

9 Wide and Ultrawide Bandgap Oxides for Photovoltaics and Photonics

Larysa Khomenkova

9.1 INTRODUCTION

Oxide-based materials and structures are becoming increasingly important in a wide range of practical fields including microelectronics, photonics, optoelectronics and magnetoelectronics, spintronics, thermoelectrics, power harvesting, and energy storage as well as for environmental application. Synthesis, fabrication, and processing of these materials, as well as transformation into particular device structures are still a challenge. Among different oxides, ZrO_2, HfO_2, La_2O_3, Y_2O_3, Al_2O_3, and the silicates $ZrSiO_4$ and $HfSiO_4$ are often considered as high-k dielectrics, being used for microelectronic devices as an alternative to SiO_2 gate materials (He, & Sun, 2012; Houssa, 2003; Kingon,Maria, & Streiffer, 2000; Robertson, 2002; Wilket al., 2001). This is because the SiO_2 layer thinner than 2 nm no longer acts as a good insulator due to direct tunneling current across. However, the replacement of SiO_2 with a thicker high-k layer, with the same equivalent capacitance or "equivalence oxide thickness," has also certain other conditions, including chemical stability in direct contact with Si substrate. This rules out both Ti and Ta oxides, which both react with Si to form silicide phase. Another key factor is to serve as the barriers for both electrons and holes. It requires that both their valence and conduction band offsets be over 1 eV. These two conditions allow the most effective oxides to be identified, considering the parameters summarized in Table 9.1 and Figure. 9.1. However, these materials have unique optical properties that give a hand to their application in solar cells, light-emitting devices, and catalysis, along with their wide microelectronic application.

Silicon oxide, SiO_2, is a covalent random network in which Si is four-fold bonded to oxygen and O is two-fold bonded to Si. Its bandgap is 9 eV (Table 9.1). Its valence band maximum consists of nonbonding oxygen *p*-states with a large effective mass. The conduction band minimum is a broad minimum of Si 3*s*-states with an effective mass of about 0.5. The ionicity of each Si–O bond is about 50%. Contrary to this, all high-k oxides are largely ionic (Lucovsky, 2001).

Aluminum oxide (Al_2O_3) is the only *s*–*p* bonded oxide with k = 9–11. It can be found in different forms (Figure 9.3) that can be both amorphous and crystalline. The Al coordination is mainly six-fold and some four-fold. Crystalline Al_2O_3 has a number of polymorphs. Sapphire (α-Al_2O_3) is hexagonal with six-fold Al. γ-Al_2O_3 is a cubic spinel structure, in which Al is both four- and six-fold coordinated, that also contains hydrogen (Busca, 2014; Houssa, 2003). The bandgap is 8.8 eV and direct. The conduction band shows a broad minimum at Γ point of Al 3*s*-states. The upper valence bands are nonbonding oxygen *p*-states. The overall valence band is 7 eV wide. There is a considerable charge transfer between Al and O states, and the valence band is mainly oxygen-like. This charge transfer is larger than that in SiO_2.

TABLE 9.1
Parameters of different oxides

Material	Gap (eV)	Electron affinity (eV)	Static dielectric constant, ε_0 (or k)	Charge neutrality level of the interface states (eV)	High-frequency dielectric constant ε_∞	Schottky pinning parameter S	CB offset with Si (eV)	Thermal stability, T_{max} (°C)
SiO_2	9	0.9	3.9	4.0	2.25	0.86	3.5	>1420
ZrO_2	5.8	2.5	12–16 (amorphous, monoclinic) 25–30 (tetragonal, cubic)	3.6	4.8	0.41	1.4	~900
HfO_2	4.5–6	2.5	12–16 (amorphous, monoclinic) 25–30 (tetragonal, cubic)	3.7	4	0.53	1.5	~430–600
Al_2O_3	8.8	1	9	5.5	3.4	0.63	2.8	~1000
$ZrSiO_4$	6.5	2.4	10–12	3.6	3.8	0.56	1.5	>1700
$HfSiO_4$	~6	2.4	10–11	3.6	3.8	0.55	1.5	>1700

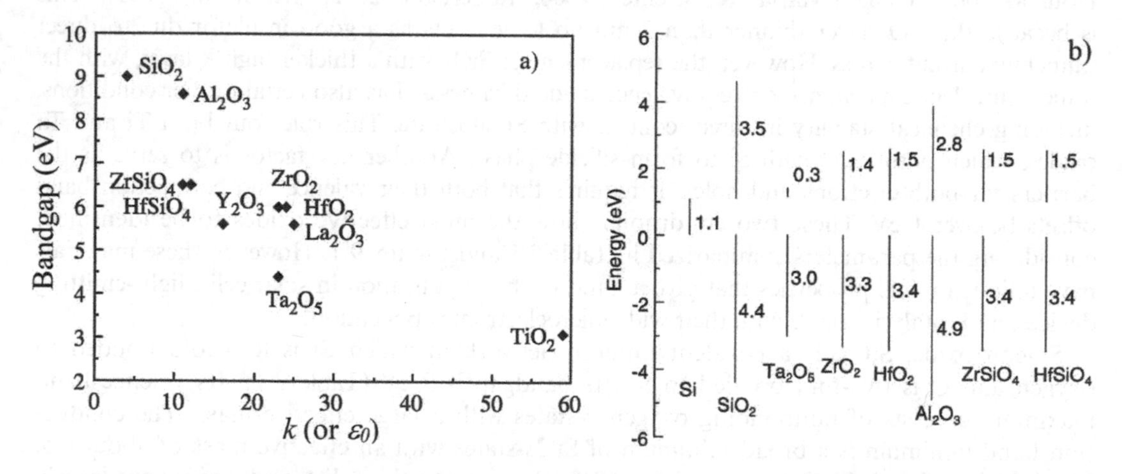

FIGURE 9.1 Variation of dielectric constant with bandgap of different oxides (a) and their corresponding band offsets on Si (b).
Adopted from (Roberston, 2002).

Among alternative to SiO_2 gate oxide, the simplest from the group IVA is zirconium oxide (ZrO_2). ZrO_2 is stable in the monoclinic structure at room temperature, it transforms to the tetragonal structure above 1170°C and to cubic one at 2200°C, but can be stabilized in both these phases at room temperature via doping with Si, Ge, or Y (Khomenkova, & Korsunska, 2019; Korsunska, & Khomenkova, 2019). In cubic ZrO_2, Zr and O have, respectively, eight O and four Zr neighbors. In monoclinic or tetragonal ZrO_2, each Zr atom has seven oxygen neighbors. This oxide has an indirect gap of 5.8–6.5 eV. The valence band is 6 eV wide, and it

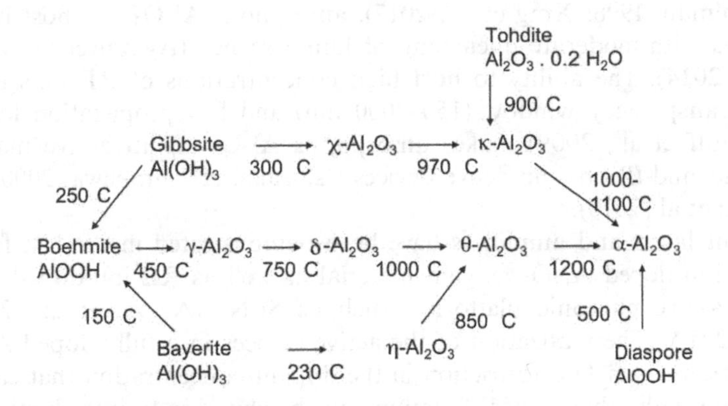

FIGURE 9.2 Schematics of the main solid-state transformations of aluminas.

Reprinted with permission from (Busca, 2014). Copyright 2014 Elsevier.

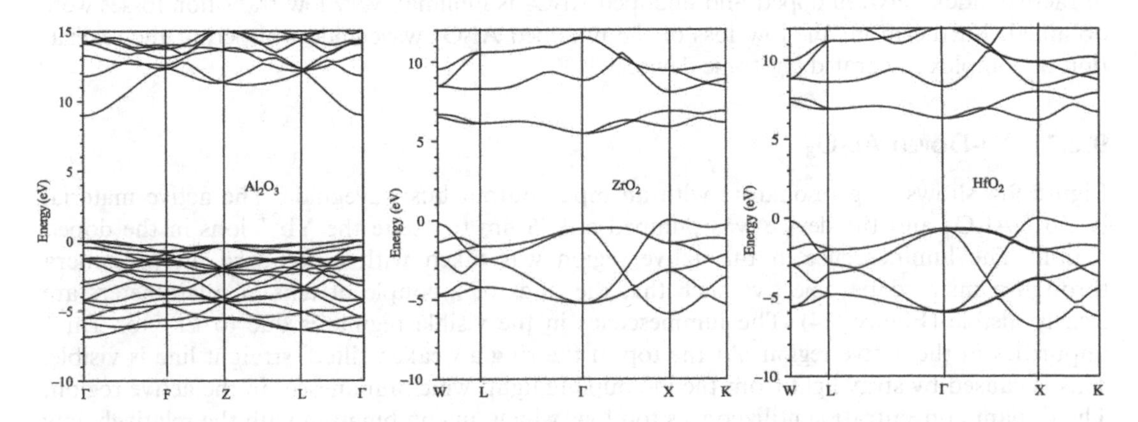

FIGURE 9.3 Band structure of Al_2O_3, ZrO_2, and HfO_2.

Adopted from (Houssa, 2003).

has a maximum at X formed from oxygen *p*-states. The conduction band minimum is a Γ_{12} state of Zr 4*d*-orbitals. A considerable charge transfer from Zr to O occurs. The valence band is strongly O *p*-states and the conduction band- Zr *d*-states, with 30% admixture (Houssa, 2003).

Hafnium oxide (HfO_2) is structurally and electronically similar to ZrO_2. It has similar bandgap, except that the crystal splitting of the Hf 5*d*-states in the conduction band is larger. Lowering the crystal symmetry of both ZrO_2 and HfO_2 results in the narrowing of their bandgaps as well as slight broadening of the valence states and intermixing of the *d*-states in the conduction band, and, in consequence, there is no longer a simple crystal field gap (Houssa, 2003). In this chapter, some recent applications of these oxides (both undoped and doped with different impurities) are considered.

9.2 AL_2O_3-BASED MATERIALS DOPED WITH RARE-EARTH IONS

Amorphous Al_2O_3 is an attractive material not only for microelectronics, but also for integrated photonics. It provides active and passive functionalities, especially, when doped with rare-earth (RE) ions. In contrast to Si, SiO_2 and Si_3N_4 that have a low solubility for RE ions (Kik, & Polman, 1998; Xing et al., 2017), amorphous Al_2O_3 can host high concentrations of RE ions with moderate quenching of luminescence (Agazzi et al., 2013; Vázquez-Córdova et al., 2014). The ability to host high concentrations of RE ions in combination with the wide transparency window (150–7000 nm) and low propagation loss (Demirtaset al., 2018; Wörhoff et al., 2009) makes amorphous Al_2O_3 an attractive material for UV, visible, near-, and mid-IR on-chip active devices (Ishizaka, & Kurokawa, 2000; Purnawirman et al., 2017; Yang et al., 2010).

Many different lasers and amplifiers have been demonstrated in the last few years using amorphous RE ion-doped Al_2O_3 as gain material as well as RE ion-doped Al_2O_3 in combination with passive photonic platforms, such as Si_3N_4 (Agazzi et al., 2013; Vázquez-Córdova et al., 2014). The realization of the active devices in a fully doped Al_2O_3 layer has two main limitations, namely a restriction in the dopant concentration that can be used and lack of integration with other optical functions on the chip due to introduction of losses via misalignment. Recently, a single-layer monolithic integration scheme was proposed to integrate active and passive regions on a wafer by using a photonics damascene method. Using this approach, only one lithography and etching step was required to fabricate all functionalities, leading to self-aligned transitions from the passive to the active sections. Since the difference of refractive index between doped and undoped Al_2O_3 is minimal, very low transition losses were obtained. Furthermore, the low loss of the undoped Al_2O_3 waveguide will permit the realization of complex integrated photonic devices.

9.2.1 Yb-Doped Al_2O_3

Figure 9.4 shows ring resonators with an input–output bus waveguide. The active material is Yb^{3+}:Al_2O_3 and the device was pumped at 976 nm to excite the Yb^{3+} ions in the doped region. The luminescence in the active region was taken with a standard photo camera through a microscope objective such that the rings of a couple of tens of micrometers are clearly visible (Figure 9.4). The luminescence in the visible regime is due to Er^{3+} or Tm^{3+} impurities in the active region. At the top of the ring a weaker, tilted, straight line is visible. This is caused by stray light from the in-coupling light, which luminesces in the active region. The dopant concentration utilized was too low, which, in combination with the relatively low

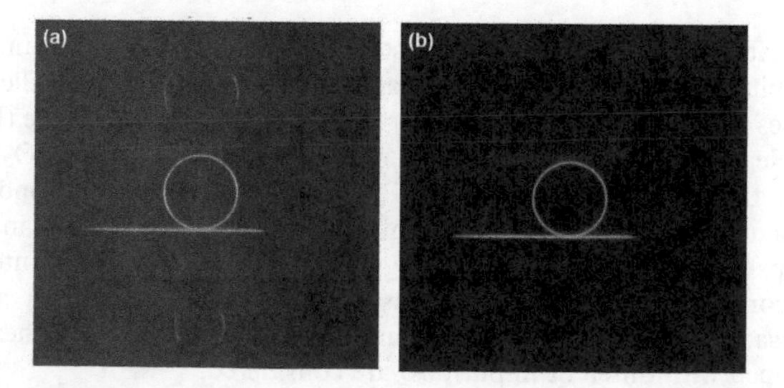

FIGURE 9.4 Photograph of (a) chip containing rings with a radius of 100 and 150 μm, showing the luminescence of the ytterbium-doped active region in the center ring. (b) Same photograph without background light. Luminescent stray light can be observed on the top of the ring.

overlap with the pump (i.e., 290 nm thick doped layer in a 400 nm total thickness of Al_2O_3, comparable with 43% of the power in the fundamental mode overlapping the active region) prevented lasing from the device. Since the fabrication is realized in a single lithography and etching step, no misalignment in that region occurs, further increasing the efficiency of the transition. Toward the center of the active region, the fraction of doped layer slightly increases and therefore the luminescence becomes stronger (van Emmerik et al., 2018).

9.2.2 Er-Doped Al_2O_3

Figure 9.5 demonstrates another example of the application of Al_2O_3 doped with Er ions (Vázquez-Córdova et al., 2014). The waveguide cross-section consisted of a 1-µm-thick by 1.5-µm-wide $Al_2O_3:Er^{3+}$ channel with ridge height of 0.35 µm. This geometry allows high

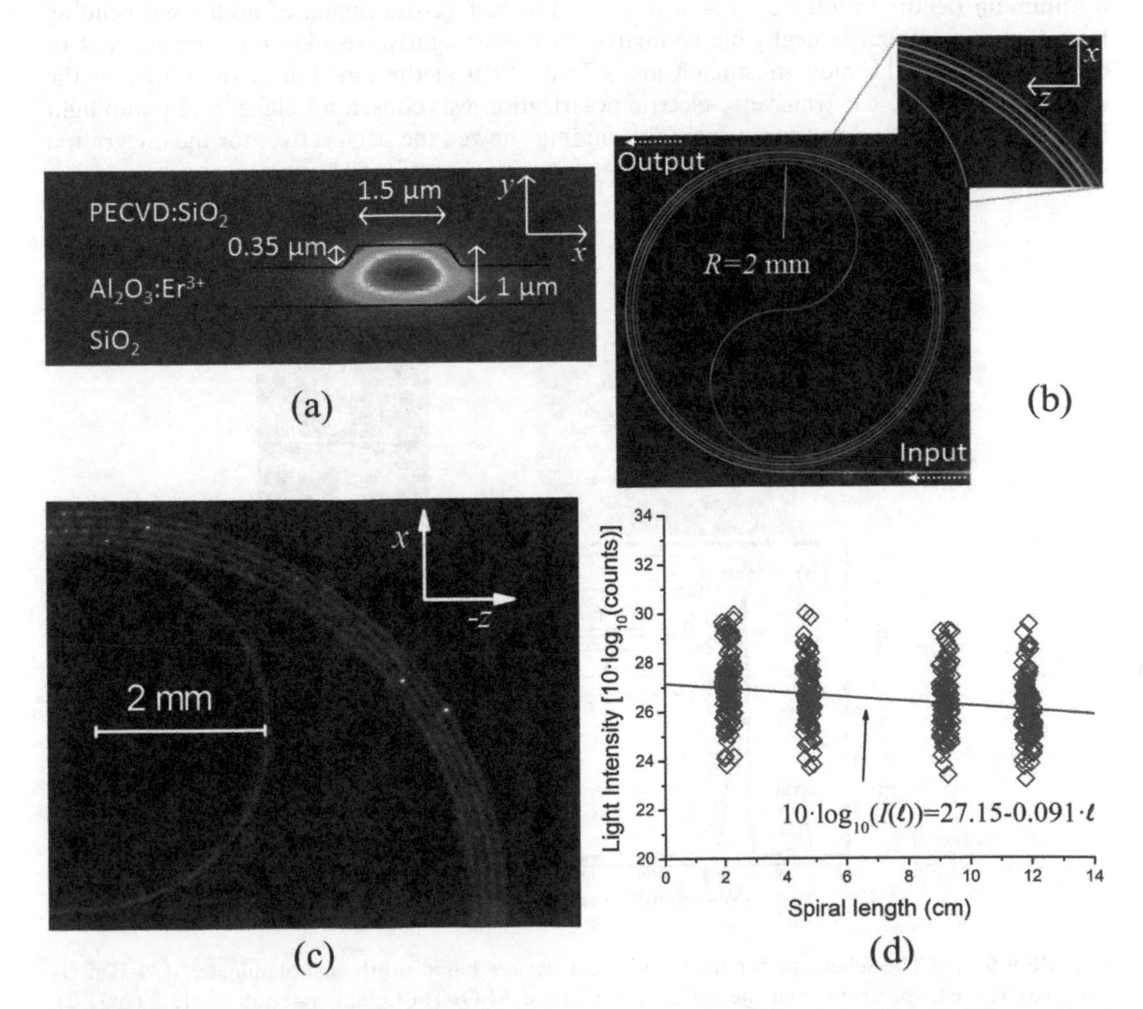

FIGURE 9.5 (a) Waveguide amplifier cross-section and simulated signal-mode profile. (b) Photograph of a pumped ($\lambda_P = 976$nm) $Al_2O_3:Er^{3+}$ spiral amplifier on a silicon chip. A close-up view of the spiral amplifier is shown in the inset. (c) Infrared image ($\lambda = 1320$ nm) of a quarter of a spiral-shaped channel waveguide. (d) Intensity distribution along the propagation direction in the spiral-shaped channel waveguide.

confinement in straight and bent sections of the waveguide. Moreover, higher signal and pump intensities were achieved in comparison with the channel geometry presented by Bradley et al. (2010). The simulated pump and signal mode fields had an overlap of ~95% and ~85% with the doped region, respectively. The refractive indices employed in the simulations were taken from previous measurements. The waveguide cross-section and the mode-field distribution at the signal wavelength of 1532 nm are displayed in Figure 9.5a.

Understanding the performance of an Er^{3+}-doped amplifier is significantly complicated by the spectroscopic processes of the Er^{3+} ion. The migration-accelerated energy transfer upconversion process $(^4I_{13/2}, ^4I_{13/2}) \rightarrow (^4I_{15/2}, ^4I_{9/2})$ induces a concentration- and excitation-dependent quenching of the $^4I_{13/2}$ amplifier level. More importantly, a fast quenching process of this level limits the optimum Er^{3+} concentration in Al_2O_3 to $(1-2)\times10^{20}cm^{-3}$. As a consequence, waveguide lengths on the order of 10 cm are desired for efficient amplifier performance. Figure 9.5b demonstrates a fabricated spiral-shaped channel waveguide with a minimum bending radius of $R = 2$ mm. For this radius, the simulated additional bending loss of $<10^{-6}$ dB/cm is negligible compared to the straight-waveguide propagation loss of ~0.1 dB/cm and the mode-mismatch loss of ~0.02 dB at the junction in the center of the spiral (Figure 9.5c, d). Transverse-electric polarization was chosen for signal and pump light in the simulations and measurements. This finding showed the perspectives for the integration of RE-doped Al_2O_3 with Si-based technology for optical communication devices.

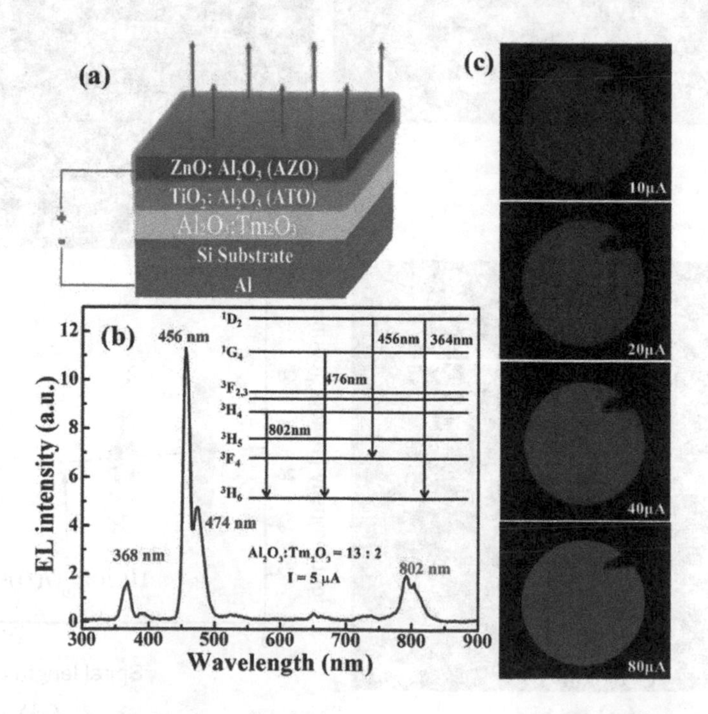

FIGURE 9.6 (a) The schematic for the luminescent devices based on the nanolaminate Al_2O_3/Tm_2O_3 films. (b) The EL spectrum from the device in which the Al_2O_3/Tm_2O_3 subcycle ratio is 13:2 (AOT-2), the inset shows the radiative transitions in the Tm^{3+} ions resulting in the EL emissions. (c) The images taken by a digital camera from this AOT-2 MOS-structured light-emitting device (MOSLED) at different injection currents.

9.2.3 Tm-Doped Al_2O_3 Materials

Aiming for the realization of compact Si-based optoelectronics, electroluminescence (EL) from RE ions has been extensively reported in many compounds, such as SiN_x, TiO_2, and ZnO. However, the efficiencies of the devices based on the aforementioned materials are far from practical utilization. One of the limitations is the large leakage current. RE-implanted SiO_2 or $SiON_x$ structures have attracted much attention due to their notable EL efficiency and silicon compatibility (Cueff et al., 2011; Jambois et al., 2009; Rebohle et al., 2014; Sun et al., 2006). However, the mismatch in the coordination structure and atomic size of silicon (tetrahedron) and RE ions (octahedron) is another limit that provokes the clustering of RE ions in the Si-based host. In comparison, similar devices based on Al_2O_3 nanofilms present much lower working voltage, and comparable efficiency while their EL performance needs more exploration (Liu et al., 2019; Yang, Li, & Sun, 2018).

Blue emission, which has the highest photon energy (2.6–2.7 eV) of the three primary colors, is of great importance in display and lighting. Tm^{3+} ions have presented efficient blue emissions in various matrixes. Recently, Liu et al. (2019) reported on the EL properties of Tm^{3+}-doped Al_2O_3 thin films grown by atomic layer deposition. Using Tm-doped Al_2O_3 might exploit the merits of both oxides to realize efficient blue EL from Tm^{3+} ions.

Figure 9.6 shows the schematic for the multilayered devices and EL spectrum from the MOSLED based on the Al_2O_3/Tm_2O_3 nanolaminate. The EL emissions mainly exhibit several peaks at the wavelengths of 368, 456, 474, and 802 nm, which originate from the radiative transitions from the $^1D_2$, $^3F_4$, $^1G_4$, and $^3H_4$ excited states to the $^3H_6$ ground state in Tm^{3+} ions, respectively (Figure.9.6b). It is noteworthy that the EL emissions at 456 and 474 nm are dominating and the blue light is easily seen by naked eyes (Figure.9.6c).

It was observed that the emission at 456 nm from Tm^{3+} ions exhibits a power density of 0.15 mW/cm^2. However, some decrease of the EL intensity and decay lifetime was observed in the samples governed by the clustering and cross-relaxation of the Tm^{3+} ions in the films with high Tm content (Liu et al., 2019). The decay lifetime for the Tm^{3+} ions under optical excitation is in the range of 0.13–1.25 μs while under electrical excitation, the decay lifetime increases to 1.13–4.02 μs. The EL is attributed to the impact excitation of the Tm^{3+} ions by hot electrons in the Al_2O_3 matrix. These results supported the development of Si-compatible RE-doped light sources by modifying the dopant structure in the nanolaminates to achieve efficient emissions.

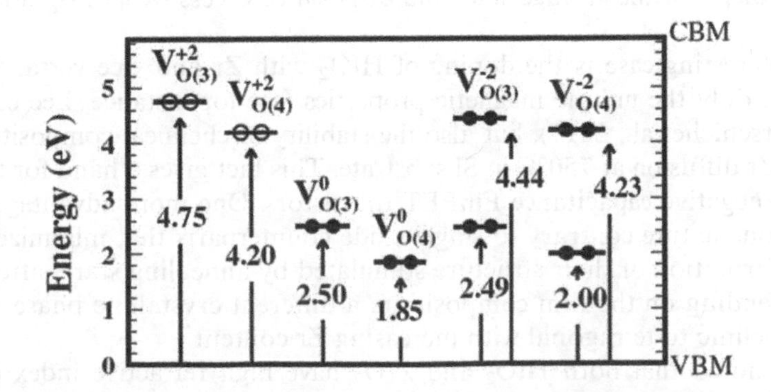

FIGURE 9.7 Schematic presentation of energy levels of different oxygen vacancies in ZrO_2 (HfO_2)-based materials.

9.3 ZRO_2- AND HFO_2-BASED MATERIALS

9.3.1 RE Free ZrO_2- and HfO_2-Based Materials and Their Solid Solutions

As it was mentioned above, ZrO_2 and HfO_2 are considered as twin oxides. Both of them have the bandgap of 5.8–6.5 eV, calculated conduction band offset on Si of about 1.5 eV and experimental one is in the range 1.4–2.2 eV that is large enough for microelectronic application. ZrO_2 and HfO_2 being doped with silicon can form different silicates. Among them, $ZrSiO_4$ and $HfSiO_4$ are glassy oxides with bandgaps of ~6.5 eV. $ZrSiO_4$ ($HfSiO_4$) consists of chains of alternate edge-sharing ZrO_4 (HfO_4) and SiO_2 tetrahedra, with additional Zr–O (Hf-O) bonds between the chains, leading to an overall six-fold Zr(Hf) coordination. The bandgap of both silicates was estimated to be about 6.5 eV with the conduction band offset of 1.5 eV and sometimes a bit higher than that for ZrO_2 (HfO_2).

Most of the properties of these materials are determined by the native defects that are different types of oxygen vacancies as it is shown for oxides in Figure 9.7. Similar defects are observed in their silicates. In most cases, intense cathode- and photoluminescence from oxygen vacancies can be found (Korsunska, & Khomenkova, 2019). However, to achieve intense light emission in specific spectral range, the doping with other ions can be performed similar to the case described for Al_2O_3.

Considering the nature of these defects as well as their charge states, Yang et al. (2018) reported on the first-principles calculations of energetically favorable defect type with the structural stability in $ZrSiO_4$ as well as on the formation energy, electronic property, and the structural deformation of point defects in $ZrSiO_4$. Similar conclusions can be made for Hf silicates also.

It was found that isolated neutral point defects are unlikely to be frequently observed. Besides, the positive charged vacancy has the lowest formation energy under five equilibrium conditions. However, when locate close to the conduction band minimum, they become more easily observed than in the O-rich environment. The charge density differences show that the charges are mainly localized on cation or anion vacant sites. Because of the Coulomb interaction, the nearest O atom of cation and anion vacant sites exhibits an opposite relaxation behavior. For the interstitial case, and with 4^+ charge state are energetically favorable, no matter the environment is O-rich or O-poor. In addition, the charge is concentrated at the interstitial sites and it is also spread between the ions along the atomic bonds. The Frenkel-pair and Schottky defects with charge states have much lower formation energy than their neutral counterpart. This finding allowed understanding of another role of ZrO_2-based materials such as catalysts or markers of different reaction. For instance, to use them for the storage state and disposal of excess weapons-grade Pu and high-actinide wastes.

One other interesting case is the doping of HfO_2 with Zr and vice versa. Such materials demonstrate not only the unique magnetic properties (see for instance, Lee et al., 2016; Shi et al., 2010; Starschichet al., 2017), but also the stability of chemical composition of the film and no Hf and Zr diffusion at 750°C in Si substrate. This fact gives a hand for the application of such films in negative capacitance FinFET transistors. One more advantage of such films is their amorphous nature contrary to single oxide counterparts that minimizes leakage current. The transformation of their structure stimulated by annealing starts after annealing at 950°C and, depending on the film composition, a different crystalline phase is formed ranging from monoclinic to tetragonal with increasing Zr content.

It should be noted that both HfO_2 and ZrO_2 have high refractive index ($n = 1.9–2.2$ at 1.95 eV, depending on preparation method), high optical transparency in the ultraviolet-infrared spectral range (bandgap is about 5.8–6.5 eV), and compactness and hardness offer optical applications as well. The phonon cutoff energy (~about 780 cm^{-1}) reduces the probability of nonradiative phonon-assisted relaxation that is attractive for doping these materials in particular with the RE elements (Khomenkova, Korsunska, Labbé, Portier, & Gourbilleau, 2019). However, in spite of the mentioned advantages, these oxides are not often addressed as a host for the RE ions.

9.3.2 RE-Doped ZrO_2 and HfO_2 and Their Silicates

Among different RE ions, trivalent erbium (Er^{3+}) is considered aiming at the application in optical communication. The trivalent neodymium (Nd^{3+}) was used in the inorganic laser materials attracting great attention of scientific researches and industrial applications. Nd^{3+} ions exhibit broad and strong absorption band around 800 nm and a very intense emission in the near-infrared luminescence range from 800 to 1430 nm associated with the $^4F_{3/2}\rightarrow{}^4I_J$ ($J = 9/2$, 11/2, 13/2) optical transitions in the 4*f* inner electronic shell of Nd^{3+} ions, following "four-level" scheme. The $^4F_{3/2}$ emitting state can be populated conveniently by the emission of low-cost commercially available laser diodes. The doping of HfO_2 and/or ZrO_2 and their silicates with RE ions allows not only to obtain specific light emission, but also to develop different types of catalysis and markers of chemical reactions.

9.3.2.1 Nd-Doped Zr Silicates

In the case of trivalent RE ions such doping results in the formation of oxygen vacancies required for RE charge compensation. These vacancies can play positive role that is the stabilization of tetragonal or cubic phases demonstrating higher dielectric permittivity and better catalytic activity. However, the appearance of numerous oxygen vacancies can result in their clustering and formation of cavities or voids causing the mechanical instability of the material. For instance, this was demonstrated for the $ZrSiO_4$ doped with Nd^{3+} ions by Li et al. (2019). It was shown that such materials are promising candidates for actinide immobilization. The series of $Zr_{1-x}Nd_xSiO_{4-x/2}$ ($x = 0$, 0.02, 0.20, and 1.0) ceramics was studied in terms of phase evolution and acidity on the chemical durability of $ZrSiO_4$-based nuclear waste forms. The single-phase $ZrSiO_4$ sample was found to have better chemical stability than that of the biphasic $Zr_{0.8}Nd_{0.2}SiO_{3.9}$ sample due to the existence of a secondary highly soluble phase ($Nd_2Si_2O_7$), which increases the contact area with leachate (Figure 9.8.)

The degradation of $Nd_2Si_2O_7$ ceramics after leaching into different leachates for 42 days at room temperature is clearly seen. Relatively slight corrosion at the grain boundaries was found for the microstructure of the samples with $x = 0.20$ leached into 1 M HNO_3 leachate (Figure 9.8h). The increase of HNO_3 concentration up to 4 M HNO_3 results in significant surface corrosion of the samples with $x = 0.2$ (Figure 9.8i). Besides, a small number of pores caused by corrosion are observed from the microstructure of the $Nd_2Si_2O_7$ ceramics after leaching into the 0.1 and 1 M HNO_3 leachates (Figure 9.8j, k). The $Nd_2Si_2O_7$ ceramic leached into the 4 M HNO_3 leachate became highly porous, increasing the area of the contact between the sample and the leachate (Figure 9.8l). At the same time, no significant changes on the surfaces of $ZrSiO_4$ and $Zr_{0.98}Nd_{0.02}SiO_{3.99}$ ceramics are observed in all leachates (Figure 9.8a–f). This observation showed that doping of $ZrSiO_4$ with Nd ions of small quantity ($x = 0.02$) results in smaller grain size and higher compactness of the ceramic along with high enough stability of the surface toward HNO_3 treatment. The increase of Nd content as well as acid concentration results in the increase of both normalized release rates of Zr and Nd in the $Zr_{1-x}Nd_xSiO_{4-x/2}$ ceramic waste forms. The lowest values for Zr and Nd were detected in $ZrSiO_4$ and $Zr_{0.98}Nd_{0.02}SiO_{3.99}$ ceramics, respectively, whereas the highest ones were reported for the $Zr_{0.8}Nd_{0.2}SiO_{3.9}$ and $Nd_2Si_2O_7$ ceramics. The difference in the normalized release rates of Zr and Nd was explained by the difference

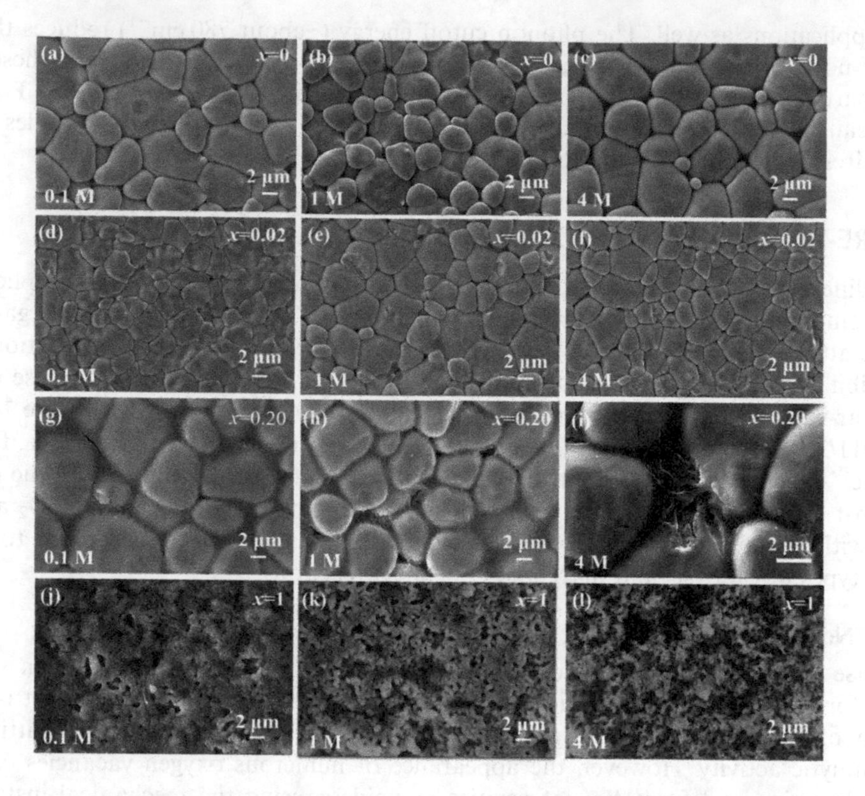

FIGURE 9.8 SEM images of $Zr_{1-x}Nd_xSiO_{4-x/2}$ (x = 0, 0.02, 0.20, and 1) ceramics after 42 days of leaching in 0.1, 1, and 4 M HNO_3 at room temperature.

Reproduced with permission from Li et al. (2019). Copyright 2019 Elsevier.

in the energies of their bonds with oxygen atoms as well as by the changes in the surfaces of the leached ceramics. Specifically, the formation of oxygen vacancies due to Nd^{3+} incorporation in $ZrSiO_4$, required by the charge compensation, results in the weakening of the lattice and degradation of ceramic surface.

9.3.2.2 RE-Doped Hf Silicates

As one can see, both $HfSiO_4$ and $ZrSiO_4$ contain high Si content. At the same time, the doping of both oxides with lower Si content has some advantages also, that is, when Si content is in the range 6–12 at%, the stabilization of tetragonal and cubic phase of both materials can be achieved at room temperature (Khomenkova, Portier, Marie, & Gourbilleau, 2011) due to the formation of Si-O covalent bonds that are flexible and allow "compensating" ionic nature of Zr-O and Hf-O bonds governed the crystallization of these phases.

When Si content reached about 15–20%, the appearance of light emission in red spectral range after annealing at high temperatures occurs (Figure 9.9). In this case, the formation of some Si clusters was observed (Khomenkova, & Korsunska, 2019; Khomenkova et al., 2019; Talbot, Roussel, Khomenkova, Gourbilleau, & Pareige, 2012) and red emission was attributed to the carrier recombination in these clusters and/or via interface cluster/host states.

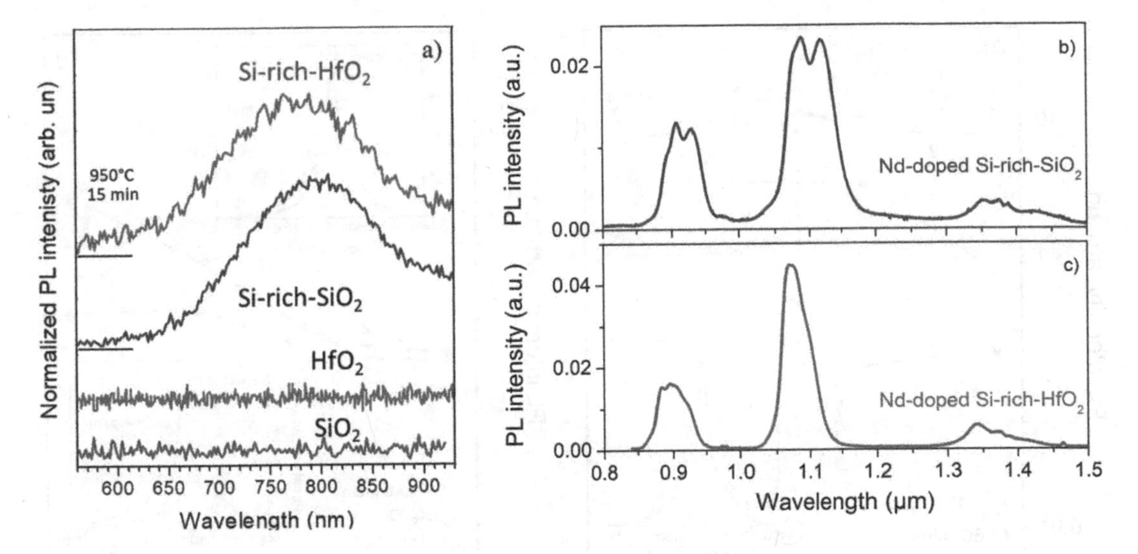

FIGURE 9.9 (a) Comparison of the PL emission in visible spectral range obtained for Si-rich HfO_2 and Si-rich SiO_2 as well as undoped oxides. (b, c) PL spectra in infrared spectral range obtained for Nd-doped Si-rich SiO_2 (b) and Nd-doped Si-rich HfO_2 (c).

It is interesting that codoping of Si-rich-HfO_2 with RE ions (Er^{3+}, Nd^{3+} or Pr^{3+}) allowed intense RE emission to be achieved (Khomenkova et al., 2019). Figure 9.9 shows the comparison of Nd^{3+} PL emission excited nonresonantly (with 488 nm wavelength) from Nd-doped Si-rich HfO_2 (Figure 9.9b) and Nd-doped Si-rich SiO_2 materials (Figure 9.9c). This intense Nd-related emission under ono-resonant excitation is explained by the spatial localization of Nd ions in the vicinity of Si clusters or Si-contain defects that take part in the energy transfer toward RE ions.

This effect was also demonstrated for Er-doped Si-rich HfO_2 (Khomenkova et al., 2019). It was shown that Er^{3+} PL emission at about 1540 nm corresponding to the $^4I_{13/2} \rightarrow {}^4I_{15/2}$ radiative transition can be obtained under resonance excitation corresponding to optical absorption in Er^{3+} ions (for instance, at 488 nm) (Figure 9.10). At the same time, nonresonant (476 nm) excitation also allowed to achieve similar PL emission, for which intensity was of the same order in amplitude (Figure 9.10). This effect was explained by the existence of two types of sensitizers, which are Si nanocluster/nanocrystals (Si-ncs) and host defects such as oxygen vacancies or their complexes. It is known that PL emission from Si nanoclusters is observed in the 700–1000 nm spectral range and that from oxygen vacancies is detected in a wide 270–600 nm spectral range. The first type of sensitizers dominates usually at blue-orange excitation, while another one is remarkable under ultraviolet-violet light excitation. However, the use of 476 and 488 nm excitations permits to minimize the contribution of host defects in the photoluminescence response. It should be noted that for Er-doped HfO_2 materials (i.e., Si-free ones), Er^{3+} PL emission was detected at a very low intensity (as a noise level) under nonresonant excitation (476 nm) that support the statement about negligible contribution of HfO_2 host defects in the excitation Er^{3+} ions under such excitation.

Taking into account the shape of Er^{3+} PL spectrum, one can conclude the Er^{3+} spatial localization in the host material as well as the crystalline nature of the surrounded matrix. This is due to the fact that when Er^{3+} ions are incorporated into a solid host, the degenerated energy levels of their 4*f* states split due to the Stark effect, and the shape of PL spectrum due to Er^{3+}optical transitions between the various Stark levels will reflect the local

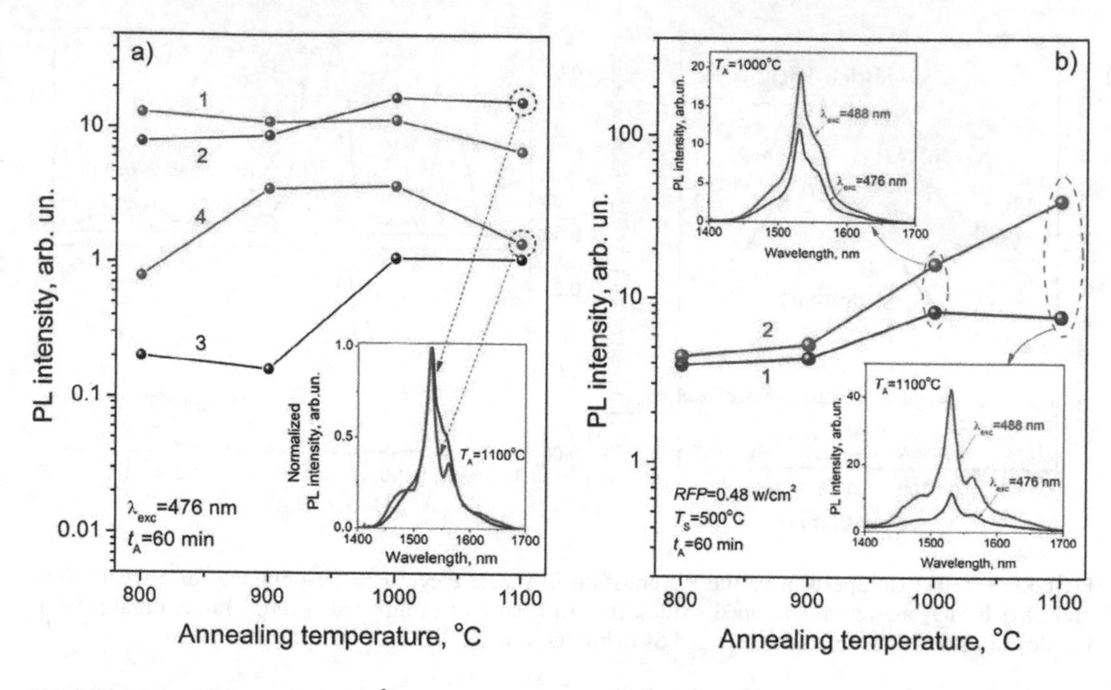

FIGURE 9.10 (a) Evolution of Er^{3+} PL intensity (at 1540 nm) for the films grown with RFP = 0.48 W/cm^2 (1,2) and 0.74 W/cm^2 (3,4) while T_S was 100 °C (1,3) or 500 °C (2,4). Inset represents normalized PL spectra for the films grown with RFP = 0.48 W/cm^2 (blue curve) and 0.74 W/cm^2 (red curve) at T_s = 500 °C after annealing at T_A = 1100 °C. Annealing time t_A = 60 min; excitation wavelength 476 nm. (b) Variation of PL intensity with T_A for the films grown with RFP = 0.48 W/cm^2 (1,2) under 476 nm (blue curve) and 488 nm (green curve) excitations. Insets represent PL spectra recorded under nonresonant (476 nm) and resonant (488 nm) excitations.

environment of the Er ion. In consequence, an appearance of sharp narrow PL peaks is expected for Er^{3+} situated in crystalline host. At room temperature, so-called "thermal" broadening of the PL spectrum is due to lattice phonons. However, for Er^{3+} ions embedded in amorphous host, for instance in SiO_2, additional factor for the spectrum widening is the fluctuation of electrical field of host matrix, that is, the fluctuation of Er–O distance. Thus, the increase of the full-width of Er^{3+}-related PL band (Figure 9.10) is due to the significant fluctuation of electric field of amorphous matrix surrounding the Er^{3+} ion in these films.

Annealing at high T_A values stimulates usually phase separation process in Si-rich HfO_2 materials via formation of HfO_2 and Si-rich phases (Si nanoclusters/nanocrystals (Si-ncs), SiO_x or SiO_2). The former (HfO_2 phase) is known to be crystallized in tetragonal or cubic phase. All Si-rich phases are still amorphous for T_A < 900 °C. The Si-ncs formation occurs at higher T_A values. Thus, an intense Er^{3+} emission under nonresonant excitation (476 nm) can occur from Er^{3+} ions located in the vicinity of Si-ncs/SiO_2 and Si-ncs/HfO_2 interfaces or inside SiO_x shell covering Si-ncs and separating them from SiO_2 and HfO_2 phases. Annealing at high temperature such as T_A = 1100 °C results in the formation of monoclinic HfO_2 phase that is accompanied by the appearance of sharp PL bands related to Er^{3+} ions. This is the evidence that these ions are situated in a regular crystal lattice, that is regular HfO_2 lattice. However, this emission is observed under nonresonant excitation (476 nm), which gives the possibility to assume an energy transfer from Si-related phase to Er^{3+} ions located in HfO_2 phase in the vicinity of the interface between Si and HfO_2 phases.

The comparison of Er^{3+} PL spectra measured under 476 and 488 nm excitations showed significant difference in PL intensity for the films annealed at high T_A. As one can see for the films grown with $RFP = 0.48$ W/cm^2 and $T_S = 500$ °C, this difference is negligible for $T_A = 800$ °C and increases up to one order of magnitude for $T_A = 1100$ °C (Figure 9.10b). The analysis of the PL spectrum shape revealed that under nonresonant (476 nm) excitation, the PL spectrum for $T_A \leq 900$ °C is determined by the emission of Er^{3+} ions located in amorphous Si-rich phase (SiO_x or Si/SiO_2 interface). However, for the film annealed at $T_A = 1100$ °C, the contribution of resonantly (488 nm) excited Er^{3+} ions situated in HfO_2 phase becomes significant and results in the sharpening of PL spectrum (Figure 9.10b). Thus, under resonant excitation (488 nm), the Er^{3+} PL emission occurs not only from Er^{3+} ions interacting with Si-ncs (i.e., located in the vicinity of Si-ncs/oxide (SiO_2 or HfO_2) interfaces), but also from Er^{3+} ions situated in crystallized HfO_2 phase.

Taking into account these results, the mechanism of RE emission in Er-doped Si-rich HfO_2 films can be proposed. Figure 9.11 shows schematic presentation of light absorption and light emission process as well as different ways for energy transfer toward RE ions. It is seen that there are several main players in these processes, namely the Er^{3+} ions and Si-ncs, as well as oxygen vacancies (so-called oxygen-deficient centers (ODC)) located in SiO_2 (Si-ODC) and HfO_2 (Hf-ODC) phases.

As it was mentioned above, under UV and visible excitations, the Si-ncs being embedded in oxide host can demonstrate bright red emission. Besides, they can be also effective sensitizers of Er^{3+} ions demonstrating infrared emission near 1540 nm. In UV-blue spectral

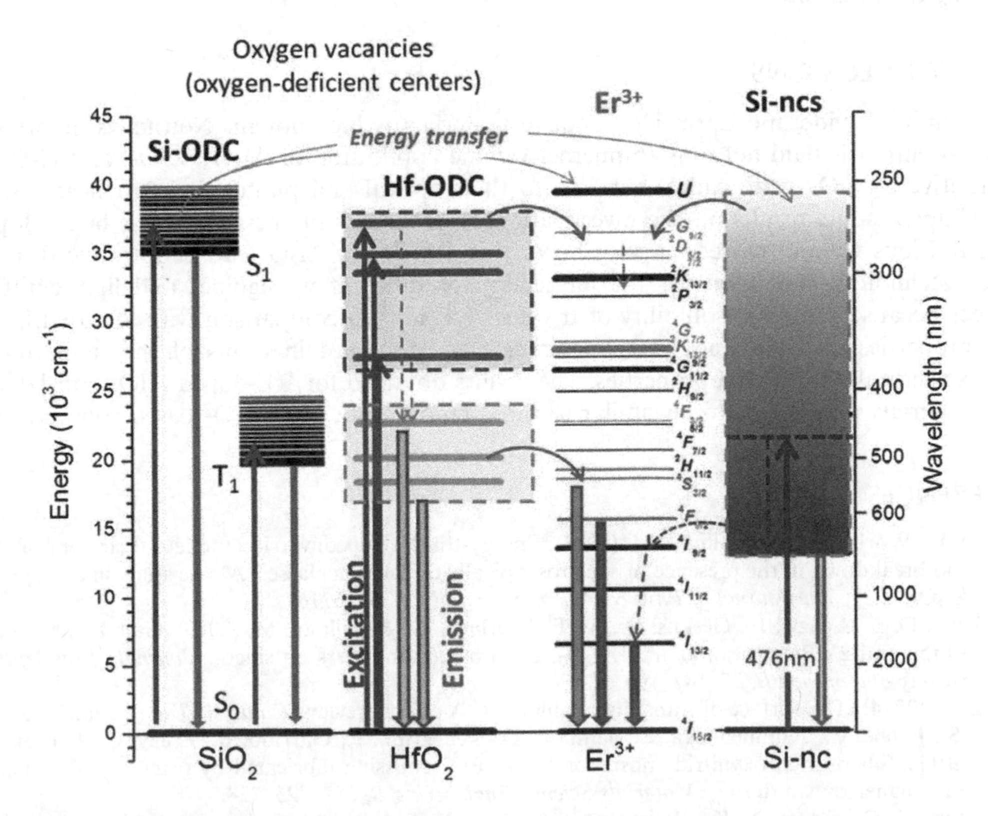

FIGURE 9.11 Schematic presentation of light-emitting mechanism in (Si,Er)-codoped HfO_2 films demonstrating excitation of Er^{3+} ions via energy transfer from host defects (oxygen vacancies or ODC) and Si-ncs. S_1 and T_1 are singlet and triplet states of Si-ODC, respectively.

range, some Er^{3+} probable transitions overlap with the energetic levels of the Si-ODC and Hf-ODC defects. Taking into account that the energy levels of these defects can form the sub-bands inside the bandgap of the host matrix, the excitation of RE ions can occur via energy transfer from these defects, as it was demonstrated for Er-doped SiO_2 and Er-doped HfO_2 materials (Cueff et al., 2011; Khomenkova et al., 2014).

Thus, the competition between PL emission from Si-ncs, Er^{3+} ions, and host defects will determine the shape of total PL spectrum and its peak position. The longer is the excitation wavelength, the smaller will be the contribution of host defects. If such excitation is nonresonant for Er^{3+} ions, the main path for light absorption will belong to Si-ncs. The carrier relaxation will occur via recombination in Si-ncs as well as via energy transfer toward Er^{3+} ions leading to Er^{3+} emission in the infrared spectral range.

The appearance of some sharp peaks in PL spectrum of Er^{3+} ions proves their location in a regular crystal lattice where the fluctuation of crystal field is low. This means that sharp PL bands are due to the Er^{3+} ions located in HfO_2 phase formed during phase separation. Thus, the sharpening of PL spectrum shape is an additional evidence of the formation of crystalline HfO_2 phase in (Si,Er) codoped HfO_2 films.

Similar results were demonstrated for (Er,Si) codoped ZrO_2 films by Rozo et al. (2008). The achievement of Er^{3+} emission under nonresonant excitation was ascribed to energy transfer from Si-ncs. However, the Si-ncs presence was not shown directly. Meanwhile, the roles of oxygen vacancies and oxygen interstitial defects as effective excitation paths for Er^{3+} ions were also addressed in that work. Such results testify that HfO_2- and ZrO_2-based materials being doped in similar way show similar PL properties and, consequently, have similar light-emitting mechanisms.

9.4 CONCLUSIONS

The family of wide and ultrawide bandgap materials is big enough. Nowadays, more and more attention is paid not only to microelectronic application of Al_2O_3, ZrO_2, and HfO_2 as alternative to SiO_2 gate oxides, but also to their optical and photovoltaic applications. In this chapter, some results on the investigation of properties of these materials being doped with RE ions were discussed. It was shown that RE-doped Al_2O_3 can be introduced in Si-based technology and used for the fabrication of different waveguides and light-emitting devices because of higher solubility of trivalent RE ions in comparison with Si-based hosts. The properties of ZrO_2 and HfO_2 materials were discussed in terms of the similarity of their structural and optical properties. The results obtained for RE-doped HfO_2 and Hf silicate materials were found to be similar to those reported for their ZrO_2-based counterparts.

REFERENCES

Agazzi, L., Wörhoff, K., & Pollnau, M. (2013). Energy-transfer-upconversion models, their applicability and breakdown in the presence of spectroscopically distinction classes: A case study in amorphous Al_2O_3: Er^{3+}. *The Journal of Physical Chemistry C, 117*, 6759–6776.

Bradley, J. D. B., Agazzi, L., Geskus, D., Ay, F., Wörhoff, K., & Pollnau, M. (2010). Gain bandwidth of 80 nm and 2 dB/cm peak gain in Al_2O_3: Er^{3+}optical amplifiers on silicon. *Journal of the Optical Society of America B, 27*, 187–196.

Busca, G. (2014). The surface of transitional aluminas: A critical review. *Catalysis Today, 226*, 2–13.

Cueff, S., Labbé, C., Khomenkova, L., Jambois, O., Pellegrino, P., Garrido, B., Frilay, C., & Rizk, R. (2011). Silicon-rich oxynitride hosts for 1.5μm Er^{3+} emission fabricated by reactive and standard RF magnetron sputtering. *Materials Science Engineering B, 177*, 725–728.

Demirtas, M. C., Odaci, N. K., Perkgoz, C., & Sevik, F. A. (2018). Low loss atomic layer deposited Al_2O_3 waveguides for applications in on-chip optical amplifiers. *IEEE Journal of Selected Topics in Quantum Electron, 24*, 1–8.

He, G., & Sun, Z. (Eds.). (2012). *High-k gate dielectrics for CMOS technology.* Weinheim, Germany: Wiley.

Houssa, M. (Ed.). (2003). *High-k gate dielectrics.* Boca Raton, FL: CRC Press.

Ishizaka, T., & Kurokawa, Y. (2000). Optical properties of rare-earth ion (Gd^{3+}, Ho^{3+}, Pr^{3+}, Sm^{3+}, Dy^{3+} and Tm^{3+}) - doped alumina films prepared by the sol–Gel method. *Journal of Luminescence, 92*, 57–63.

Jambois, O., Berencen, Y., Hijazi, K., Wojdak, M., Kenyon, A.J., Gourbilleau, F., Rizk, R., & Garrido, B. (2009). Current transport and electroluminescence mechanisms in thin SiO_2 films containing Si nanocluster-sensitized erbium ions. *Journal of Applied Physics, 106*, 063526.

Khomenkova, L., An, Y.-T., Khomenkov, D., Portier, X., Labbe, C., & Gourbilleau, F. (2014). Spectroscopic and structural investigation of undoped and Er^{3+} doped hafnium silicate layers. *Physica B, 453*, 100–106.

Khomenkova, L., & Korsunska, N. (2019). Solid state composites and multilayers produced by magnetron sputtering. Chapter 2. In R. Savkina & L. Khomenkova (Eds.), *Solid state composites and hybrid systems* (pp.152–185). Boca Raton, FL: CRC Press.

Khomenkova, L., Korsunska, N., Labbé, C., Portier, X., & Gourbilleau, F. (2019). The peculiarities of structural and optical properties of HfO_2-based films co-doped with silicon and erbium. *Applied Surface Science, 471*, 521–527.

Khomenkova, L., Portier, X., Marie, P., & Gourbilleau, F. (2011). Hafnium silicate dielectrics fabricated by RF magnetron sputtering. *Journal of Non-Crystalline Solids, 357*, 1860–1865.

Kik, P. G., & Polman, A. (1998). Erbium-doped optical-waveguide amplifiers on silicon. *MRS Bulletin, 23*, 48–54.

Kingon, A., Maria, J., & Streiffer, S. (2000). Alternative dielectrics to silicon dioxide for memory and logic devices. *Nature, 406*, 1032–1038.

Korsunska, N., & Khomenkova, L. (2019). Multifunctional zirconia-based nanocomposites. Chapter 2. In R. Savkina & L. Khomenkova (Eds.), *Solid state composites and hybrid systems* (pp. 28–57). Boca Raton, FL: CRC Press.

Lee, M. H., Fan, S.-T., Tang, C.-H., Chen, P.-G., Chou, Y.-C., Chen, H.-H., … Liu, C. W. (2016). Physical thickness 1.x nm ferroelectric $HfZrO_x$ negative capacitance FETs. *Proc of IEEE Int Electron Dev Meeting (IEDM)*, San Francisco, CA, pp. 12.1.1–12.1.4.

Li, S., Liu, J., Yang, X., Ding, Y., Zhu, L., Liu, B., … Duan, T. (2019). Effect of phase evolution and acidity on the chemical stability of $Zr_{1-x}Nd_xSiO_{4-x/2}$ ceramics. *Ceramincs International, 45*, 3052–3058.

Liu, Y., Liu, Y., Ouyang, Z., Yang, L., Yang, Y., & Sun, J. (2019). Blue electroluminescent Al_2O_3/Tm_2O_3 nanolaminate films fabricated by atomic layer deposition on Silicon. *Nanomaterials, 9*, 413.

Lucovsky, G. (2001). Transition from thermally grown gate dielectrics to deposited gate dielectrics for advanced silicon devices: A classification scheme based on bond iconicity. *Journal of Vacuum Science and Technology A, 19*, 1553–1561.

Purnawirman, N., Li, N., Magden, E. S., Singh, G., Singh, N., Baldycheva, A., … Watts, M. R. (2017). Ultra-narrow-linewidth Al_2O_3: Er^{3+}lasers with a wavelength-insensitive waveguide design on a wafer-scale silicon nitride platform. *Optics Express, 25*, 13705–13713.

Rebohle, L., Braun, M., Wutzler, R., Liu, B., Sun, J. M., Helm, M., & Skorupa, W. (2014). Strong electroluminescence from SiO_2-Tb_2O_3-Al_2O_3 mixed layers fabricated by atomic layer deposition. *Applied Physics Letters, 104*, 251113.

Robertson, J. (2002). Band structures and band offsets of high K dielectrics on Si. *Applied Surface Science, 190*, 2–10.

Rozo, C., Jaque, D., Fonseca, L. F., & Solé, J. G. (2008). Luminescence of rare earth-doped Si-ZrO_2 co-sputtered films. *Journal of Luminescence, 128*, 1197–1204.

Shi, X., Tielens, H., Takeoka, Sh., Nakabayashi, T., Nyns, L., Adelmann, C., … van Elshocht, S. (2010). Development of ALD $HfZrO_x$ with TDEAH, TDEAZ and H_2O. *ECS Transactions, 27*, 699–704.

Starschich, S., Schenk, T., Schroder, U., & Boettger, U. (2017). Ferroelectric and piezoelectric properties of $Hf_{1-x}Zr_xO_2$ and pure ZrO_2 films. *Applied Physics Letters, 110*, 182905.

Sun, J. M., Prucnal, S., Skorupa, W., Dekorsy, T., Müchlich, A., & Helm, M. (2006). Electroluminescence properties of the Gd^{3+} ultraviolet luminescent centers in SiO_2 gate oxide layers. *Journal of Applied Physics, 99*, 103102.

Talbot, E., Roussel, M., Khomenkova, L., Gourbilleau, F., & Pareige, P. (2012). Atomic scale microstructures of high-k HfSiO thin films fabricated by magnetron sputtering. *Materials Science and Engineering B*, *177*, 717–720.

van Emmerik, C. I., Dijkstra, M., de Goede, M., Chang, L., Mu, J., & Garcia-Blanco, S. (2018). Single-layer active-passive Al_2O_3 photonic integration platform. *Optical Materials Express*, *8*, 3049–3054.

Vázquez-Córdova, S. A., Dijkstra, M., Bernhardi, E. H., Ay, F., Wörhoff, K., Herek, J. L., …, & Pollnau, M. (2014). Erbium-doped spiral amplifiers with 20 dB of net gain on silicon. *Optics Express*, *22*, 25993–26004.

Wilk, G. D., Wallace, R. M., & Anthony, J. M. (2001). High-κ gate dielectrics: Current status and materials properties considerations. *Journal of Applied Physics*, *89*, 5243–5275.

Wörhoff, K., Bradley, J. D. B., Ay, F., Geskus, D., Blauwendraat, T. P., & Pollnau, M. (2009). Reliablc low-cost fabrication of low-loss Al_2O_3: Er^{3+}waveguides with 5.4-dB optical gain. *IEEE Journal of Quantum Electronincs*, *45*(5), 454–461.

Xing, P., Chen, G. F. R., Zhao, X., Ng, D. K. T., Tan, M. C., & Tan, D. T. H. (2017). Silicon rich nitride ring resonators for rare-earth doped telecommunications-band amplifiers pumped at the O-band. *Scientific Reports*, *7*, 9101.

Yang, J., van Dalfsen, K., Wörhoff, K., Ay, F., & Pollnau, M. (2010). High-gain Al_2O_3:Nd^{3+}channel waveguide amplifiers at 880 nm, 1060 nm, and 1330 nm. *Applied Physics B*, *101*, 119–127.

Yang, X.-Y., Wang, S., Lu, Y., Bi, P., Zhang, P., Sh, H. Y., & Yi, T. D. (2018). Structures and energetics of point defects with charge states in zircon: A first-principles study. *Journal of Alloys and Compounds*, *759*, 60–69.

Yang, Y., Li, N., & Sun, J. M. (2018). Intense electroluminescence from Al_2O_3/Tb_2O_3 nanolaminate films fabricated by atomic layer deposition on silicon. *Optics Express*, *26*, 9344–9352.

Index

R

S

T

U

V

X

W

Z